CZECHOSLOVAKIA AT THE WORLD'S FAIRS
BEHIND THE FAÇADE

CZECHOSLOVAKIA AT THE WORLD'S FAIRS

BEHIND THE FAÇADE

MARTA FILIPOVÁ

CEU PRESS

CENTRAL EUROPEAN UNIVERSITY PRESS

BUDAPEST–VIENNA–NEW YORK

Published in 2024 by
Central European University Press

Nádor utca 9, H-1051 Budapest, Hungary
Tel: +36-1-327-3138 or 327-3000
E-mail: *ceupress@press.ceu.edu*
Website: *www.ceupress.com*

This publication is part of a project that has received funding from the European Research Council (EDC) under the European Union's Horizon 2020 research and innovation programme (grant agreement no. 786314).

European Research Council
Established by the European Commission

On the cover: Czechs in traditional dress – In front of building. New York World's Fair 1939–1940 records, Manuscripts and Archives Division, New York Public Library.

ISBN 978-963-386-766-2 (hardback)
ISBN 978-963-386-767-9 (ebook)

Library of Congress Cataloging-in-Publication Data

Names: Filipová, Marta, author.
Title: Czechoslovakia at the world's fairs : behind the façade / Marta Filipová.
Description: Budapest ; New York : Central European University Press, 2024. | Includes bibliographical references and index.
Identifiers: LCCN 2024035611 (print) | LCCN 2024035612 (ebook) | ISBN 9789633867662 (hardback) | ISBN 9789633867679 (adobe pdf)
Subjects: LCSH: Exhibition buildings. | Architecture and state--Czechoslovakia. | Exhibitions--History--20th century. |
Czechoslovakia--Exhibitions. | BISAC: DESIGN / General Classification: LCC T395.5.C95 F55 2024 (print) | LCC T395.5.C95 (ebook) | DDC 607/.3409437--dc23/eng/20241009
LC record available at https://lccn.loc.gov/2024035611
LC ebook record available at https://lccn.loc.gov/2024035612

Contents

List of Figures · vii

Acknowledgments · xi

Introduction · 1

A new state and a new nation? · 5
The small Czechoslovak nation · 7
Selective hierarchies · 10
Book structure · 12

1. Continuities and Raptures: Exhibiting a Modern State · 17

Grounds of exhibitions · 20
Establishing an exhibition model: *Rio de Janeiro 1922* · 23
The politics of a national pavilion: *Paris 1925* · 29
The hesitant exhibit: *Philadelphia 1926* · 34
Exhibiting during the crises: *Barcelona 1929* · 38
Recovery: *Chicago 1933* · 42
Colonial company: *Brussels 1935* · 47
Flow of display: *Paris 1937* · 51
End of an era: *New York and San Francisco, 1939–1940* · 55
The final rupture · 60
Conclusion · 65

2. Exhibition Spaces · 67

Ephemerality of design · 69
Contemplating exhibition spaces · 73
Three-dimensional graphic design: Ladislav Sutnar · 78
Designing a fairy tale · 84
Journey through the pavilion · 87
Czechoslovakia on display · 90
Display techniques · 94
Reception · 97
Conclusion · 100

3. Art and Design: The Limits of Modernism · 103

First international exhibitions · 107
Searching for a representative style · 109
The state in arts and crafts · 111
Crafts and tapestry · 116
What is modern art and design · 120
Whose modernism · 121
What modernism? · 124
The odd one out · 129
Modernity across the ocean—Art in America · 132
Conclusion · 136

4. Commodifying the State · 139

Economy and the new state · 142
Modern folk industries: Detva · 145
Detva on display · 149
Nationalized consumption: Pilsner beer · 153
Economies and nationalities · 156
Truly Bohemian glass? · 159
Exhibiting glass: Moser · 161
Building a company, exhibition and the state: Baťa · 166
Business with pleasure · 169
Conclusion: the limits of loyalty · 174

5. People of the Fair · 177

"Czechoslovaks" in the USA · 180
A painting under the arm · 182
Czechs and Slovaks of Chicago · 184
Where is my homeland · 189
The diaspora at the fair · 191
"Native" experience · 197
Gendered fairs · 200
National days · 205
Conclusion · 211

6. Race Science, and Ethnicity: Molding the Ideal Czechoslovak · 215

Establishing hierarchies · 217
What are the Czechoslovaks · 222
Eugenics at exhibitions and congresses · 224
Eugenics in Czechoslovakia · 227
Race and ethnicity displays · 230
Visualizing race and humanity · 234
The ideal man, the ideal Czechoslovak · 237
Hygiene and health · 242
Conclusion · 245

7. Conclusion · 247

Modernity and modernism · 252
Classifications and hierarchies · 254
Agency · 256
Trajectories and developments · 257
Crumbling façade · 257

Bibliography · 259

Index · 283

List of Figures

1. Antonín Strnadel and Josef Novák, collage of Czechoslovakia, Czechoslovak pavilion, New York world's fair, 1939, Ladislav Sutnar Papers, Getty Research Institute, Los Angeles (2013.M.6)
2. Rio de Janeiro, The Centennial International Exhibition, 1922. Museu Paulista of the University of São Paulo, Pub Wiki Commons Public Domain - Museu Paulista (USP) Collection.
3. Pavel Janák, Czechoslovak pavilion in Rio de Janeiro, 1922, f. 85 Janák, Archive of Architecture and Civil Engineering, National Technical Museum, Prague.
4. View of the Czechoslovak pavilion interior in Brussels, 1935, Baťa fund, sign XV, inv. no. 165, Moravian Regional Archive in Brno, branch Zlín.
5. International Exhibition of the Decorative Arts – General view, Paris, 1925, postcard. Author's collection.
6. Josef Gočár, Drawing of the Czechoslovak pavilion in Paris, 1925, f.14 Gočár, Archive of Architecture and Civil Engineering, National Technical Museum, Prague.
7. Edwin Olsen, Blake Clark, and Houghton Mifflin Company, "A kite view of Philadelphia and the Sesqui-Centennial International Exposition," map. Boston: Houghton Mifflin Co., 1926. Digital Commonwealth, https://ark.digitalcommonwealth.org/ark:/50959/0r96fm962.
8. Ladislav Machoň, The Czechoslovak pavilion in Philadelphia, 1926. f. 26 Machoň, Archive of Architecture and Civil Engineering, National Technical Museum, Prague.
9. Panoramic view of the Barcelona Exposition, Spain. 1929. Records of the Bureau of Foreign and Domestic Commerce (151-FC-106-67), National Archives and Records Administration, Records of the Bureau of Foreign and Domestic Commerce (151-FC-106-67) National Archives and Records Administration. Public Domain.
10. Vintage 1933 Chicago World's Fair Postcard, A Century Of Progress International Exposition - Looking South Over The World's Fair Grounds. Attribution licence (CC BY 2.0) Joe Haupt, Flickr.
11. Kamil Roškot, The Czechoslovak pavilion for the Chicago world's fair, 1933. f. 39 Roškot, Archive of Architecture and Civil Engineering, National Technical Museum, Prague.
12. Interior of the Czechoslovak pavilion, Chicago, 1933. Private collection. Courtesy of Marcy Cameron.
13. View of the The Universal and International Exposition in Brussels, 1935, postcard. Wiki Commons, public domain.
14. The Czechoslovak pavilion in Brussels, 1935, The Royal Institute of British Architects, record control no.: P013009, Call/Ref. no.:55122-55122/23, Call/Ref. no.: ACN5-ACN7.

15. Site map of the Paris World's Fair 1937. Wiki Commons Public Domain.
16. View of the Czechoslovak pavilion, Paris, 1937. *Pestrý týden* 12, no. 35, August 28, 1937, 19. Moravian Regional Library, Brno.
17. Treasure Island, Magic City of Golden Gate International Exposition, 1939. Steve Shook, Creative Commons Attribution 2.0.
18. Czechoslovak exhibit at the Golden Gate Fair, San Francisco, The Royal Institute of British Architect, Record control no. P013165.
19. Inscription Czecho-Slovakia, design for the pavilion, New York, 1939. f. 39 Roškot, Archive of Architecture and Civil Engineering, National Technical Museum, Prague.
20. Foldout from the General Motors brochure, "Things to see at Highways and Horizons," 1939. takomabibelot, Creative Commons Attribution 2.0.
21. Kamil Roškot, Design of the Czechoslovak pavilion, 1939. f. 39 Roškot, Archive of Architecture and Civil Engineering, National Technical Museum, Prague.
22. Views of the Paris Exhibition, 1937, The Royal Institute of British Architect, Record control no.: P019346, Call/Ref. no.: AP57/13-AP57/81A.
23. Pavel Janák, Floor plan of the Czechoslovak pavilion in Rio de Janeiro, 1922, f. 85 Janák, Archive of Architecture and Civil Engineering, National Technical Museum, Prague.
24. Ladislav Sutnar, Prager Presse exhibit, The Press pavilion, Paris, 1937. II Division, 1918-39, box 589, file 4, inv.no. 93703/III-3/37 of July 12, 1937, Ministry of Foreign Affairs Archive, Prague.
25. Ground plan of the Czechoslovak pavilion, Paris, 1937, *Stavba* XIII (1936-7): 177. Moravian Regional Library, Brno.
26. View of the Industries Hall, Czechoslovak pavilion, Paris, 1937. *Pestrý týden,* 28 August 1937, 14. Moravian Regional Library, Brno.
27. View of the Hall of Honor, Czechoslovak pavilion, Paris, 1937. *Exposition Paris 1937, Arts et techniques* no. 14, July 1937, Museum of Decorative Arts, Prague.
28. View of the Baťa exhibit, Czechoslovak pavilion, Paris, 1937, *Pestrý týden* 12, no. 35, 28 August 1937, 16, Moravian Regional Library, Brno.
29. Antonín Heythum, Glass section, Czechoslovak pavilion, Paris, 1937. *Pestrý týden* 12, no. 35, 28 August 1937, 18, Moravian Regional Library, Brno.
30. Czechoslovak pavilion, New York World's Fair, 1939, Ladislav Sutnar Papers, Getty Research Institute, Los Angeles (2013.M.6).
31. Otakar Švec, Motocyclist, 1924, bronze, 114x226x96cm, inv.no. 1382, Photograph © National Gallery Prague 2024.
32. Josef Gočár, Interior of the Czechoslovak pavilion in Paris, 1925, f.14 Gočár, Archive of Architecture and Civil Engineering, National Technical Museum, Prague.
33. Josef Gočár, interior of the Czechoslovak pavilion in Paris with František Kysela's tapestries, 1925, f.14 Gočár, Archive of Architecture and Civil Engineering, National Technical Museum, Prague.
34. František Kysela, Weaving, tapestry, 1924, Museum of Decorative Arts, Prague.
35. Emilie Paličková Milde, Little Sun, lace doily, 1925. Museum of Decorative Arts, Prague.
36. Zdeněk Pešánek, Spa Fountain, 1936, author's model replica from 1959. The Benedikt Rejt Gallery, Louny.
37. Ľudovít Fulla, Song and Work, 1934–35, painting. Slovak National Gallery, OP10.

38. Martin Benka, Countryside of Terchova, 1936, painting, Slovak National Museum in Martin.
39. Lace doilies, *Detva. Československý ľudový umelecký priemysel, úč. spol.* (Bratislava: Detva, 1929).
40. Example of Detva products on display, *Detva. Československý ľudový umelecký priemysel, úč. spol.* (Bratislava: Detva, 1929).
41. Colorful embroidered pillow cases, *Detva. Československý ľudový umelecký priemysel, úč. spol.* (Bratislava: Detva, 1929).
42. View of the Pilsner restaurant, Czechoslovak pavilion, New York, 1939-40, Archive of the Czech and Slovak Association, University of Chicago.
43. Advertising of the Glass Company Moser. Wikimedia Commons, public domain.
44. Glass display at the Czechoslovak pavilion in 1937, *Exposition Paris 1937, Arts et techniques,* no. 14, July 1937, Museum of Decorative Arts, Prague.
45. Vase – Fan by Lotte P. Moser, 1925, purple-red royalit glass, height 21.1cm, The Museum of Decorative Arts, Prague inv. no. 97168.
46. Design for the Czechoslovak pavilion in New York, f.60 exhibitions, Archive of Architecture and Civil Engineering, National Technical Museum, Prague.
47. Exhibit Profile of Czechoslovak Person for the Czechoslovak pavilion in New York, f. 60 Exhibitions, Archive of Architecture and Civil Engineering, National Technical Museum, Prague.
48. Interior. Czechoslovak exposition, Sesqui-Centennial Exposition, 1926. City of Philadelphia, Department of Records, City Archives, A1474, box 17, 232-4.8.1 Folder 32, Dep. of foreign participation Czechoslovakia.
49. *Chicago on Parade* pamphlet, Century of Progress Collection, box 5.13, Chicago Public Library.
50. Two singing waitresses at the Czechoslovak restaurant, Century of Progress Scrapbooks, June 26 – July 22, 1933, box 1, folder 3, Chicago Public Library.
51. Sokol gathering at Century of Progress, Chicago, 1933, Archive of the Czech and Slovak Association, University of Chicago.
52. Pageant "Oh my Country," Century of Progress, Chicago, 1933. Archive of the Czech and Slovak Association, University of Chicago.
53. Horse parade at Czechoslovak national day, Century of Progress, Archive of the Czech and Slovak Association, University of Chicago.
54. Women selling beer in Pilsner restaurant, New York World's Fair 1939–1940 records, Manuscripts and Archives Division, New York Public Library.
55. "The Relation of Eugenics to Other Sciences" from A decade of progress in Eugenics. Scientific papers of the Third International Congress of Eugenics (Baltimore: The Williams & Watkins company, 1934. based on a paper by Dr. Harry H. Laughlin. Wellcome Trust, Photo number: L0032341.
56. Malvina Hoffman, Tony Sansone, A Man from New York (The Nordic Type). Author's photograph.
57. Louise Lentz Woodruff, *Science Leading the Youth,* 1933. Chicago History Museum.
58. The Transparent Woman in the Museum of Hygiene, Dresden, 1947, df_roeneg_0000182_001. Wikimedia Commons Attribution 4.0: Deutsche Fotothek.
59. Zlín, houses for Baťa workers. Postcard, Author's collection.
60. View of the Baťa pavilion, Brussels, 1935. Fond Baťa, sign XV, inv. no. 165, Moravian Regional Archive in Brno, branch Zlín.

61. Czechoslovakia Participation - Czechs in traditional dress - In front of building. New York World's Fair 1939-1940 records, Manuscripts and Archives Division, New York Public Library.
62. World's Fair. Czech mural I, Gottscho-Schleisner Collection, Library of Congress, Prints and Photographs Division.
63. World's Fair. Window panels with illustrations and text on the profitsharing system in the shoe industry III, Gottscho-Schleisner Collection, Library of Congress, Prints and Photographs Division.
64. World's Fair. Window panels with illustrations and text on the profitsharing system in the shoe industry I, Gottscho-Schleisner Collection, Library of Congress, Prints and Photographs Division.
65. World's Fair. Window panels with illustrations and text on the profitsharing system in the shoe industry II, Gottscho-Schleisner Collection, Library of Congress, Prints and Photographs Division.
66. World's Fair. Interior of Czech Building, Gottscho-Schleisner Collection, Library of Congress, Prints and Photographs Division.

Acknowledgements

This book would not be possible without the help of a number of people and institutions who supported me at different stages of research, writing, rewriting and editing. First of all, I am grateful to my colleagues with whom I was fortunate to work on the project "Continuity and Rupture: Art and Architecture in Central Europe, 1919–1939." Our focus was on the visual arts in interwar Central Europe and on, as the project's title suggests, patterns of continuities that were carried from the Habsburg Monarchy into the successor states. We placed them against the better known historical, artistic, economic, and social ruptures that lead to the establishment of the new political elites. Matthew Rampley, Julia Secklehner and Nóra Veszprémy, alongside Christian Drobe, helped me shape this book's focus, patiently read and commented on my various drafts. It goes without saying that "Continuity and Rupture" would not be possible without the funding from the European Research Council that allowed the concentrated and ambitious research. Our treasured project administrator, Jana Hájková, played a vital role in this because she went above and beyond her job description to support us. I am also thankful to the Departments of Art History in Birmingham and in Brno that hosted the project.

This book is grounded in extensive archival research. I am very grateful to all the archivists and librarians I met along the way who were always extremely helpful. This includes the archivists at the Archives of Architecture and Civil Engineering of the National Technical Museum, Robert Vejvoda and Martin Šámal, Tomáš Klusoň from the archives of the Ministry of Foreign Affairs, Jakub Doležal from the Archives of the President of the Republic Office, colleagues in the National Archives in Prague and Bratislava as well as those in the Moravian Regional Archive in Brno and Zlín.

I also encountered kind response and assistance during my research abroad. In Chicago, the archivists at the Special Collection Division of the Chicago Public Library and their colleagues at the Abakanowitz Research Center of Chicago History Museum played a key role in identifying further material on the Chicago world's fair in 1933. In relation to other world's fairs in the United States, I was also met by extremely helpful approach of Emily M. Orr from the Cooper Hewitt, the staff at the Barncroft Library and the Environmental Design Archives at the University of California, Berkeley, at the San Francisco Public Library and New York Public Library, and San Francisco Museum of Modern Art.

While the Czech and Slovak diaspora in the USA is still thriving, the historic—and written—memory is scattered in many locations. I am therefore thankful to those that keep it alive and accessible. Megan Browndorf of the Archive of the Czech and Slovak Association held at the libraries of the University of Chicago was my great source of material and knowledge. The National Czech and Slovak Museum and Library at Cedar Rapids in Iowa holds a rich collection, and the librarian David Muhlena and the director, Cecilia Rokusek, provided me with insight into the local expat communities and their sources. I am also grateful to George Drost for his expansive knowledge and network, Dawn Clewett of the Czech and Slovak Heritage Library at Riverside, Chicago, to Marcy Gerwing, nee Cameron, Peggy Hazard and Jaroslav Kříž, of the Antonín Marek Municipal Library in Turnov for sharing their expertise with me.

In many other locations, librarians and archivists were indispensable. Valeria Carullo and her colleagues at the RIBA British Architectural Library in London and the staff at the library of the Bibliothèque de l'Institut national d'histoire de l'art in Paris were incredibly helpful. I was also fortunate to benefit from the knowledge of Barbara Dudás, who assisted with the translations from Hungarian, and of Miroslav Kleban's of the East Slovak Gallery in Košice. There have, indeed, been many others who informed and shaped by ideas and direction, and consisted of many colleagues, readers, reviewers and members of audiences in my lectures and conferences. I would also like to express my gratitude to the publisher, especially to Linda Kunos and to the copy editor, Christopher Cox.

And last but not least, my friends and family, not to mention the dog, helped me to maintain my mental health. And while the writing process can be lonely, the journey to a book is a joyful collaboration with great many great people who enrich the result. I am forever thankful to all of them.

INTRODUCTION

What lies behind the sleek façade of national pavilions at world's fairs? Looking behind the veneer of official presentations of states reveals who and what helps to shape what is displayed and why. Any state that decides to invest in a national pavilion at a world's fair pursues a concrete, rational set of goals ranging from establishing new business relationships to proclaiming its political and ideological stance to the allies and enemies. How the pavilions are constructed and conceived, read and received is, nevertheless, subject to various circumstances that do not always align with the original intent. The political, cultural and social environment, in which the pavilions appear, together with the agency of various exhibition participants construct national pavilions and their meaning. This book is concerned with world's fairs, which is a term I use as a shorthand for the largest exhibition events between the two World Wars. The interwar period was a time marked by important changes that included political restructuring in Europe, the Great Depression, new threats of war, and rising nationalisms. All of these had a major impact on the logic and purpose of world's fairs and how they were organized.

Ever since the Great Exhibition of 1851, world's fairs had been staged on an enormous scale and were visited by millions of people.[1] For participating states, they provided a significant opportunity to gain visibility, strengthen political

1 Some more recent contributions on the subject include David Raizman and Ethan Robey, eds., *Expanding Nationalisms at World's Fairs: Identity, Diversity and Exchange, 1851–1915* (London: Routledge, 2017); Joseph Leerssen and Eric Storm, eds., *World Fairs and the Global Moulding of National Identities: International Exhibitions as Cultural Platforms, 1851–1958* (Leiden: Brill, 2022); Robert W. Rydell and Laura Burd Schiavo, eds., *Designing Tomorrow: America's World's Fairs of the 1930s* (New Haven: Yale University Press, 2010).

and economic alliances, and find new trading partnerships, embedding a distinct image of the state in the minds of others. This was especially the case for the new states that were created out of the wreckage of Tsarist Russia, the Ottoman Empire, imperial Germany and Austria-Hungary during and after the First World War. In this respect, one of the keenest participants in international exhibitions and world's fairs between the wars was Czechoslovakia. Created on October 28, 1918, it was a new political entity and one of the successor states to the Habsburg monarchy. In many respects, Czechoslovakia stands out amongst the countries created out of Austria-Hungary. It saw itself on the winning side of the First World War and it remained a more-or-less stable democracy until 1938. This relative political stability helped its diplomatic relations, especially with the USA, France and the United Kingdom. Czechoslovakia also participated in all major exhibitions of the interwar period.

Within the specific environment of world's fairs, states—including Czechoslovakia—promoted a sense of their national identity through their pavilion architecture, exhibits and accompanying events. At the same time, participating countries endeavoured to fit in and embraced the shared vision of modernity that international exhibitions promoted. Yet alongside showcasing themselves as modern and progressive, states such as Czechoslovakia also bolstered with legitimacy by emphasising their longstanding cultural traditions and history. This inevitably led to the creation of a tension between the competing principles on which their participation in world's fairs were based. Various studies of national participations at world's fairs have highlighted the numerous contradictions in the displays that foregrounded national exceptionalism while also trying to demonstrate their place in the shared modernity.[2]

If we recall Zygmunt Bauman's thesis that order and chaos are the twin pillars of modernity, we can view world's fairs as reflecting the chaos and order of the modern world.[3] The exhibitions, via their initiators, organizers and exhibitors, gathered the seeming chaos of the present, sorted it, classified it and explained it. The seeming order they brought to the disarray of competing ideas that enter the exhibition space was, however, idealised as well as ambiguous. One of the purposes of this book is to interrogate what was selected to be or-

---

2 For example, Christine Romlid, "Visualizing Sweden at the 1937 World Fair in Paris," in Holger Weiss, ed., *Locating the Global Spaces, Networks and Interactions from the Seventeenth to the Twentieth Century* (De Gruyter Oldenbourg, 2020); Mauricio Tenorio-Trillo, *Mexico at the World's Fairs: Crafting a Modern Nation* (Berkeley: University of California Press, 1996), Patricia A. Morton, *Hybrid Modernities: Architecture and Representation at the 1931 Colonial Exposition, Paris* (Cambridge: MIT Press, 2000) and others.

3 Zygmunt Bauman, *Modernity and Ambivalence* (Oxford: Polity Press, 2007), 4.

dered and what exhibition organizers (in the widest sense of the expression) sought to achieve with that selection.

Interwar world's fairs embraced modern ideas of progress, capitalism, individualism, and secularization, reflecting the interests of states, as well as the processes of urbanization and the growing interconnectedness of the world by new means of communication. The visions presented at world's fairs built an idealised interpretation of the future and of how it could be planned and constructed. World's fairs were therefore both the products of modernity and its agents. As Joseph Leerssen has recently argued in relation to world's fairs, "modernity turns the world into a panorama."[4] The panoramic view of the globe offered classifications and explanations in a concise place and time.[5] Yet who was responsible and what motivated them? This question is seldom addressed directly, and the concern about who was responsible for selecting, ordering, classifying and displaying the world is one of the key issues I address in this book. I acknowledge that the state, in my case Czechoslovakia, was the primary agent in the organization of world's fair, represented by what could be called the *exhibition elites* that consisted of government officials, business leaders, ambassadors, as well as architects and designers. However, I argue that their effort was complemented, challenged and modified by other actors too. They ranged from female performers and casual workers to scientific experts, all of whom promoted their own understanding of what they represented or promoted in these contexts.

Agency has become a concern of recent world's fairs scholarship, which has turned away from focusing solely on those representatives of the state who were, at least in principle, in control. World's fairs as physical and mental spaces were co-constructed or disrupted by participating countries and audiences, as well as workers and performers of various ethnic backgrounds. A new perspective on the topic has been provided by for example approaches from gender studies that turned long needed attention to the multifaceted roles of women in world's fairs.[6] Restoring their presence and highlighting their absence has enriched the understanding of world's fairs as intricate mechanisms reflective of societal structures. Attention to questions of gender is a distinguishing feature of *Czechoslova-*

4 Leerssen and Storm, *World Fairs*, 37. Cf also Thomas Richards, *The Commodity Culture of Victorian England: Advertising and Spectacle, 1851–1914* (Stanford University Press, 1990).

5 Michel Foucault, *Discipline and Punish, The Birth of the Prison* (London: Allen Lane, 1977); Walter Benjamin, *The Arcades Project* (Harvard University Press, 1999); Tony Bennett, "The Exhibitionary Complex," *new formations* 4, no. 1 (1998): 73–102.

6 Tracey Jean Boisseau and Abigain M. Markwyn, eds., *Gendering the Fair: Histories of Women and Gender at World's Fairs* (Illinois University Press, 2010); Myriam Boussahba-Bravard and Rebecca Rogers, eds., *Women in International and Universal Exhibitions, 1876–1937* (London: Routledge, 2018).

kia at the World's Fairs, too, which argues that the noticeable absence of women amongst the exhibition elites reflected the general situation in Czechoslovakia.

The topic of great exhibitions and world's fairs in the interwar era continues to generate new studies because of the ability of these events to reflect the vast complexity of the modern world in a single space, that keep offering new material to consider.[7] Yet Czechoslovakia and its participation in interwar world's fair has seldom been a subject of sustained research in this connection. The reasons for this are numerous, but they undoubtedly are related to the fact that it was often regarded as a state of little political and economic significance in the global scheme of things. And while some literature has been devoted to Czechoslovak presence at expos after the Second World War, especially in Brussels 1958 and Montreal 1967, the interwar period remained understudied.[8]

The careful examination of the different aspects of Czechoslovak participation offered here is based on in-depth archival research and reveals not only one state's trajectory through the interwar world and world's fairs but also sheds light on the important political, economic, social, and cultural circumstances that shaped these events. The attempts by successive Czechoslovak governments to be part of every large exhibition of the interwar period indicate that the state had the ambition to become a recognised player in global affairs. I argue that this was done by two methods: on the one hand, the state put strong emphasis on presenting itself as a new entity that took a democratic direction. On the other hand, it benefitted from the industrial and cultural legacies of the Habsburg empire it could build on. The selective inclusions and exclusions of objects, people and ideas for display in Czechoslovak pavilions therefore demonstrate the extent to which national presentations, as well as nations, reflected

7 Martin Wörner, *Vergnügung und Belehrung: Volkskultur auf den Weltausstellungen 1851–1900* (Münster: Waxmann, 1999); Peter H. Hoffenberg, *An Empire on Display: English, Indian and Australian Exhibitions from the Crystal Palace to the Great War* (Berkeley: University of California Press, 2001); Sven Schuster, *Die Inszenierung der Nation. Das Kaiserreich Brasilienim Zeitalterder Weltausstellungen* (Frankfurt am Main: Peter Lang, 2015); M. Elizabeth Boone, *"The Spanish Element in Our Nationality," Spain and America at the World's Fairs and Centennial Celebrations, 1876–1915* (University Park, PA: Penn State University Press, 2021); Robert W. Rydell, *All the World's a Fair: Visions of Empire at American International Expositions, 1876–1916* (Chicago: Chicago University Press, 1984).

8 Pavlína Morganová, Terezie Nekvindová, and Dagmar Svatošová, *Výstava jako médium: české umění 1957–1999* (Prague: AVU, 2020); Martina Mertová and Rostislav Švácha, *Byli jsme světoví!: Expo Brusel, Montreal, Ósaka z archivu architekta Miroslava Řepy: pohledy do sbírek Muzea umění Olomouc = We were world-class! : Expo Brussels, Montreal and Ósaka from the archive of the architect Miroslav Řepa: collection insights* (Olomouc: Muzeum umění Olomouc, 2022); Cathleen M. Giustino, "Industrial Design and the Czechoslovak Pavilion at EXPO '58: Artistic Autonomy, Party Control and Cold War Common Ground," *Journal of Contemporary History* 47, no. 1 (2012): 185–212; Daniela Kramerová and Terezie Nekvindová, *Automat na výstavu: československý pavilon na Expo 67 v Montrealu = The automatic exhibition: the Czechoslovak pavilion at Expo 67 in Montreal* (Cheb: Galerie výtvarného umění, 2017).

the careful crafting of narratives designed to underpin the construction and promotion of a putative identity.

A new state and a new nation?

Czechoslovakia was formed as an independent state by the Declaration of Independence of the Czechoslovak Nation by its Provisional Government, drafted in the USA by the politician and philosopher Tomáš Garrigue Masaryk (1850–1937) and published on October 18, 1918. It created a union of Czechs and Slovaks in a single state with a view that Slovaks would eventually gain autonomy.[9] The Czech and Slovak National Council, formed during the war and working from Paris, formed the basis of the first Czechoslovak government, Masaryk became the President and was supported by his close aids in France—Edvard Beneš as foreign minister and Milan Rastislav Štefánik as minister of war.[10] The territory of the new state was composed of Bohemia, Moravia, Slovakia and Sub-Carpathian Ruthenia. The latter region joined Czechoslovakia following negotiations that started in 1917 in the USA between Masaryk, the American President Woodrow Wilson and Rusyn-Americans in a temporary solution. The largely rural territory of Sub-Carpathian Ruthenia remained under direct Czechoslovak governance until 1924 when the Ruthenians gained representation (however small) in the national parliament. Autonomy of the region, promised in the so-called Uzhhorod memorandum in 1919, however, was not achieved until late 1938.

Apart from the Ruthenians, many other minorities, most importantly Germans, Hungarians, Poles, and Jews, lived in the new Czechoslovak state. Germans comprised of some 23% of the inhabitants of the new state, and Hungarians about 12%.[11] The idea of a single Czechoslovak nation and people was therefore devised by politicians and backed by historians already during the war. It was a conscious strategy of Czech and Slovak diplomats to be presented under a single identity so as not to confuse the politicians of the Alliance, who were expected to be unfamiliar with the history and complex ethnic composition of

9 Pavel Kosatík, *Slovenské století* (Prague: Torst, 2021), 85–86.

10 Brent Mueggenberg, *The Czecho-Slovak Struggle for Independence, 1914–1920* (Jefferson, North Carolina: McFarland, 2014), 206.

11 Zdeněk Kárník, "Výsledky sčítání lidu 1910 a 1021. Němci v Českých zemích, jejich počet a relativní čísla," in *České země v éře První republiky 1918–1938. I. Vznik, budování a zlatá léta republiky (1918–1929)* (Prague: Libri, 2000), 89.

Central Europe.[12] Czech—and to a lesser extent Slovak—politicians, journalists and scholars would refer to the "Czechoslovak nation" that formed a Slavic majority, and it was implicitly set against the substantial population of Germans and Hungarians. Apart from a shared "Czechoslovak" culture, they pointed to the similarities of the Czech and Slovak languages and identified shared historical moments in Czech and Slovak history and culture.[13]

However, there was no consistent or binding definition of any nation, whether Czechoslovak, Czech, or Slovak.[14] The Constitution of 1920 only codified the idea of a Czechoslovak language when it included a section on national, religious and ethnic minorities, securing their rights.[15] In practice, this meant that Czech and Slovak were seen as two branches of the same Czechoslovak language and were used in the Czech and Slovak regions respectively. At the same time, the concept of Czechoslovakism relied on the Czech idea of national self-determination and the conception of a Czech nation, which predated the vision of creating an independent state. The ultimate decision for a joint Czechoslovak state was a solution to the realization of the limited size of each individual group that sought autonomy or independence from the Habsburg Monarchy.

The union of the ethnic groups in Czechoslovakia was not to last. Slovak, German, Hungarian and Polish nationalism within the republic and outside of its territory intensified during the late 1920s and eventually caused a weakening of the state's internal and international position. The Second Czech-Slovak Republic, created after the Munich Agreement in September 1938, which ceded the border regions of the Sudetenland to Germany, introduced more than a hyphen in the name of the state to indicate Slovak emancipation. It was also a start to the eventual breakup of the country the following year into the Protectorate of Bohemia and Moravia, on the one hand, and the Slovak Republic on the other hand. Sub-Carpathian Ruthenia, with a large Hungarian minority, fell under Hungarian administration.

---

12 Ladislav Holý, *The Little Czech and the Great Czech Nation. National Identity and the Post-Communist Social Transformation* (Cambridge: Cambridge University Press 1996), 95.

13 For example, Albert Pražák, *Československý národ* (Prague: Akademie, 1925); Leopold Weigner, *Lidové umění československé* (Prague: Josef Vilímek, 1917); Renata Tyršová, *Svéráz v zemích československých* (Plzeň: Český deník, 1921); Zdeněk Wirth ed., *Československé umění* (Prague: Vesmír, 1926).

14 Milan Ducháček, "Čechoslovakismus v prvním poločase ČSR: státotvorný koncept nebo floskule?," in *Čechoslovakismus*, edited by Michal Kopeček, Jan Mervart, Adam Hudek (Prague: NLN, 2019), 158.

15 Act 122/1920 Coll. article 129 of the Constitution, on the principles of the language right in the Czechoslovak republic, passed on 29 February 1920 and valid as of 6 March 1920, published in the Code no. 26/1920, 268.

The simplifications of the ethnic composition in the notion of the Czechoslovak nation had a direct consequence in the selective way Czechoslovakia was presented at world's fairs. As I argue throughout the book, *Czechoslovak* in the national displays was often synonymous with Czech and the industrial heartlands of Bohemia. This was grounded in the latter's economic and political history going back to the Habsburg monarchy. Bohemia, along with Upper and Lower Austria, and Vienna, had been a highly industrialised region and it enjoyed a boom of new industries from the 1890s onwards. Its private enterprises and agriculture were on the rise and the region started developing national capital in the form of its own (Czech) banks and financial institutions that could compete with Austro-German ones.[16]

After the creation of Czechoslovakia, Bohemia continued to be undeniably more industrialised, with about 90% of all Czechoslovak industrial production concentrated here.[17] Slovakia and Sub-Carpathian Ruthenia, on the other hand, remained mainly agricultural. In the narrative constructed for the world's fairs, the predominantly rural territories in the east of the state would be commonly reduced to a rather simplified display of the areas as embedded in agriculture and folk culture. Proportionately, Slovakia was usually harshly underrepresented.[18] Moreover, Czechoslovak displays also hardly ever featured other minorities and it was only Germans who enjoyed representation through various trading organizations and businesses that participated in the pavilions.

The small Czechoslovak nation

The way many politicians, writers and artists envisaged Czechoslovakia can be demonstrated by a short text authored by the writer Karel Čapek (1890-1938). Extracts of the article were used in Czechoslovak national presentations at the International Exposition of Art and Technology in Modern Life in Paris in 1937 and the World of Tomorrow in New York in 1939.[19] The acclaimed writer de-

16 Eduard Kubů and Jaroslav Pátek, "Základní charakteristika výchozí situace a vývoje československé ekonomiky v meziválečném období," in *Mýtus a realita hospodářské vyspělosti Československa mezi světovými válkami*, eds. Eduard Kubů and Jaroslav Pátek (Prague: Karolinum, 2000), 10.

17 Kubů and Pátek, "Základní charakteristika," 16.

18 Pavol Komora, *Hospodárske a všeobecné výstavy 1842–1940* (Bratislava: Slovenské národné muzeum, 2017), 233.

19 Karel Čapek, "Hle, Československo," *Srdce Evropy* I, unpag. (1936), reprinted in *Od člověka k člověku* III. Prague: Československý spisovatel; B. Soumar, ed., *Tchecoslovaquie à l'Exposition Internationale des Arts et Techniques dans la Vie Moderne Paris 1937* (Prague: Melantrich, 1937).

scribed the state as a land of contradictions between modernity, civilization and industrialization on the one hand, and primitive naivety and traditions on the other. This, indeed, was the putative divide between west and east, between Bohemia and Slovakia respectively. Čapek also defined Czechoslovakia as a land at the intersection of history, a site of political, religious and cultural crossroads, and an island. Throughout modern history, Czechoslovakia has often been referred to as an "island of democracy and freedoms," as well as an "island of order and calm," as Čapek calls it in his text.[20]

Such accounts contributed to the myth of a democratic, liberal and peaceful state, tolerant to its minorities and not eliciting conflicts. Nevertheless, some contemporary historians and philosophers like Josef Pekař and Emanuel Rádl and more recently historians, such as Andrea Orzoff and others, have challenged the traditional nationalist rhetoric of Czechs which saw them as suffering under Habsburg rule, eventually liberating themselves, their language and culture in 1918 in an embrace of democratic values.[21]

This myth-making was promoted not only by the so-called "Castle," the term used for the informal group of institutions and allies of T. G. Masaryk and the foreign minister Edvard Beneš, but also more widely and popularly in literature, the arts, and—of course—exhibitions.[22] As a close collaborator and friend of Masaryk, Čapek was an ardent advocate of Czechoslovakia as a small, but capable state with a liberal democracy and a "hardworking and valiant people, whose risorgimento is not yet complete, and who still remain capable of astonishing the world."[23] It is one of the aims of this book to look beyond these myths created by politicians, writers, and artists as well as those involved in constructing the Czechoslovak national displays.

20 Karel Čapek, "La visage de la Tchécoslovaquie," in *Tchecoslovaquie à l'Exposition*, 15–21.

21 For example, Josef Pekař, *Smysl českých dějin* (Prague: v.n., 1929); Emanuel Rádl, *O smysl našich dějin předpoklady k diskusi o této otázce* (Prague: Čin, 1925); Andrea Orzoff, *Battle for the Castle: The Myth of Czechoslovakia in Europe, 1914–1948* (Oxford: Oxford University Press, 2009); Mary Heimann, *Czechoslovakia, The State that Failed* (Yale University Press, 1997); Ladislav Holý, *The Little Czech and the Great Czech Nation: national identity and the post-communist transformation of society* (Cambridge: Cambridge University Press 1996); Robert Pynsent, *Questions of Identity: Czech and Slovak Ideas of Nationality and Personality* (Budapest: Central European University Press, 1994); Ondřej Slačálek "The Postcolonial Hypothesis. Notes on the Czech 'Central' European Identity," *Annual of Language & Politics & Politics of Identity* 10, no. 10 (2016): 27–44.

22 Michal Kopeček, "Czechoslovak Interwar Democracy and its Critical Introspections," *Journal of Modern European History / Zeitschrift für moderne europäische Geschichte / Revue d'histoire européenne contemporaine* 17, No. 1 (February 2019): 7–15. See also Karel Čapek, *O demokracii, novinách a českých poměrech*, ed. Ivan Klíma (Prague: Academia, 2003); Karel Čapek, *Talks with T.G. Masaryk*, transl. by M. H. Heim and D. Round (North Haven: Catbird Press, 1995).

23 Čapek, *Tchecoslovaquie*, 21.

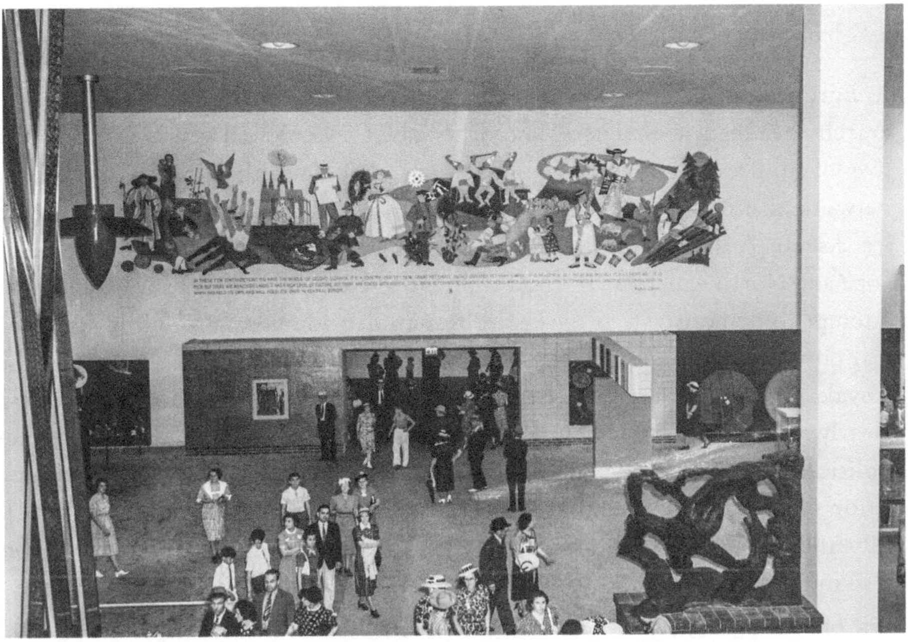

Fig. 1. Antonín Strnadel and Josef Novák, collage of Czechoslovakia, Czechoslovak pavilion, New York world's fair, 1939.

When Čapek's quotation about the small but civilised and cultured nation appeared in New York, it was underneath a collage which visualised many of the stereotypes via which Czechoslovakia was presented [fig. 1].[24] The collage showed the "typical" cultural phenomena of the entire country extending from the western border with Germany to the eastern regions of the Sub-Carpathian Ruthenia. The westernmost regions were represented by an inhabitant of Chodsko clad in a heavy coat—the Chods were the historic protectors of the borders with Germany. A group of Sokol gymnasts followed while Prague was represented by the castle with middle-class inhabitants. Halfway through the image, more rural scenes appeared with mountainous and agricultural regions of eastern Moravia and Slovakia. These were symbolised by peasants either attending to animals or engaged in dancing or playing a musical instrument. The easternmost Sub-Carpathian Ruthenia was depicted as a land of forests where timber was mined. Both the collage and the quotation extended the simplified reading of the country's character using the west-east hierarchies and stereotypes about cultured Bohemia and primitive Slovakia, so often replicated in exhibitions.

24 The work was designed by the artists and professors of the School of Decorative Arts in Prague, Antonín Strnadel and Josef Novák.

Selective hierarchies

In Europe and the USA, world's fairs and colonial expositions established a hierarchy between "westerners" and others, be they colonial subjects, representatives of cultures unknown in the "west," or indigenous inhabitants.[25] Indeed, the network also included international and universal exhibitions in South America, Asia or Australia, but for Czechoslovakia, most crucial venues were in Europe and North America. The hierarchies presented here were motivated by the attempt to demonstrate progress in human development, economic and political power, and the superiority of western culture in general. As the Czechoslovak government entered these established classifications as a newcomer, it clearly associated itself with the western paradigm. It had already made its key political allies during the First World War when negotiating the future composition of Central Europe and subscribed to their version of democracy, modernity and attitudes to minorities. Alliances were forged not only in politics but also in the fields of the visual arts and science, all crucial areas when constructing a nation, national vision or a national pavilion.

Through their exhibition elites, world's fairs as well as national pavilions tried to influence what should be selected for display and what should be excluded. In this regard, Bauman's comparison of state control with gardening offers a way of reading such practice.[26] In a garden, one can control what should be planted, let thrive or what should be cut back, and uprooted. Bauman has applied this to state apparatuses (or any institution of power) and their control of, e.g., population and society. He has focused on the extremist consequences of such an approach leading to the Second World War genocide, but he also pointed out the global nature of the science of human heredity and the art of human breeding, in other words, eugenics. "Ordering—planning the order and executing it—is essentially a rational activity, attuned to the principles of modern science, and,

25 Zeynep Çelik, *Displaying the Orient: Architecture of Islam at Nineteenth-Century World's Fairs* (Berkeley: University of California Press, 1992); Sadiah Qureshi, *Peoples on Parade: Exhibitions, Empire and Anthropology in Nineteenth-Century Britain* (Chicago: Chicago University Press, 2011); Nicolas Bancel et al. eds. *Zoos humains. Autemps des exhibitions humaines* (Paris: La Découverte, 2002); Dana S. Hale, *Races on Display: French Representations of Colonized Peoples, 1886–1940* (Bloomington: Indiana University Press, 2008); Matthew F. Bokovoy, *The San Diego World's Fairs and Southwestern Memory, 1880–1940* (Albuquerque: University of New Mexico Press, 2005); Nancy J. Parezo and Don D. Fowler, *Anthropology Goes to the Fair: The 1904 Louisiana Purchase Exposition* (Lincoln: University of Nebraska Press, 2007); James Gilbert, *Whose Fair? Experience, Memory and the History of the Great St. Louis Exposition* (Chicago: Chicago University Press, 2009).

26 Zygmunt Bauman, *Modernity and the Holocaust* (Cambridge: Polity Press, 1989).

more generally, to the spirit of modernity."[27] As I argue in this book, the national pavilion, just like Bauman's "garden," is planned and designed but also interfered with from many angles.

In *Czechoslovakia at the World's Fairs*, I also build on the assertions of scholars like Marius Turda and Roger Griffin who have proposed, in the wake of Bauman, that many of the states of central and eastern Europe informed their sense of nationhood with ideas derived from eugenics.[28] Czechoslovakia was no exception, and it brought these ideas to its national pavilions. In the early 1930s, eugenics as a science was implemented in many countries across the globe. Legislation on sterilization, racial laws, foundation of research and medical institutes were packaged as driven towards welfare and better societies. Creating comparative hierarchies between different groups of people, whether social or ethnic, was also common. Colonial exhibitions were particularly full of displays that stressed the need for retaining or expanding colonies in the name of a civilizing mission.

In Europe, exhibitions like the Ibero-American Exposition in Seville in 1929, the Empire Exhibition in London in 1924–25, and the Colonial Exhibition in Paris in 1931, included displays of improvement to local infrastructures, education and healthcare in the colonies. Re-creations and re-imaginations of native dwellings, places of worship and whole villages were commonplace often inhabited by indigenous people recruited to perform various tasks and entertain the visitors. Clad in what was conceived as native dress, their primitivism was exaggerated. These practices were not reserved only for colonial subjects; the exoticization of domestic rural people and their cultures also regularly appeared in exhibitions and world's fairs. They were driven by attempts to display the picturesque quality and cultural diversity of the country in question as well as different levels of culture and civilization.[29] Such an approach was also upheld in the Czechoslovak pavilions, including their adjacent restaurants, where female attendants would be clad in folk dresses in order to provide an aura of exoti-

---

27 Zygmunt Bauman, *Modernity and Ambivalence* (Oxford: Polity Press, 1991), 38.

28 Roger Griffin, "Tunnel Visions and Mysterious Trees: Modernist Projects of National and Racial Regeneration: 1880–1939," in *Blood and Homeland: Eugenics and Racial Nationalism in Central and Southeastern Europe, 1900–1940*, eds. Marius Turda and Paul J. Weindling (Budapest–New York: CEU Press, 2007), 418; Marius Turda, "Eugenics and Biopolitics, 1933–1940," in *Modernism and Eugenics*, 117 and other texts.

29 Eric Storm, "The Transnational Construction of National Identities: A Classification of National Pavilions at World Fairs," in *World Fairs and the Global Moulding of National Identities: International Exhibitions as Cultural Platforms, 1851–1958*, eds. Joseph Leerssen and Eric Storm (Leiden: Brill, 2022), 53–83; Dagnosław Demski and Dominika Czarnecka, eds. *Staged Otherness: Ethnic Shows in Central and Eastern Europe 1850–1939* (Budapest–New York: Central European University Press, 2022).

cized tradition, inviting a further examination of the relationship between gender and primitivism.

Book structure

Czechoslovakia was, indeed, not an exploited colony that would be displayed by means of a pretence of a reconstructed village with indigenous inhabitants and staged behaviour. Nor was it one of the largest political or economic powers that would showcase its possessions and greatness at world's fairs. Nevertheless, it was an ambitious country that invested impressive amounts of resources into taking part in all world's fairs of the interwar period, while selecting and ordering the key aspects of its identity for display. It occupied a space between the unknown and exotic, usually reserved for non-European participants, and the modern and progressive.

Competition in this space was harsh because many other new countries endeavoured to offer their products, cultures and identities. These were new entities like Poland, Austria, Hungary, Yugoslavia, or Ireland, to mention just a few examples from Europe. Austria, Hungary, and Yugoslavia, specifically, are my main comparative examples, as they shared many similar starting points after the war. Their political development, however, took a different course from Czechoslovakia; Austrofascism developed in Austria and Miklós Horthy introduced an authoritarian regime in Hungary. In the Kingdom of Yugoslavia, which also consisted of diverse ethnic groups like Czechoslovakia, leaders also moved away from democracy toward a dictatorial regime.

Czechoslovakia at the World's Fairs interrogates how the process of political, societal and economic modernization that Czechoslovakia underwent was displayed at world's fairs and specific international exhibitions, from the visual representation in architecture and arts to the displays of modern technology and scientific visions of a modern nation. Most of the chapters that follow do not follow a chronological development of Czechoslovak participation at world's fairs and international exhibitions but instead are focused thematically. The exception is Chapter One, "Continuities and Ruptures: Interwar Displays," where I trace the consequences that the breakdown of the Habsburg empire caused for successor states like Czechoslovakia and their representation at world's fairs. Although Czechoslovakia only came into existence in 1918, its participation in world fairs was informed by a long history of trade and national exhibitions in Bohemia and, more widely, Austria-Hungary that had been taking place since the mid-eighteenth century. On the background of the political, economic and social circumstances of the interwar period, I examine the foundations and contents of these exhibitions, looking for the patterns of continuities and ruptures amongst them. The historical overview of interwar world's fairs and the individual national pavilions sets out the discussion of the trajectory Czechoslovakia followed in the short span of not even twenty years.

The main issues that Czechoslovak representations faced included practical problems of space organization and selection of objects. Chapter Two, "Exhibition Spaces," explores the physical designs of the national pavilions that reflected political, commercial, and artistic needs of the time and aimed at creating exhibitionary order. The concerns of architects and designers of how to entice visitors in its physical and mental spaces of display and keep them there became a pressing issue for everyone. While they were limited by the interests of the state and the exhibitors, designers applied complex theories about space distribution, visitors' psychology and information communication, as well as inspiration from other disciplines like graphic or stage design. They were inevitably responsible for the most visible aspects of the national presentations.

Chapter Three, "Art and Design: The Limits of Modernism," focuses on the visual arts displayed in Czechoslovak pavilions and on the fluid understanding of what *modern* means. The artistic displays throughout the interwar period were driven by the question of what kind of art should represent the state. Embracing modernity and progress was important for the Czechoslovak organizers and while architecture of the national pavilion was the first and most visible feature of the official presentation, the modernity of the display was also projected through fine art and design. Yet as I argue in this chapter, *modern* often did not stand for radical or avant-garde. The attempt to find the right balance between international and national forms of visual expression informed the selection of artworks. Alongside its modernity, the state tried to proclaim its historical legitimacy and traditions through its visual arts, most visibly at the 1925 Exposition in Paris where crafts and decorative tendencies had a prominent place. It also found an important medium in the inclusion of folk art and culture, which were a fitting vehicle for displaying the country's distinctiveness as well as a means of including Slovakia in the displays.

In Chapter Four, "Commodifying the State," the book examines what *Czechoslovak* meant in relation to commercial goods. In the volatile environment of the interwar period, marked especially by postwar shortages, the Great Depression and German expansionism, many large and small companies also used Czechoslovak pavilions to promote their products to find new markets. The analysis of four case studies of Detva textiles, Moser glass, Pilsner beer and Baťa shoes shows that the content of the national presentation was largely shaped by business elites in which the showcasing "national" products was intertwined with business interests. These interests were, however, not always aligned with those of the state, as in the case of the shoe company Baťa, which proposed its own vision of the national pavilion and its focus for the world's fair in New York. The involvement of the four companies ultimately demonstrates how they navigated the political and economic challenges of the interwar period.

Apart from businesses and the government, the content of the pavilions was created by numerous individuals and interest groups. In Chapter Five, "People of the Fair," I turn the focus away from the exhibition elites to actors whose interventions received smaller exposure but helped to shape the content of the pavilions. These less visible, yet formative actors were not a uniform but a diverse group, including members of émigré communities, especially in the USA, the workers who helped to design and build the displays, and the performers who participated here. American Czechs and Slovaks participated not only in the physical construction of the national pavilions but also in the mental formation of the idea of the Czechoslovak nation. The idea of Czechoslovakia was born abroad at the end of the First World War and émigré communities especially in the USA retained influential views of the new political entity and identity. They were often involved in constructing the ideological content of the national pavilions, shaping their material as well as ideological content.

I use the discussion of the diaspora to further consider the topic of inclusion and exclusion at world's fairs. I draw attention to those who were employed by the exhibitions, as well as to visitors and journalists, for example. Their views of the way the state and the nation should be presented revealed alternative views of Czechoslovakia to the official narratives. Many of the employees in Czechoslovak, and indeed other pavilions, were women. In the Czechoslovak case, however, women predominantly occupied service positions; they were waitresses and sales attendants, performers and objects of display through their products or bodies, and they attended as visitors and observers. In a few cases, women were designers and exhibitors, but in general, their presence in Czechoslovak pavilions beyond the services like sales and waitressing was marginal. Set against the debates on gender and emancipation held in Czechoslovakia at the time, I argue that the exclusions of women from the Czechoslovak representations contradict the image of Czechoslovak democracy constructed in the fairs.

Alongside gender and class, the ethnic diversity of Czechoslovakia was a complex matter that exhibition organizers did *not* readily display at world's fairs. The Czechoslovak majority dominated at the expense of the minorities of Germans and Hungarians, as well as Jews, Poles, Ruthenians, and Roma. In Chapter

Six, "Race, Science and Ethnicity: Molding the Ideal Czechoslovak," I interrogate the ethnic construction of the unified Czechoslovak nation and put it into the context of international discussions about race and ethnicity during the interwar period. Fairs were often used to impress messages about race upon visitors; these included a condoning of colonial expansionism, white supremacy and racial hierarchy. Although Czechoslovakia had no colonies, these issues were still relevant to it when one considers the conscious ethnic and racial hierarchies of the displays.[30] The chapter scrutinises especially the relation between the efforts to present Czechoslovakia as a young and healthy nation based on theories of social Darwinism and eugenics that were widespread in the 1920s and 1930s.

In the Czechoslovak pavilions, health establishments were promoted next to depictions of the latest housing projects and social care facilities that the state, and in some cases private enterprises, built not only in the name of modernization and progress but also future improved society. Throughout chapter and the book, I argue that the concerns with presenting the nation as young and progressive were linked to the interest in the betterment of *humanity* in the first half of the 20th century, or, in other words, eugenics.[31] Such alignment with eugenics brought dangerous ideas into Czechoslovak politics, scholarship, and, indeed, exhibitions in the form of justification of the biological, social and cultural hierarchies between people. By examining concrete examples, I reveal that as part of the international eugenic community, Czechoslovakia brought these ideas either explicitly or implicitly to its presentations at international exhibitions.

The engagement in eugenics of many scholars and politicians in Czechoslovakia lead to a selective presentation of material and facts. One of the main concerns of *Czechoslovakia at the World's Fairs* is the way information and ideas were shown and visualised. Diagrams and photographs explicitly or implicitly linked to eugenics are one example, display of heavy machinery, modern art, touristic destinations, or food and drink as Czechoslovak another. National presentations become specific forms of communication where the repetition of the same narrative, i.e., display of the same or similar message in successive national pavilions, endorsed and normalised such modes of representation.[32]

30 Mark A. Brandon, *The Perils of Race-Thinking: A Portrait of Aleš Hrdlička* (Budapest–New York: Central European University Press, 2023).

31 Marius Turda, *Modernism and Eugenics* (Basingstoke: Palgrave Macmillan, 2010); Marius Turda, ed. *The History of East-Central European Eugenics. Sources and Commentaries* (London – New York: Bloomsbury Academic, 2015).

32 Jean Baudrillard, *Simulacra and Simulation* (New York: Semiotext(e), 1983), Susan Sontag, *On Photography* (London: Penguin Books, 1977).

How exceptional was Czechoslovakia in its efforts to be at every single world's fair with its products, arts and visions? The Conclusion, and the entire book, claim that behind the primary concern with establishing trade relations at the world's fairs was a complex array of motivations about what kind of Czechoslovakia should be displayed, who should represent it and how. These drives created a discrepancy between the official portrayal of the state and the agendas of individuals, companies and organizations. The organizers, consisting of the political, business and cultural elites, used the means of visual diplomacy to validate Czechoslovakia as a new, modern, democratic state and partner for the external audiences. Interests of various other actors from outside of these closed circles of the exhibition elites disrupted the seemingly perfect picture. The questions of inclusions and exclusions whether of race, gender or class affected many individuals and groups who took part in the world's fairs. Czechoslovakia with its insistence on showcasing democratic principles throughout the entire interwar period makes a great example of the volatility that was taking place behind the official façade of its pavilions. The attempt to place Czechoslovakia through its pavilions in the world order becomes symptomatic of the efforts of many other countries and new states who were not the largest players in world's fairs.

Chapter One

CONTINUITIES AND RUPTURES

EXHIBITING A MODERN STATE

From their early days in the 1840s and 1850s, international exhibitions developed into complex events with the participation of many countries. The increasing size of exhibitions, as well as the need to distinguish between various national economies and cultures, lead to the construction of national pavilions. The *Exposition Universelle* in Paris of 1867 was the first event that saw such distribution of nations into separate buildings.[1] Such division instilled order at the exhibition ground and was adopted by most later fairs. Clear national differentiation became particularly important at specific moments linked to a political rupture, such as a redrawing of state frontiers after world wars or the dissolution of empires. Soon after Czechoslovakia was formed in 1918, it started using world's fairs to find its own trading opportunities. These events were also used to project, and to an extent consolidate, the country's image of a new state to the rest of the world.

As an independent state, Czechoslovakia took to the international world's fairs stage quickly. Unlike any other successor state of the Habsburg Empire, Czechoslovakia participated in all major interwar world's fairs and international exhibitions between 1920 and 1940; this was around twenty events in

Europe, and North and South America. For comparison, Austria took part in five interwar exhibitions (Paris 1925 and 1937, Philadelphia 1926, Barcelona 1929 and Brussels 1935), and Hungary in six (Paris 1925 and 1937, Philadelphia 1926, Brussels, New York and San Francisco 1939–40). Poland was also present in Paris 1925 and 1937, in Brussels, and New York; while The Kingdom of Serbs,

1 Irena R. Makaryk, *April in Paris: Theatricality, Modernism and Politics at the 1925 Art Deco Expo* (Toronto: Toronto University Press, 2018).

Croats and Slovenes (from 1929 the Kingdom of Yugoslavia) appeared in both Paris exhibitions, Barcelona, and New York.[2] And finally, Germany, one of the most important trading partners and competitors for Czechoslovakia, built its official pavilions only in Barcelona and Paris 1937.

Why was Czechoslovakia exceptional, taking part in so many exhibitions abroad? There are several reasons which are related, on the one hand, to the ability of the new state to continue various practices from Austria-Hungary rather than having to invent them anew. Exhibiting abroad was one of them. On the other hand, Czechoslovakia broke off ties with the monarchy and was able to present itself as a new democratic and forward-looking state. The gravity the Czechoslovak state saw in the participation could therefore be read as a result of the relative stability of Czechoslovak politics and economy and as a sign of the confidence of the exhibition elites that put forward these costly presentations.

As mentioned in the Introduction, many historians, politicians and writers, including Karel Čapek, called Czechoslovakia the "island of democracy."[3] This suggestion of political isolation stands out especially in comparison with neighboring Austria, Hungary and Germany who had to grapple with fascism and extreme nationalism. All the countries formed out of the Habsburg Monarchy saw the collapse of their democratic systems during the interwar period.[4] Except for Czechoslovakia. Indeed, the state had its own severe problems, which included growing ethnic conflicts and autonomy claims from Slovaks, Germans and Ruthenians, yet it managed to retain its democracy until the Second World War.[5]

The multi-ethnic composition of Czechoslovakia was a defining feature of the state that linked it with the Habsburg monarchy.[6] The fall of Austria-Hungary was a political turning point for central Europe, yet many other continuities underlie the transition from an imperial subject to an independent democracy.

---

2 Aleksandra Stamenković, "Yugoslav Pavilions at International Exhibitions in Artistic and Political Discourse 1918–1941," *Istorija 20. veka* 40, no. 2 (2022): 301–322.

3 Karel Čapek, "Hle, Československo", *Srdce Evropy* I, unpag. (1936), reprinted in *Od člověka k člověku* III. Prague: Československý spisovatel; Karel Čapek, "La visage de la Tchécoslovaquie," in *Tchécoslovaquie à l'Exposition Internationale des Arts et Techniques dans la Vie Moderne Paris 1937*, ed. B. Soumar (Prague: Melantrich, 1937), 15–21.

4 Sabrina P. Ramet, *Interwar East-Central Europe, 1918–1941* (Abingdon, Oxon – New York: Routledge, 2020); Joseph Rothschild, *East Central Europe Between the Two World Wars* (London: Routledge, 2015).

5 Mark Cornwall and R. J. W. Evans, eds., *Czechoslovakia in a Nationalist and Fascist Europe 1918–1948* (Oxford: Oxford University Press, 2007); Mary Heimann, *Czechoslovakia, The State that Failed* (Yale University Press, 1997); Andrea Orzoff, *Battle for the Castle: The Myth of Czechoslovakia in Europe, 1914–1948* (Oxford: Oxford University Press, 2009); Zdeněk Kárník, *České země v éře První republiky 1918–1938. I. Vznik, budování a zlatá léta republiky (1918–1929)* (Prague: Libri, 2000).

6 Pieter M. Judson, "Epilogue: The New Empires," *The Habsburg Empire: A New History* (Cambridge, Massachusetts: The Belknap Press of Harvard University Press, 2016), 442–52.

As much as the dissolution of the monarchy was not one definable moment in time but rather a series of events that led to the collapse of the empire, Czechoslovakia emerged as a protruding negotiation of the state's identity, status and independence. Interwar Czechoslovakia inherited and continued many political processes and strategies from Austria-Hungary, while various bureaucratic practices and Austrian laws were retained in the Czechoslovak system until the early 1930s and only gradually replaced.[7]

Despite the gradual emancipation, the economic relations and bonds within Austria-Hungary were tight and the political restructuring of Central Europe after the war had serious consequences for all parts of the former monarchy, including Czechoslovakia. While in Austria-Hungary, the Czech lands and Slovakia followed different trajectories, these needed to be united into a joint legal, economic and social systems after 1918. The first years of the new republic were therefore marked by rapid economic changes, which I discuss in more detail in Chapter Two. In this regard, Czechoslovakia was quite successful and the quick introduction of economic measures by the state, such as new tariffs and currency, or repatriation of capital led to relative financial stability of the new state.[8] Modernization became key for building the new state and its marketing at international exhibitions. Apart from the financial reforms, improvements of infrastructure in the state, such as building of roads and railways, as well as electrification of the country played a key role in connecting the different parts of the country in the name of progress. Yet, this progress and modernization did not come equally to all parts and reached the eastern parts of Slovakia and Sub-Carpathian Ruthenia, with their large rural areas, at a much slower pace.

I explore the topics of inequalities, modernization, as well as economic and political influences throughout the book in relation to the specific issues of representation of minorities, arts and businesses. First, however, it is important to

7 Orzoff, *Battle*, 129; Vít Vlnas, "Mýty a kýče první republiky." *Nová Přítomnost* 8 (1991): 28–29; Robert Pynsent, *Questions of Identity: Czech and Slovak Ideas of Nationality and Personality* (Budapest: CEU Press, 1994); Claire Morelon, "State Legitimacy and Continuity," in *Embers of Empire: Continuity and Rupture in the Habsburg Successor States after 1918*, eds. Paul Miller and Claire Morelon (New York–Oxford: Berghahn Books, 2017), 43–63; Kárník, *České země* I, 106–112; Julia Secklehner, "Crossing Borders and Period Boundaries in Central European Art: The Work of Anna Lesznai (ca. 1910–1930)," in *Rethinking Period Boundaries: New Approaches to Continuity and Discontinuity in Modern European History and Culture*, eds. Lucian George and Jade McGlynn, 119–148 (Oldenbourg: De Gruyter, 2022), doi:10.1515/9783110636000-005; Nóra Veszprémi, "Universalizing The Local? Museums, Centres And Peripheries," *Revista da História da Arte. Instituto de História da Arte* 9, no. 2 (2021): 99–108.

8 Eduard Kubů and Jaroslav Pátek, "Základní charakteristika výchozí situace a vývoje československé ekonomiky v meziválečném období," in *Mýtus a realita hospodářské vyspělosti Československa mezi světovými válkami*, eds. Eduard Kubů and Jaroslav Pátek (Prague: Karolinum, 2000), 13.

outline in a more historical sequence the continuities and ruptures of the exhibiting efforts which informed the individual displays. While there were obvious precedents of exhibition participation and organization from Austria-Hungary that Czechoslovakia could build on, the new state also needed to provide a clear message about its uniqueness and new orientation. Continuities and ruptures can also be detected between the different exhibitions in the interwar period in which Czechoslovakia took part as each participation in a world's fair saw a different set of debates about the content and rationale of the national exhibits.

Grounds of exhibitions

After the Dual monarchy was created in 1867, the Czech lands (the territory of Bohemia, Moravia and a part of Silesia) were governed from Vienna, while Hungarians administered Upper Hungary (the future Slovakia) and Sub-Carpathian Ruthenia. Austria, or Cisleithania, and Hungary, or Transleithania, were both participants at exhibitions abroad and they both organized their own large domestic exhibitions. The Weltausstellung in Vienna in 1873 became an immense display of science and industry, which was aimed at showing the Habsburg monarchy as a great political and economic power.[9] The Millennium Exhibition in Budapest in 1896 celebrated 1,000 years since the arrival of the Magyar tribes led by Arpad to the Carpathian basin.[10] This event was made an integral part of the great national past, which was linked to the present and to achievements in economy, culture, and education.[11] The Magyars were presented as the protectors of the various minorities in Greater Hungary in a way similar to how Austrians had been framed as the guardians of the Cisleithanian ethnicities in the 1873 Weltausstellung.[12]

9 Jutta Pemsel, *Die Wiener Weltausstellung von 1873: das gründerzeitliche Wien am Wendepunkt* (Vienna and Cologne: Böhlau, 1989); Matthew Rampley, "Peasants in Vienna. Ethnographic at the 1873 World's Fair," *Austrian History Yearbook* 42 (2011): 110–132; Martin Wörner, *Vergnügung und Belehrung: Volkskultur auf den Weltausstellungen, 1851–1900* (Münster–New York: Waxmann, 1999).

10 Samuel D. Albert, "The National for Itself: The 1896 Hungarian Millennium and the 1906 Romanian National General Exhibition," in *Cultures of International Exhibitions*, ed. Marta Filipová (Farnham: Ashgate Publishing, 2015), 113–136; Bálint Varga, "Anchoring a Millennium-Old Past in the Hungarian Minds," *The Monumental Nation: Magyar Nationalism and Symbolic Politics in Fin-de-siècle Hungary* (New York – Oxford: De Gruyter Berghahn 2016), 62–112.

11 Pavol Komora, *Hospodárske a všeobecné výstavy 1842–1940* (Bratislava: Slovenské národné muzeum, 2017), 123–132.

12 Matthew Rampley, "Peasants in Vienna. Ethnographic at the 1873 World's Fair," *Austrian History Yearbook* 42 (2011): 110–132; Matthew Rampley, "Vernacular Cultures and National Identities: The

The two parts of the Dual Monarchy also appeared separately at various world's fairs—for instance Hungary was represented at Paris 1900 and in St. Louis in 1904. In Paris in 1900, Austria also built a Bosnia and Hercegovina pavilion for a region, which had recently fallen under its administration, and both states contributed to the Panama Pacific Exposition in 1915. Bohemia and Slovakia, or rather Upper Hungary, most often appeared as integral parts of the two crown lands, although occasionally Bohemia was also represented separately before 1918. Objects labelled as Bohemian appeared, for example, at the Great Exhibition in London in 1851, the Paris Expositions Universelles, the Louisiana Purchase Exposition of 1904, and as far as Launceston, Tasmania, where a Bohemian section formed part of the International Exhibition in 1891–92.

Yet "Bohemian" at these occasions related to a geographical region rather than to a specific ethnic or national group. Moreover, many businesses had branches across Cisleithania irrespective of ethnic and linguistic borders. Before the war, companies from Bohemia exported abroad mainly via foreign companies and this situation carried into the post-war period.[13] When shortages of commodities immediately after the war occurred, new countries like Czechoslovakia felt the pressure to actively promote their foreign trade. World's fairs were therefore a good opportunity to foster new relationships, find new partners and introduce the existence of new states.

Participation in world's fairs was organized on the governmental level and for each fair, a cross-ministerial committee was formed. It consisted of representatives of the ministries most involved in exhibitions, determined by the nature of the exhibits, potential gains and invested finances: in the first place it was the Ministry of Trade, the Ministry of Foreign Affairs, the Ministry of Public Works and the Ministry of Education. The committees oversaw the selection and preparations of exhibits, as well as the focus of the representation. In 1927, the Ministry of Public Works and the Ministry of Education initiated an attempt to create a permanent committee that would secure continuity between the individual entries in world's fairs, but such a body was never formed. The discrepancies between the various ministries and their views of the principal purpose of the national presentation were just too large.[14] Instead, a new

---

Politics of Folk Art," *The Vienna School of Art History. Empire and the Politics of Scholarship, 1847–1918* (University Park: Pennsylvania University Press, 2013).

13 Ladislav Turnovský, "Československo na světové výstavě v Rio de Janeiro," *Národní listy*, February 10, 1923, 6.

14 "Návrh na utvoření stálé meziministerské komise pro organisaci státní účasti na výstavách v cizině," July 26, 1927, Organisace Československé účasti na výstavách v cizině, kD2519/7, APRO

committee would be composed for each Czechoslovak entry into a world's fair to oversee the organization under new political and economic circumstances.

The elementary debate between ministries revolved around the question of whether a national pavilion should have an economic or cultural focus. These two aims were often understood to be at odds.[15] The first aim, attached to the Ministries of Education and Public Works, promoted the state broadly under the rationale that the world needed to be informed about the cultural, social and economic development of the Czechoslovak state. The promotion, or rather propaganda, of the country was not only political and economic but also cultural. It incorporated various works of art and design that indicated the country's modern artistic trajectory taken as a sign of its independence. Such motivation can also be related to more general educational efforts of world's fairs and their exhibits, presented often as "folk universities" and effective educational venues for visitors of various backgrounds.[16] National representation became a tool of diplomacy and soft power which placed emphasis on a country's attractive features related to culture, history and politics.[17]

The other aim of participation in a world's fair, according to the governmental proposal, was to generate trade and economic benefits. This was, after all, the original purpose of trade fairs and was particularly important for political newcomers like Czechoslovakia that had to restructure their economies and find new trading partners. This focus was promoted by the Ministry of Trade which worked closely with trade organizations, industries, chambers of commerce and large and small business representatives to prepare commercially oriented displays. The duality between the representative and trade functions of the pavilions informed many decisions about the content from the very start. This also impacted an array of features of the displays, from the choice of architects and designers to the selection of objects. The lack of a permanent committee nevertheless did not affect the frequency and enthusiasm with which Czechoslovakia entered interwar world's fairs.

The overall economic climate and the financial standing of the state also played a role in the ability of Czechoslovakia to take part in the interwar world's fairs. Several domestic monetary reforms, introduced from 1919 by the Minister

15 "Nástin organisace československých exposic na výstavách v cizině," 1927, Organisace Československé účasti na výstavách v cizině, kD2519/7, 1–9, APRO .

16 J. Klečka, "Naše výstava," *Czechs and Slovaks. World's Fair Memorial of the Czechoslovak Group* (Chicago, 1933), 100.

17 Joseph S. Nye, "Soft Power," *Foreign Policy* 80 (1990): 153–171; cf. Bartosz Dziewanowski-Stefańczyk, "World's Fairs as Tools of Diplomacy: Interwar Poland," in *World Fairs and the Global Moulding*, eds. Leerssen and Storm, 300–328.

of Finance, Alois Rašín, strengthened the Czechoslovak currency. They created favorable currency exchange conditions for Czechoslovakia which reduced construction or transportation costs. At the same time, though, with a strong crown, Czechoslovak products, displayed in the national pavilion as export goods, became rather expensive for foreign markets.[18] These economic circumstances underlay the considerations of those involved in the exhibition planning, yet there were other motivations which influenced the decisions about participation. Becoming a part of the network of countries, for example, that regularly exhibited in world's fairs, a network that included France, the United Kingdom and USA, seems to be more important than generating trade.

Establishing an exhibition model: *Rio de Janeiro 1922*

Czechoslovakia took part in smaller exhibitions in the early 1920s, which included The International trade fair in Lyon and the Exhibition of Czechoslovak Folk Art in Paris that both took place in 1920. France was an important military and political ally for Czechoslovakia during the war, a topic further developed in Chapter Three, and exhibiting here so soon after the end of the conflict was motivated by further strengthening these ties. The first properly planned participation of Czechoslovakia at an international fair was at The Centennial International Exhibition in Rio de Janeiro in 1922, held between September 7, 1922 and March 23, 1923. The Czechoslovak pavilion built here established a few practices that were followed and developed in later exhibitions [fig. 2].

Originally planned as a national exhibition to commemorate one hundred years of Brazilian independence from the Portuguese, the event was opened to international exhibitors. The fair changed the physical landscape of the city as many Portuguese-era buildings were demolished in the name of modernization to make way for the pavilions and a progressive identity of the state.[19] The Brazilian organizers used the exhibition as an opportunity to promote the republican government, economic recovery and future trade, in other words as a vehicle to display the state's modernity.[20] And it was a vision with which Czechoslovakia could easily associate. Even though it showcased its products for reasons that

18 Turnovský, "Československo," 6.

19 Livia Rezende, "Nature and the Brazilian State at the Independence Centennial International Exhibition in Rio de Janeiro, 1922," in *Cultures of International Exhibitions*, ed. Marta Filipová, 163–182, Bontempi and Sobe, "Exposição Internacional," 229.

20 Rezende, "Nature and the Brazilian State," 167.

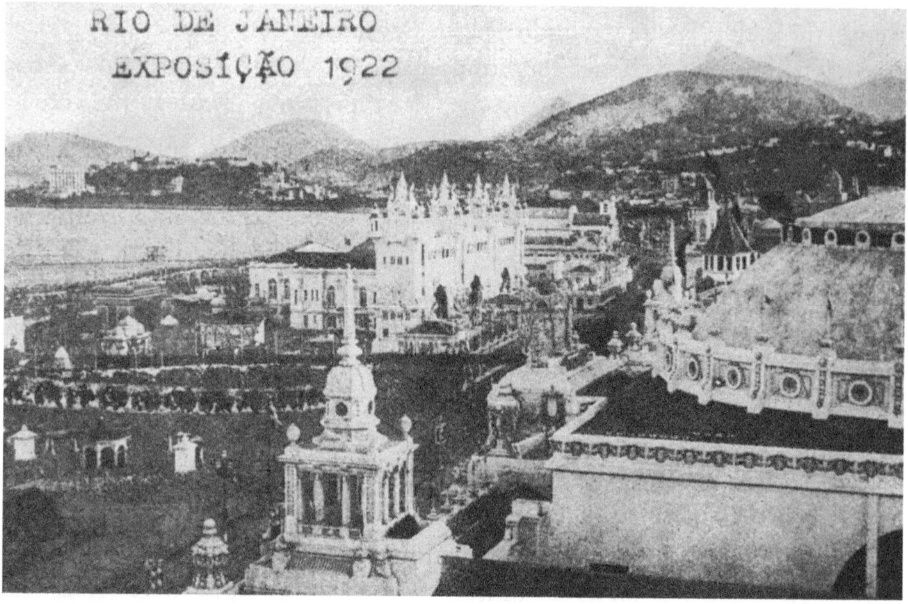

Fig. 2. Rio de Janeiro, The Centennial International Exhibition, 1922.

the government officially advertised as purely economic, the political motivations were implicitly evident.

This was also the first exhibition where Ladislav Turnovský of the Ministry of Trade was appointed the commissioner general of the Czechoslovak representation. Turnovský would become a commissioner at many other future pavilions where he oversaw their construction and content and therefore had a significant influence on them. He emphasized the unique position of Czechoslovakia in Rio de Janeiro as the "only state from the entire Central Europe," which had a place next to a limited number of international participants from the United States, Argentina, Mexico, Chile, Japan, Portugal, Denmark, Italy, Norway, Sweden, Belgium, Great Britain, and France that also accepted the invitation from the Brazilian government.[21] This was a rather small number of states in comparison with later interwar fairs, and therefore it could be seen as a marker of political and economic stability of the participating states so soon after the end of the war.

The main trade competitors in this case were countries known to be trading in South America, i.e., Japan, Italy, Belgium, and France. They already had commercial representation of their industries and financial support from banks

21 Turnovský, "Československo na světové výstavě," 2.

and their subsidiaries in the region. Expansion to new markets of the Americas and the conscious creation of political alliances with the western powers—especially France, USA, and Great Britain—were the key motivations for Czechoslovakia to appear at this fair. Germany did not participate at the Rio de Janeiro exhibition, which was important for Czechoslovakia as Germany continued to be seen as one of the largest international trade competitors for the new state.[22]

The official priorities for the participation were therefore the Ministry of Trade's economic interests and the attempt to give "the impression of noble demonstration of the political and economic independence."[23] Yet already here, the different agendas of the different ministries meant that for example, the Ministry of Education tried to emphasize *cultural* representation of the state. This was visible in the architecture of the national pavilion, immediately the most noticeable symbol of Czechoslovakia.[24] The pavilion was designed by the Czech architect Pavel Janák (1882–1956), who was approached to modify the original plans of another architect.[25] Janák's building was a rectangular structure with a sloping roof with most emphasis on the decoration of the façade. Janák made

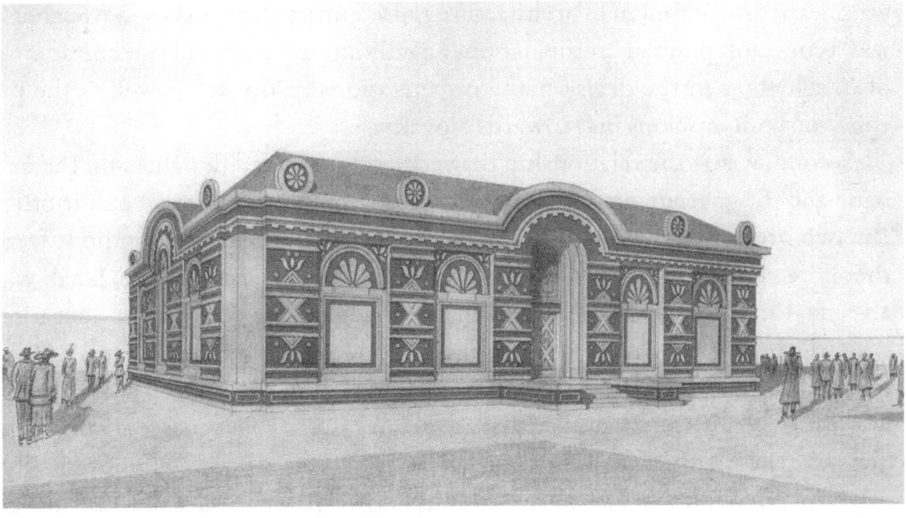

Fig.3. Pavel Janák, Czechoslovak pavilion in Rio de Janeiro, 1922

22 "From the Ministry of Trade to the Presidium of the Ministerial Committee," March 25, 1922, 5, 22II Rio de Janeiro, Jubilejní výstava v Rio de Janeiro no. 35490/22, *Výstava v Rio de Janeiro*, no. 35490/22, APRO.
23 "From the Ministry of Trade," 5.
24 "From the Ministry of Trade," 5.
25 In some documents, the name is misspelled as Pitlík.

distinctive references to folk culture, with motives derived from, for example, floral ornaments and bright coloring used in folk decoration in villages across Czechoslovakia [fig. 3]. The selected language of the building followed the principles of the so-called national style that was being formulated in Czechoslovakia as the official representative style at the time and that combined the legacy of Cubism with Slavic traditions.[26] In Rio de Janeiro, it consisted mainly of decorative elements in the form of abstracted ornament in the contrasting colors of red and white.

The reason for Janák's replacement of the earlier design by Josef Pytlík demonstrates the importance of representative architecture of national pavilions. Pytlík was an architect of Czech descent, who settled and worked in Brazil.[27] The governmental officials in charge of the pavilion, however, found that the original proposal did not "correspond to today's level and direction of Czech architecture"[28] and the design was labelled "reactionary in that it follows fake theatrical and pompous exhibition architecture," more fitting for exhibitions around 1900.[29] At least three questions are raised by these comments that I further develop throughout the book. First, the reference to *Czech* architecture was meant as a synonym of architecture representing Czechoslovakia. Such use was a common practice in the national pavilions and reflected the dominance of Czech elites in the decision and organizational processes as well as the unconscious or conscious bias towards Slovakia.

Second, it was the relationship between architects and designers on the one hand and the governmental officials responsible for the exhibitions on the other. The two groups were often personally connected and national pavilions were therefore closely linked with a specific group of people. In this case, Janák was an acquaintance of Václav Vilém Štech, a secretary at the Ministry of Education who was intimately involved in the preparation of many Czechoslovak exhibitions abroad and who suggested Janák redesign the Czechoslovak pavilion. Štech was also an art historian and critic and the architect's colleagues from organizations like the Association of the Czechoslovak Werkbund (Svaz československého díla) and The School of Decorative Arts in Prague. After 1918, artists and design-

26 Vendula Hnídková, *Národní styl, kultura a politika* (Prague: VŠUP, 2013), 184; Marie Benešová, *Pavel Janák* (Prague: Nakladatelství československých výtvarných umělců, 1959), 18.

27 "From the Ministry of Trade," 3.

28 "Úřad pro zahraniční obchod v Praze – Protokol," The Office for Foreign Trade, February 4, 1922 22II Rio de Janeiro Jubilejní, "To the Ministry of Education," February 4, 1922, The Office of Foreign Trade, *Výstava v Rio de Janeiro*, no. 13765/22, Archive of the Presidential Office and "From the Ministry of Trade," 3.

29 "Exhibition in Rio de Janeiro," The Ministry of Education, January 23, 1922, no. 7037, APRO.

ers working at the School were frequently assigned to supply the visual identity for the state that included the designs of the state emblem, stamps, banknotes, official posters, or diplomas.[30] For these links with the governing structures, the School, its teachers and students had an outstanding place in national pavilions and presentations at world's fairs and became part of the exhibition elite.

Finally, the governmental statement about the kind of architecture needed in a national pavilion also reveals the acknowledgement of the requirement of such buildings to be reflective of the contemporary course both in architecture and politics. Janák's Czechoslovak pavilion in Rio de Janeiro was the only pavilion in the history of Czechoslovak interwar participation where the architecture explicitly quoted elements of vernacular architecture, which could be seen as traditionalist and outdated. However, in doing so, it echoed the contemporary trend of searching for a Czechoslovak modern national style, in this case, found in Slavic colors and forms.

While the exterior tried to embrace contemporary vernacularism translated into the national style, the interior was organized as a standard exhibition room which consisted of a large space divided by tables, cabinets and partitions. To equip the interior, the Ministry of Trade contacted the Central Association of Czechoslovak Industrialists and several trade chambers in Czechoslovakia, while some companies applied to be exhibited directly. This pattern would become commonplace in future exhibitions which were a mixture of invited exhibitors and those that asked to have their work or products included. In Rio de Janeiro, around 80 companies showed a variety of smaller products of the glass industry (like Bohemian crystal), ceramics, music instruments, clothes, hops, pencils and carpets, alongside agricultural machinery and vehicles, most probably located outside the national pavilion.[31]

One of the main features in the interior was a stained-glass window showing the Czechoslovak emblem which was designed by František Kysela (1881–1941), an artist and instructor at the School of Decorative Arts in Prague. The official section was in the center of the pavilion and promoted, for example, agriculture, the spa industry, including an exhibit of Czechoslovak folklore, diagrams with statistics about the production capacities of the state, and various promotional materials. How exactly the objects were distributed and displayed is, nev-

30 Pavla Pečinková, "Chapters from the History, 1895–1946," in *Vysoká škola uměleckoprůmyslová v Praze/ Academy of Arts, Architecture and Design in Prague, 1885–2005*, eds. Martina Pachmanová and Markéta Pražáková (Prague: VŠUP, 2005), 45.

31 "Čs. Pavilion," 9 and Turnovský, "Československo na světové výstavě," 1923, 6; Levy, *A exposição*, 147.

ertheless, not clear because of the lack of evidence in the case of this and many other national pavilions. Textual documentation from archives, catalogues and newspapers often does not mention the exact distribution of exhibits and the number of photographs from interiors is small, especially in some cases like the 1922 exhibition or the world's fairs in Chicago in 1933. A vital source is drawings of ground plans, cuts and sections as well as blueprints, where they survived.[32] The plans were, nevertheless, often modified according to local circumstances, as I discuss in the following chapter. Occasionally, written accounts of personal visits to pavilions provide a good additional source which enhances our understanding of the interiors.[33]

The very few images from the pavilion in Rio de Janeiro reveal shelves full of porcelain running past the walls and glass cabinets with decorative objects while the bust of President T. G. Masaryk dominated the space amid greenery.[34] The subsequent Czechoslovak pavilions followed a similar pattern of space distribution and object arrangement. They routinely comprised a large entrance hall with a painting of Prague and a centrally located bust of the President [fig. 4]. Although Masaryk was not directly involved in the Czechoslovak exhibitions, it is clear from the amount of archival documentation of the Presidential Office that he was aware of the organization process and the eventual successes of the exhibits.

In Rio de Janeiro, the bust was made of bronze by the sculptor Jan Štursa (1880–1925) and it was reproduced from a marble original placed in the parliament in Prague. Copies were cast for various Czechoslovak pavilions in the interwar period as well as for many locations across Czechoslovakia. Masaryk's presence through a prominently placed sculpture in the Czechoslovak pavilions or sections was highly symbolic because foreign allies would recognize him as the representative of the state he helped to form during the war. Even after he stepped down from his presidential role in 1935, it was his likeness rather than that of his successor Edvard Beneš that would appear more prominently in the national pavilions.

Masaryk also played the role of the country's protector, which was a narrative not dissimilar to that given to the Habsburg monarch Franz Joseph, whose sculptural likeness had adorned various Austrian and Bohemian exhibitions before 1918. For instance, at the Jubilee Exhibition in Prague in 1891 which fo-

32 The main location of the plans is the National Technical Museum in Prague, whose storage was flooded in 2002 when many artifacts in the depository were damaged or destroyed.

33 Speaking of the world's fair in Chicago, Ladislav Janík, for example, reported on the green in the antechamber and blue-grey in the main parts, "Na výstavišti," *Svornost*, June 16, 1933, 3.

34 *Světozor* 23, no. 5 (1922/23): 113.

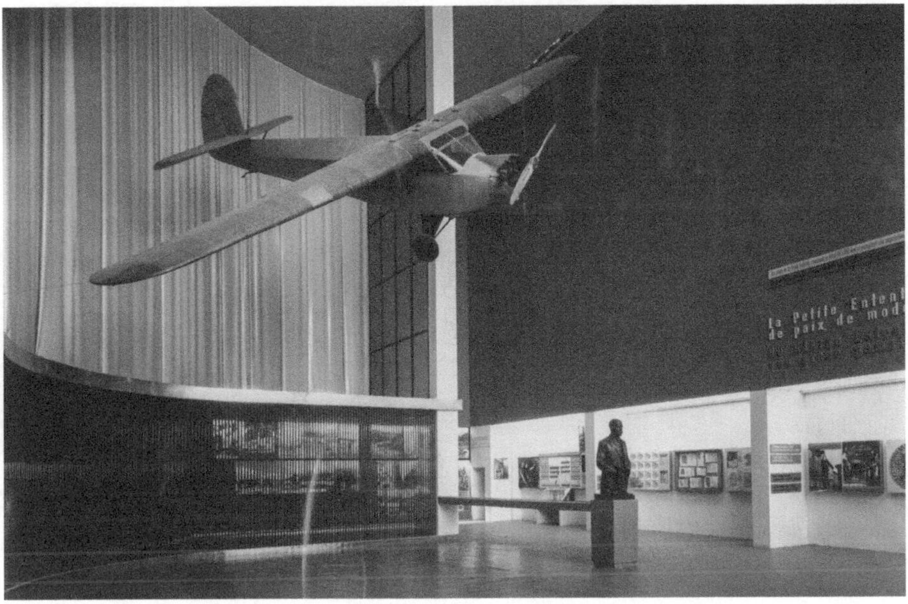

Fig. 4. View of the Czechoslovak pavilion interior in Brussels, 1935.

cused on the arts and industries of Bohemia, Franz Joseph was described as the "bright, fatherly and kind head of state" and his statue dominated the main pavilion.[35] Like the monarch, Masaryk was also regularly referred to as a caring, charitable, and modest paternal figure.[36] As an aged leader of the new state, he provided a certain familiarity of a figurehead known from the monarchy.[37] Apart from the presidential bust, other features, like the basic content, were shared across the pavilions in the interwar period and reiterated a similar narrative. As such, visitors were reassured that even though Czechoslovakia was a new nation-state, it was built on familiar and recognizable traits.

The politics of a national pavilion: *Paris 1925*

For the first time in interwar exhibitions history, the pavilion in Rio de Janeiro embraced the idea of a single Czechoslovak nation, home of the Czechoslovak people, even though the state consisted of many different ethnic groups.

35 Rudolf Jaroslav Kronbauer, *Naše Jubilejní výstava* (Prague: Josef Vilímek, 1892), 279.
36 Jiří Rak, "Stařičký mocnář a tatíček Masaryk," in *19. století v nás: modely, instituce a reprezentace, které přetrvaly*, ed. by Milan Řepa, 267–271 (Prague: Historický ústav 2008), 267.
37 Morelon, "State Legitimacy."

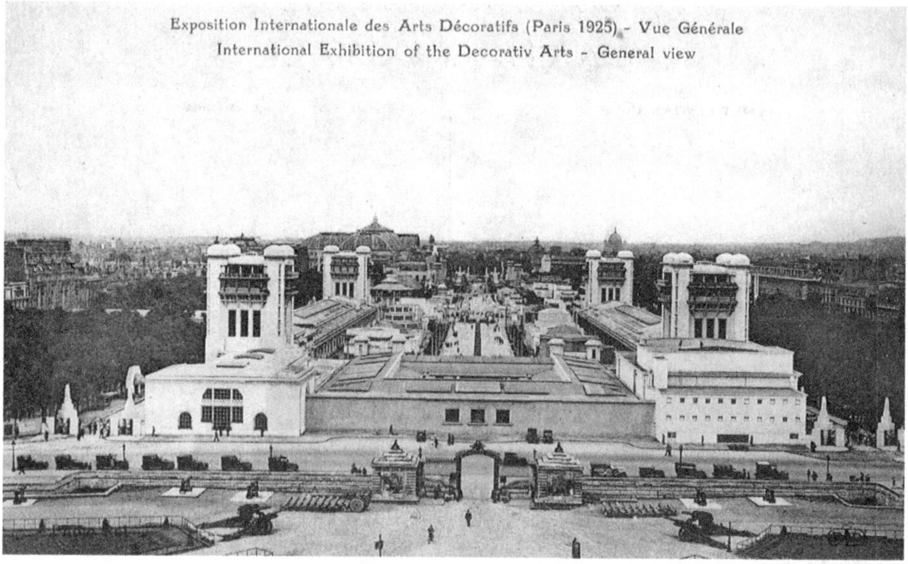

Fig. 5. The International Exhibition of the Decorative Arts – General view, Paris, 1925, postcard.

Czechoslovakism is a running theme of this book because the use of this artificial identity raises the question of how the state and nation were defined in exhibitions. The most visible promoter of the political program of Czechoslovakism, which emphasized (and more or less invented) the joint historic roots of Czechs and Slovaks, was the government and it gave prominent space to such narrative in the national presentations.

Czechoslovak displays were very much orchestrated from Prague and heavily oriented towards audiences and partners in the west. The inclusion of different aspects of the entire state was therefore an issue often discussed during the preparatory works. Already the next significant Czechoslovak pavilion at an international exhibition, The International Exhibition of Decorative Arts and Modern Industries (Exposition internationale des arts décoratifs et industriels modernes) in Paris in 1925, revealed that the Prague-based authorities recognized the tension between the two ethnic parts of binational Czechoslovakia, yet prioritized Czech exhibits.

The 1925 Exhibition was not a world's fair per se, but it holds a key place in exhibition histories as an event that showed the international and national "divisions that continued to grow in the decades to come"[38] [fig. 5]. The first large post-war European exposition was primarily a pastiche of spectacular visual

38 Makaryk, *April in Paris*, 23.

information and competing artistic directions that demonstrated the link between art and design on the one hand and the political motivations of the participating exhibitors. The official aim of the Exhibition was to promote a new visual language that would be identifiably French. France, as the organizer, sought to reclaim its place among the world's leading producers of applied arts, having faced strong competition from Great Britain, Germany, and America in the last few decades.[39]

The individuals and groups involved in organizing the exhibition, however, did not unanimously agree on how this could be achieved and had different interpretations of what *modern* decorative arts and their production meant. While some, like the prominent French designer René Lalique or the writer and critic Paul Géraldy, argued for the return to the exclusivity of art and French traditions, others called for a complete break with such heritage.[40] Such direction was suggested by, for example, Le Corbusier's model home Esprit Nouveau or Robert Mallet-Stevens's Pavilion du Tourisme, both buildings of rectangular shapes and minimal decoration. They, nevertheless, were one of the few lone examples of international modernism allowed within the confines of the Exposition, while for instance the Bauhaus and De Stijl were banned from appearing there.

The Exposition took place from April 28, 1925 until October 25, 1925 and its 131 pavilions were located along the Seine between Les Invalides and the Grand and Petit Palais and included boutique shops on the Alexander III Bridge.[41] Czechoslovakia was one of the twenty foreign states that built their national pavilion here, alongside Austria, the United Kingdom, the Soviet Union, Poland, Sweden, Yugoslavia and Italy, who represented the main trade competitors for Czechoslovakia.[42] The largest trade rival for Czechoslovakia, Germany, was absent.

The Czechoslovak participation extended across several locations as was typical of other fairs. Apart from the national pavilion, there was a substantial display in the Grand Palais focused on education and a section on modern hous-

39 Philip Whalen, "Paris 1925," 236–43, in *Encyclopedia of World's Fairs and Expositions*, eds. by John E. Findling and Kimberley D. Pelle (Jefferson N.C.: McFarland, 2008), 237.

40 Paul Géraldy, "L'Architecture vivante," in *L'illustration, special section on L'exposition internationale des Arts décoratifs et industriels modernes* (April 15, 1925), n.p.; Gabriel Mourey, "L'Exposition des arts décoratifs et industriel de 1925," in L'Amour de l'art (August 15, 1925); Waldemar George, "L'Exposition des Arts Décoratifs et Industriels de 1925: les tendances générales," *L'Amour de l'art* (1925): 283.

41 Philip Whalen, "Paris 1925," in *Encyclopaedia*, 238.

42 The other countries were Denmark, Finland, Belgium, Japan, Holland, Estonia, Latvia, Denmark, Turkey, Luxemburg, Greece and Spain.

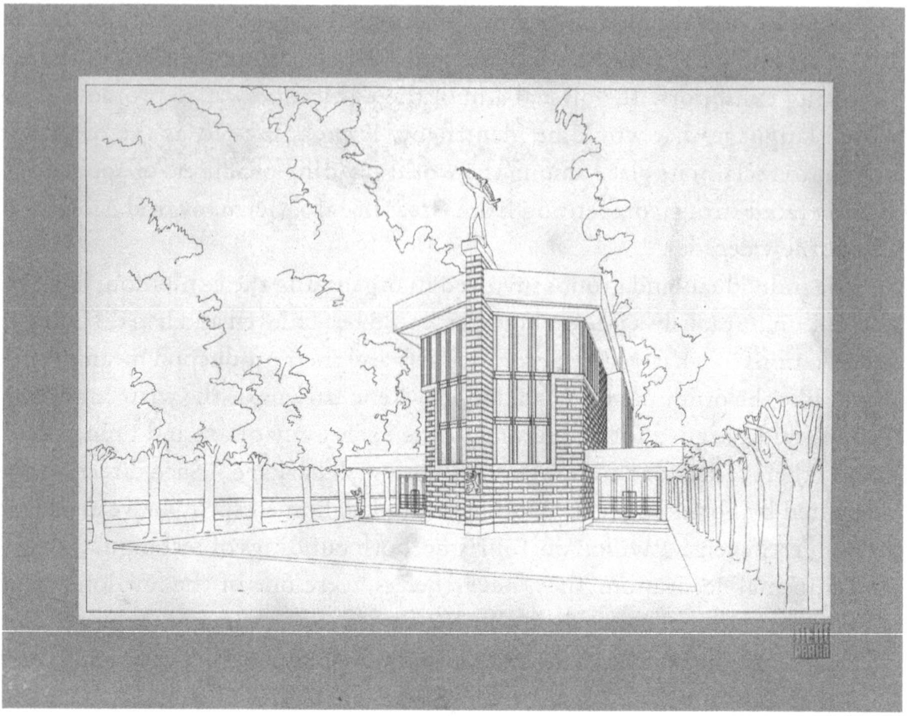

Fig. 6. Josef Gočár, Drawing of the Czechoslovak pavilion in Paris, 1925.

ing at the Galleries of the Esplanade des Invalides.[43] The Czechoslovak pavilion was located on the Grand Palais side of the river in the proximity of French regional pavilions and those of the Basque country and Poland. It was designed by the architect Josef Gočár (1880–1945) who was assisted by the architect Adolf Benš (1894–1982). Their design came out of an anonymized open competition, which was a practice that developed gradually over the interwar period. While the national pavilion in Rio de Janeiro was assigned to a named architect, subsequent buildings came out of competitions of several entries by a single architect or a group of collaborators. A cross-ministerial committee would publish a call for submission based on the specifics set out by the host country and the allocated budget. A select committee would assess the competition entries, comment on them and often propose changes and modifications [fig. 6].

43 The organization was managed by František Hodáč, professor of political economy at the Technical University in Prague who was also the secretary general of the Association of the Czechoslovak Industrialists. The Ministry of Education to the Ministerial Committee, February 4, 1925, *Mezinárodní výstava dekorativního umění v Paříži 1925*, D1441/25, APRO.

In Paris, the rectangular building had two floors and was constructed from concrete, brick, and glass. The exterior featured the state emblem by the sculptor Otto Gutfreund and a sculpture of *Victory* by Jan Štursa whose bust of the President stood in the main hall. As I offer a more focused discussion of the pavilion's artistic content in Chapter Three, in the context of searching for an appropriate modern expression, I will only mention here that many instructors and students at The School of Decorative Arts participated heavily on the content under the direction of the exhibition commissioner Štech. This meant the inclusion of sculptors like Gutfreund, who taught at the school, or the designers that had already appeared in Rio de Janeiro, like Kysela and Janák and were linked with the institution. The work of other professors, including Josef Drahoňovský's (1877–1938) glass and Emilie Paličková Milde's (1892–1973) lace, would also appear regularly in future national pavilions.[44]

The architect Otakar Novotný (1880–1959), also of the School of Decorative Arts, was responsible for the national display located in the Grand Palais and he distributed it across ten rooms on the ground floor which represented Czechoslovak trade and industries.[45] Some of the rooms were conceived as offices of the trading associations and chambers behind the official participation. Art and design education at a secondary level was presented here too; there was a room devoted to the Prague Academy of Fine Arts with an interior by Gočár and one devoted to the School of Decorative Arts by Janák, both rooms displaying work by staff and students.[46] The rooms were equipped with furniture and decorations by the designers who participated in the main national pavilion. Showcases of manufactories like the furniture maker Thonet Mundus, glass and ceramics producer Moser, bijou from the town of Jablonec nad Nisou in northern Bohemia and many others displayed the products of companies that already had commercial representation in Paris. A small section of Slovak and Sub-Carpathian decorative folk art was included here to incorporate the regions into the Czechoslovak narrative, a topic I develop in more detail in Chapter Three.

Another small Slovak display appeared in the Gallery of the Esplanade des Invalides as part of the Modern housing section based on the project by the architect Václav Ložek (1892–1951). It showed an interior with decorations derived from folk culture surrounded by murals painted by Kysela.[47] Other rooms followed with examples of different types of habitation: a bedroom equipped with

44 "Pavillon national," 21.
45 "Pavillon national," 25.
46 "Pavillon national," 35.
47 "Pavillon national," 39.

modernist furniture of the Brno based company of Jan Vaněk, an ordinary apartment, two rooms for single men (the so-called *svobodárna* in Czech) and a dining room, all again with decorations by designers and companies that had been represented across the other Czechoslovak displays, including the textiles by Teinitzerová and Paličková Milde.

One of the less prominent structures was the architect Bohuslav Fuchs's contribution to the Pastry and Charcuterie pavilion at the Galleries of the Esplanade des Invalides.[48] Food outlets would soon become an indispensable part of Czechoslovak presentations and an important place where national ideology was transmitted through food, drink, attire, music and gender. Again, I expand upon these connections in more detail in the chapters to come, starting with the Parisian food pavilion. Czechoslovak restaurants and cafés were a regular occurrence next to, or in, national pavilions. Generally, they would not serve complicated dishes but instead focus on quick snacks like Prague ham and popular drinks. Pilsner beer would therefore always feature in the role of a national drink and play the role of a key commodity and marker of nationality.

The hesitant exhibit: *Philadelphia 1926*

While food and drink provided one continuous attribute of Czechoslovak representations, the main emphasis lay elsewhere, namely on the trading and the representational benefits an exhibition could bring. The dividing line between the expected commercial gains and the ideological representation of the state was not usually clearly cut and this became prominent at the next entry in a world's fair that Czechoslovakia underwent only a year after Paris. The Sesqui-Centennial International Exposition in Philadelphia in 1926 is not one of the best-known or researched events of the world's fairs history. In terms of visitor numbers and profit, it was a failure. The poor attendance driven partly by very rainy weather for most of the time of the fair. Financially, it ended up with a huge loss and it left the organizer, the Sesqui-Centennial Exhibition Association, bankrupt and in need of bailing out by the city government.[49] In the trajectory of Czechoslovak interwar entries in world's fairs, however, it played a crucial role in moving the establishing narrative of national displays forward.

48 "Pavillon national," 42.
49 David Glassberg, "Philadelphia, 1926," *Encyclopaedia*, 246.

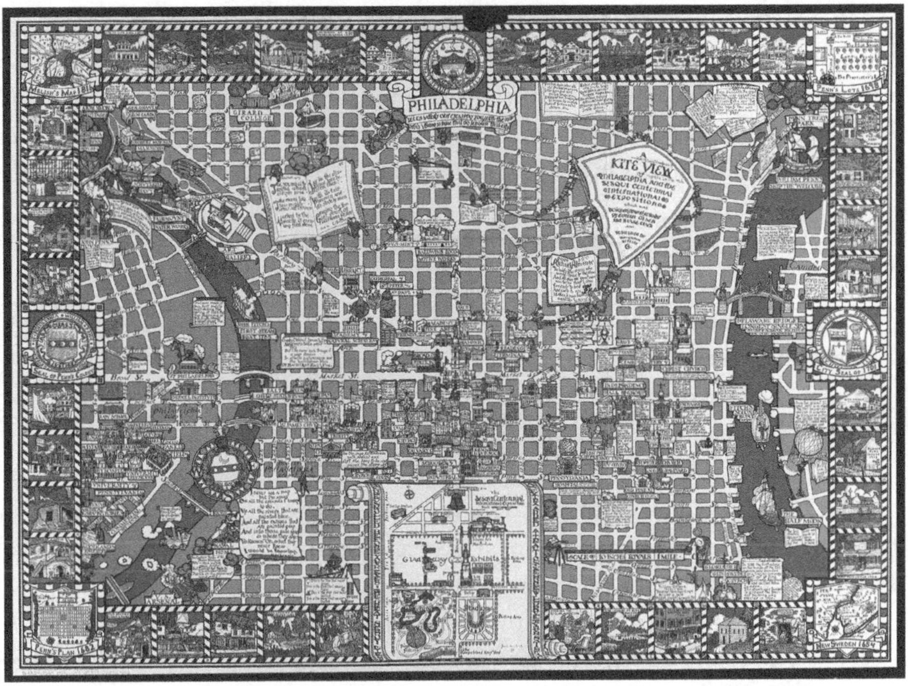

Fig.7. Edwin Olsen, Blake Clark, and Houghton Mifflin Company, "A kite view of Philadelphia and the Sesqui-Centennial International Exposition," map. Boston: Houghton Mifflin Co., 1926.

The grounds were in south Philadelphia in and around League Island Park, which is today's FDR Park [fig. 7]. The Exhibition lasted from May 31 until November 30, 1926, and the three main pavilions of the Palace of Liberal Arts and Manufacturers, the Palace of Agriculture and Foreign Exhibits, and the Palace of US Government, Machinery and Transportation, were complemented by a variety of smaller halls mostly built in colonial architecture style, a large stadium and various entertainment zones. Forty-three foreign nations participated in the fair and seven, including Argentina, Cuba, Czechoslovakia, Japan, Persia, Spain, and Tunisia, built their pavilions.[50] Apart from the national building, Czechoslovakia also built an exhibit in the Palace of Agriculture and Foreign Exhibits alongside Hungary that was represented here semi-officially. Austria, Germany, Poland or Yugoslavia had no official exhibits here and Czechoslovakia was therefore once again one of the few countries that invested its effort into such a venture.[51]

50 E. L. Austin and Odell Hauser, *The Sesqui-Centennial International Exposition* (New York: Arno Press, 1976), 81.

51 There were, nevertheless, Polish, Austrian and Hungarian days and a German week. Austin and Hauser, *The Sesqui-Centennial*, 377.

The participation of Czechoslovakia at Philadelphia was aimed at strengthening the links between the USA and the new state. Celebrating its 150th anniversary from the signing of the Declaration of Independence, the USA and Philadelphia, especially, were particularly significant, as the Czechoslovak president summed up in his message sent to the city mayor W. Freeland Kendrick:

> Hearty greetings on the opening day of the Sesquicentennial exposition in which Czechoslovakia shall have its modest share in gratitude for the help the United States rendered to our nation during its struggle for independence in the world war. In 1918 in Philadelphia we have declared our independence on the same memorial place where George Washington declared Independence in 1776.[52]

The USA served as an inspiration and a useful model for the forming Central European republic in some of its aspects, for instance in the presidential system, republican state, and universal suffrage.[53]

In Philadelphia, the Czechoslovak organizers emphasized the affiliation between the USA and Czechoslovakia not only in the historic role the city played in the birth of the state but also in the proposed content for display. The pavilion was designed by the Czech architect Ladislav Machoň (1888–1973) as an L-shaped, single-floor building. The individual rooms were again lain out in a museum-like manner with glass cabinets and pictures on walls. On paper, the pavilion's space was divided into the promotional hall, the Prague room, the national liberation hall, and a hall dedicated to the American Czechs. [fig. 8]

The liberation hall was planned to contain photographs of resistance during the First World War and related memorabilia. The sculpture of President Masaryk by Otto Gutfreund featured here. Maps, diagrams and images of the Czechoslovak contribution to the Allies' victory surrounded the statue together with a small exhibit of the Czechoslovak legionnaires. These volunteer soldiers fought in Russia, Italy, and France in the First World War alongside the Allies and in some cases settled in the USA to form a vocal part of the diaspora.

---

52 Letter from T. G. Masaryk to the mayor of Philadelphia of May 29, 1926, Světová výstava ve Filadelfii D2982/26, APRO.

53 Andrea Orzoff, "Interwar Democracy and the League of Nations," in *The Oxford Handbook of European History, 1914–1945*, ed. Nicholas Doumanis (Oxford: Oxford University Press 2014–2016), 263; Peter Bugge, "A Nation Allied with History. Czech Ideas of Democracy, 1890–1948," in *Democracy in Modern Europe: A Conceptual History*, eds. Jussi Kurunmäki, Jeppe Nevers, Henk te Velde (New York–Oxford: Berghahn Books, 2018), 208.

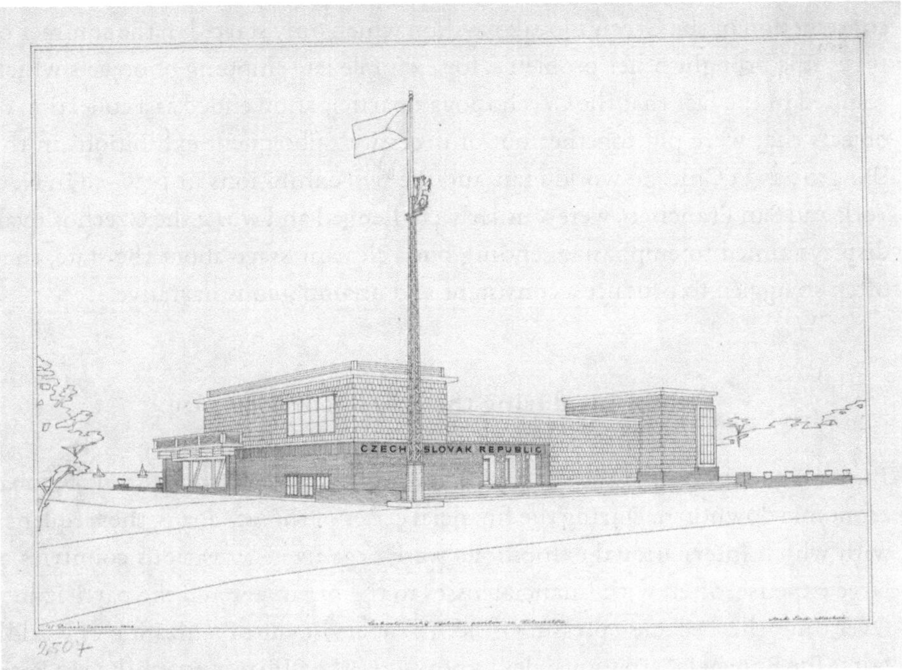

Fig. 8. Ladislav Machoň, The Czechoslovak pavilion in Philadelphia, 1926.

The room dedicated to Prague contained artifacts related to the Sokol gymnastic organization and to the capital which was represented by a large painting by the post-impressionist painter Jaroslav Šetelík (1881–1955). Vistas of Prague would become another constant in Czechoslovak pavilions occasionally accompanied by a view of the Slovak Tatra mountains in a juxtaposition of Czech history and Slovak nature. Alongside Masaryk's bust, the paintings formed a group of objects that provided visual continuity in national pavilions throughout the interwar period.

Many of these objects were included as planned but there was always a number that were added at a later stage for assorted reasons. This was the case in Philadelphia as much as in other world's fairs. One of the causes for the changes in the 1926 exhibition was the belated official decision about the state's participation. Several ministries of the Czechoslovak government (foreign affairs, public works, finance, industry, commerce and trade, as well as education) were involved in the national pavilion and other sites where Czechoslovakia appeared.

It can be said with a little exaggeration that the protruding discussions and hesitations about a Czechoslovak representation became another attribute of interwar exhibition efforts. In the case of Philadelphia, they seriously delayed the construction of the Czechoslovak pavilion which only started in the summer of 1925. This brought other problems, for example late shipping of objects which resulted in the fact that the Czechoslovak participation ended as a collection of objects that were put together out of necessity. Subsequent exhibitions in the USA, the 1933 Chicago world's fair and the two exhibitions in 1939–40 in New York and San Francisco, were similarly challenged and while the Czechoslovak displays aimed to emphasize sending out a clear message about the state, they often struggled to produce a consistent and unambiguous narrative.

Exhibiting during the crises: *Barcelona 1929*

The ambiguous result of the Philadelphia world's fair followed a period of global economic downturn. During the financial crises of the late 1920s, the frequency with which international exhibitions were organized by various countries at large expense, often with financial losses to the organizer and the participants. Over time, this became a pressing issue. To supervise the organization of world's fairs, The Bureau International des Expositions (the BIE) was established in Paris in 1928 to regulate the frequency and location of these events. The BIE convention was signed by 39 states, including Czechoslovakia, who agreed on the classification of general exhibitions, introduced a rotation system and the need to oversee the quality of the exhibits.[54] One of the founding member states of the BIE was Spain that was responsible for the first exhibition to adhere to the rules.

The Exposición International de Barcelona (The International Exhibition of Barcelona) of 1929 was the first exhibition seriously affected by the global economic crisis prompted by the Stock Market crash on Wall Street in New York. While the USA, which halved its manufacture between 1929 and 1932, was the most affected country, other economies across the world suffered too. Until then, Czechoslovakia had enjoyed relatively successful financial emancipation. During the first decade of the state's existence, its exports and import had been increasing, reaching their peak in 1929. Compared to other successor states of the Habsburg monarchy like Austria, Hungary, Yugoslavia, and Romania, the results of Czechoslovak foreign trade and economy were much better.[55]

---

54 "Our history," *Bureau International des Expositions*, https://www.bie-paris.org/site/en/about-the-bie/our-history. Accessed on July 2, 2023.

55 Zdeněk Kárník, *České země v éře První republiky. Vznik, budování a zlatá léta republiky (1918–1929)* (Prague: Libri, 2003), 436.

The impact of the global crisis was felt later in Czechoslovakia than for instance in the USA and Germany and the drop in international trade was much more drastic.[56] During the Depression years in Czechoslovakia, industrial and consumer goods production plunged, export slowed down, unemployment in many areas increased and the influence of banks over local industries only grew.[57] The worst economic decline took place in March 1933; compared to 1929 industrial production almost halved, glassmaking was down to 48% and ceramics manufacture to 41%.[58] Slovakia and Sub-Carpathian Ruthenia suffered from the extended crisis of agriculture but Slovak industries like steelworks, chemical industries and metallurgy managed to bounce back more quickly than those in Bohemia.[59]

The situation in the Bohemian and Moravian border regions, with a proportionally large number of German inhabitants, helped exacerbate the political tensions between the different national groups. The Sudetenland relied heavily on local industries, mainly producing textiles, glass and porcelain, got hit not only by the economic crisis but also by the German blockade of the borders. The regions between Germany and Czechoslovakia also had a great number of tourist attractions and spas, which similarly suffered from the economic decline. Places like Karlovy Vary and Mariánské lázně would habitually feature in the Czechoslovak exhibits, firmly related to the narrative of the Czechoslovak state.

In this difficult economic and political atmosphere, the cost of the Czechoslovak exhibit in Barcelona reached about 700 million crowns (130 million pesos), which was a substantial amount at the time.[60] The Exhibition took place between May 19 and January 15, 1930 and followed the steps of the 1888 Universal Exhibition in the same city.[61] It was also held simultaneously with the Ibero-American Exhibition in Seville which saw the participation of nearly all countries from the Americas, as well as some Spanish regions and cities. Under the protection of the prime minister and dictator Primo de Rivera, the Barcelona Exhibition focused on industry, art and sport as well as the revitalization

56 Eduard Kubů, "Zahraniční obchod," in *Mýtus a realita hospodářské vyspělosti Československa mezi světovými válkami*, ed. Eduard Kubů and Jaroslav Pátek (Prague: Karolinum, 2000), 209.

57 Zdeněk Kárník, *České země v éře První republiky (1918–1938) II. Československo a české země v krizi a v ohrožení (1930–1935)* (Prague: Libri, 2002), 36.

58 Kárník, *České země II*, 36.

59 Kárník, *České země II*, 46.

60 Jaroslav Pechar, "Mezinárodní výstava v Barceloně," *Právo lidu*, April 19, 1929, Newspaper Excerpts archive, Oh8, MFA Archive.

61 Marina Muñoz Torreblanca, "Barcelona's Universal Exhibition of 1888: An Atypical Case of a Great Exhibition," *Cultures of International Exhibitions 1840–1940. Great Exhibitions in the Margins*, ed. Marta Filipová (Farnham: Ashgate Publishing, 2015), 45–68.

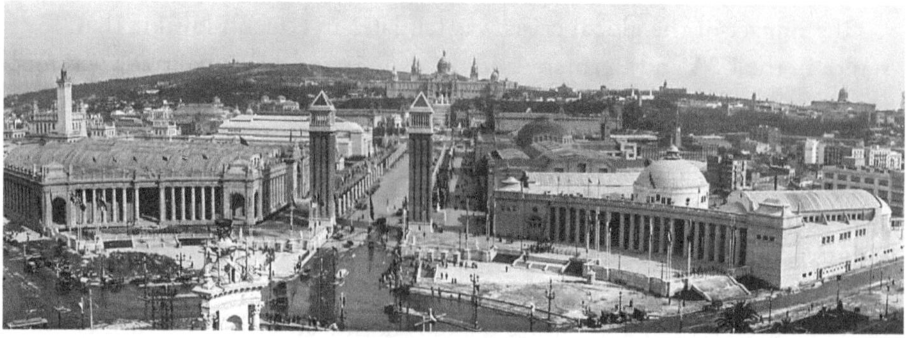

Fig. 9. Panoramic view of the Barcelona Exposition, Spain. 1929.

of the city.[62] Twenty-nine nations participated officially here, fourteen, including Hungary and Austria, with national pavilions while the USA and Japan took part semi-officially[63] [fig. 9].

One of the greatest attractions—apart from the notorious German pavilion designed by Mies van der Rohe and Lily Reich—was the Spanish village, a replica of a mediaeval pueblo with "real" natives. Here, "[w]ith a view to the national and international audience, Spain's past was staged as glorious and at the same time homely, creating an idealized retreat within the progressive industrialization."[64] Recreations of villages and historic towns in the modern settings of world's fairs retained their popularity for decades and emphasized not only local heritage but also the juxtaposition with the modernity and sanitation of the contemporary world.[65]

The Exhibition did not make a significant mark in the history of Czechoslovak exhibition activities. No national pavilion was built here, and Czechoslovakia was only represented in one of the large exhibition palaces. The Czechoslovak exhibit is also one of the least documented displays in archival sources, photographic documentation and reports. Various newspaper articles published about the exhibit, nevertheless, give a sense of what the Czechoslovak section

62 *Exposición International de Barcelona 1929, guía official* (Rudolf Mosse Ibérica, 1929), Duque de Berwick y Alba, *Catálogo Histórico y Bibliográfico de la Exposición International de Barcelona* (Madrid, 1931).

63 John E. Findling, "Barcelona 1929/30," in *Encyclopaedia of World's Fairs and Expositions*, eds. John E. Findling and Kimberley D. Pelle (Jefferson N.C.: McFarland, 2008), 251–53.

64 Anke Wunderwald, "Das Poble Espanyol der Weltausstellung 1929. Barcelona als Imagination nationaler Einheit," in Cornelia Jöchner, Christin Nezik, Gáspár Salamon, Anke Wunderwald, *Museale Architekturdörfer 1880–1930: Das Eigene in transnationalen Verflechtungen* (Dresden: Sondstein, 2023), 233.

65 Wilson Smith, "Old London, Old Edinburg: Constructing Historic Cities," in *Cultures of International Exhibitions 1840–1940. Great Exhibitions in the Margins*, ed. Marta Filipová (Farnham: Ashgate Publishing, 2015), 203–230.

looked like. It was located in the Palau Meridional, or the Central Palace, and opened on July 15, 1929.

The exhibit was designed by Richard Klenka (1873–1954), a professor of architecture, and overseen by the general commissioner Turnovský.[66] Selection of objects that by now had become customary was located here. They included a centrally located bust of Masaryk was accompanied by portraits of the kings of Bohemia, and numerous charts and maps about the development of the region over the last ten centuries.[67] The Exhibition took place one year after the first decade of the state's existence, which was commemorated in the display of historic documents on the foundation of the state. Although a young state, Czechoslovakia was also presented as an entity built on a kingdom going back hundreds of years. This was visible in, for example, the memorabilia related to the millennium of Saint Wenceslaus, the 10th century Duke of Bohemian and a Czech patron saint, that fell on 1929.

Over one hundred paintings were dispatched to the exhibition including the now canonical representations of Czechoslovakia of the capital of Prague by Šetelík and a vista of the Slovak mountains by the Czech landscape painter Otakar Štáfl (1884–1945). They were complemented by a commercial and industrial display which consisted of a range of products of heavy and light industries, decorative arts and design.[68] Barcelona was the first world's fair, in which the designer Ladislav Sutnar (1897–1976) made a significant mark with his exhibition design. This will be the subject of my discussion in the next chapter. Apart from a praised map of Central Europe, showing train and air links from Prague, Sutnar installed the section of specialized schools on behalf of the Ministry of Education. It featured the Association of the Czechoslovak Werkbund, in which Sutnar was involved, and showcased its design dedicated for sale. It aimed to "capture Czechoslovak work in the area of new, utility items of high quality and valuable artistic industry through the most typical examples."[69] This statement by Sutnar summarized an important point about such displays. The *typicality* of the design and art production, or any other production, was a characteristic feature of national pavilions. And while the exhibits deemed as typi-

66 "La sección de Checoeslovaquia," *La Veu di Catalnuya, Las noticias*, July 16, 1929, Newspaper Excerpts Archive, MFA, oh8.

67 "Inaguració de la Secció Txechoslovaca," *La publicitat*, July 16, 1929, Newspaper Excerpts Archive, MFA, oh8.

68 This included lace by Emilie Paličková Milde which was awarded grand prix. Ladislav Sutnar, "Státní odborné školy na mezinárodní výstavě v Barceloně 1929," *Horizont*, no. 25–26, (1930): 48.

69 Sutnar, "Státní odborné školy," 48.

cal were selected for the display, the fact of their inclusion also contributed to their representative status.

The decisions on what commercial items and goods to display were also driven by predictions of what markets in the given location and the potential partners would welcome. Czechoslovak exhibition commissioners reported on exhibits of other countries so that the Ministry of Trade could consider the information for the next large exhibition. After Barcelona, the next large world's fair in which Czechoslovakia took part, was held in Chicago in 1933. As the export industry of Czechoslovak goods plummeted during the Depression, participation in Chicago became even more important despite the high cost.

Recovery: *Chicago 1933*

The Century of Progress world's fair opened on May 27 and closed on November 12, 1933 after being extended by two weeks from the official closure. It was also renewed the following year to run between June 1 and October 31, 1934.[70] The exhibition was accompanied by the end of the Great Depression and the ensuing recovery measures, with one key event in its course being the abolition of prohibition. This event helped to boost visitors' attendance, alcohol consumption and revenue generated by the fair [fig. 10].

Many fairs were organized to commemorate specific events or earlier exhibitions and the Chicago world's fair recalled the 1893 World's Columbian Exposition that took place in the same city in its southern parts. The very idea of the exhibition, located centrally by Lake Michigan, originated in the mid-1920s while the theme of progress was added later when the emphasis on renewal became even more pertinent during the Great Depression years. To finance the world's fair, A Century of Progress corporation was founded in 1928 as a not-for-profit organization with membership fees and golden notes issued to guarantors. The Chicago organizers proudly proclaimed that the fair was put together with no cost to the taxpayer and in the end, ended in a profit of 688,000 dollars.[71] And unlike many previous exhibitions in the USA, for instance, the

---

70 John E. Findling, "Chicago 1933–1934," *Encyclopaedia*, 268–277; John E. Findling, *Chicago's Great World's Fairs* (Manchester – New York: Manchester University Press, 1994); Cheryl R. Ganz, *The 1933 Chicago World's Fair: A Century of Progress* (Urbana, Chicago, Ill.: University of Illinois Press, 2012); Marcy Cameron, The Dual Identity of the Czechoslovak Pavilion at the 1933 Chicago World's Fair. A Master Thesis at the School of Architecture, The University of Virginia, 2012.

71 Findling, "Chicago 1933–1934," 276.

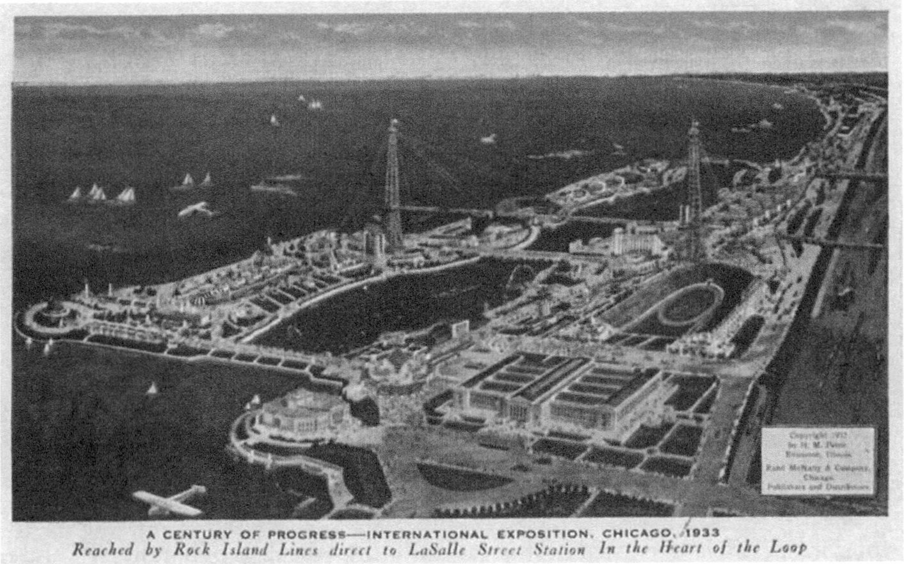

Fig.10. Vintage 1933 Chicago World's Fair Postcard, A Century of Progress International Exposition - Looking South over The World's Fair Grounds.

fairs in Chicago of 1893, St. Louis of 1904 and San Francisco of 1915, The Century of Progress was not subsidized by the local or federal government.

Several features of the fair were aimed at making the visitors forget the hardships of ordinary life. In the fair ground, they could encounter things out of the ordinary, whether it was the Sky Ride transporter bridge or the nude dancer Sally Rand. Still, many critics saw the very existence of such a costly enterprise as an unnecessary extravaganza. Like numerous other public and private organizations, the city of Chicago went bankrupt in 1929 and recovery was slow. The plans for the exhibition therefore received substantial criticism, especially from left-leaning parties, organizations and individuals who stressed the poor economic situation of people outside the exhibition grounds as well as poor working conditions at the fair.

This is a topic I discuss further in Chapter Five in relation to casual workers who were key to the success of any world's fair. They consisted of hostesses, waitresses and performers, as well as builders, craftspeople, and guards, which were positions often taken up by members of the local diaspora. Twenty countries built their national pavilions here and many became showcases of modernism as well as key locations for their respective diasporas. This was the case not only of Czechoslovakia but for example of Yugoslavia, Italy, and Ukraine too, for whom the national pavilion became a vital representation of homeland.

With regard to the architecture of the pavilions, the grounds comprised of temporary buildings mainly made of a steel frame clad with wallboard. Very few windows were included not only to avoid disrupting the facades but also to allow better regulation of the interior lighting. Illumination was key on the outside too, where electrical light in different colors were used at night. Science and its progress had their dedicated pavilions in the Hall of Travel and Transportation, Hall of Science, and Hall of Social Science, which were built next to large company buildings that included enterprises of for example Ford, General Motors or the Electrical Group. Their spectacular displays, that for instance involved suspended cars, contributed to the visualization of progress.

The entertainment features varied from attractions like the already mentioned Sky Ride, half a kilometer long, overground transportation system, to the sections with various reconstructions of American and non-American past or present in sections like Old Europe and the European village. Reimagined Europe, for example, consisted of "typical" buildings, displays of handicrafts and regional costumes, or the sale of traditional food and drink. It aimed at depicting the diversity of people in America and their ancestry. The plan was to present the traditional European structures in juxtaposition with the modernity of Chicago and bring in more visitors who could explore their own European roots.[72]

A single old European village, as it was designed, did not materialize, though, for mostly financial reasons. Moreover, many European countries objected to being presented in a village but rather opted for individualized modern pavilions. Ethnic and ethnographic reconstructions were therefore also left to individual countries and to local entrepreneurs. For instance, the Belgian village, "a faithful reproduction of an actual Old-World City," and the Streets of Paris which were constructed in 1933 turned out to be very popular. They were followed the next year, for example, by the Swiss, Irish, English, and Dutch villages, and the German pub, the Old Heidelberg Inn, advertised as "the most popular restaurant at the world's fair grounds."[73] Alongside the Oriental, Colonial or Mexican Villages, they were therefore driven by profit rather than authenticity, but their popularity also signaled which exhibits visitors found most entertaining.

The Czechoslovak pavilion at the Century of Progress was officially opened on June 17, 1933. The pavilion was an elongated, mostly one-floor building by the architect Kamil Roškot (1886–1945) and the engineer Bedřich Sirotek

72 Ganz, *The 1933 Chicago*, 128.

73 Advertisement in "Century of Progress Scrapbook, June 3–June 29, 1933," Century of Progress Collection (COP) Box 1, folder 2, Chicago Public Library.

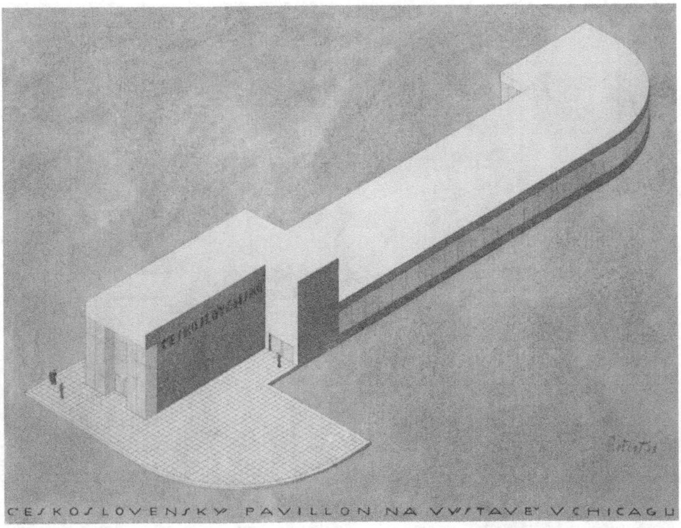

Fig.11. Kamil Roškot, The Czechoslovak pavilion for the Chicago world's fair, 1933.

(1899–1974) [fig. 11]. The higher side of the entrance to the pavilion contained the reception hall and had a glass front, which was followed by a sequence of exhibition rooms. The design was a compromise between the involved ministries and was affected by limited finances, little time necessary for the decisions and installations, and the local circumstances. They included the already mentioned keen involvement of the diaspora, which planned their own exhibit in the pavilion. Turnovský, the exhibition commissioner, was in charge of the interior mood and distribution of exhibits.[74] A major role was played by Bohuslav Soumar whom Czech-language papers in Chicago mentioned in relation to staffing and dissatisfaction of the diaspora with the management of the pavilion, all topics explored deeper in Chapter Five.[75]

The exact distribution of objects in the pavilion, apart from a few works, is unclear. The interior of the pavilion housed the now-established tropes of Czechoslovak representation—Masaryk's statue and large panoramas of Prague by Šetelík and the High Tatras by Štáfl [fig. 12]. However, in the case of the Czechoslovak pavilion in Chicago, there is a curious lack of visual evidence that would provide a good sense of the appearance of the interior displays. Apart from a couple of photographs and building plans, the content can only be reconstructed from descriptions in articles and reports by Czechoslovak officials. Official

74 Halada and Hlavačka, *Světové výstavy*, 145.

75 Critical articles of Soumar Josef Falta, "Pan Propagandista," *New Yorské Listy*, October 23, 1945, 4, "Okolo článku o čsl. pavilonu na světové výstavě," *Spravedlnost*, October 13, 1933, 8.

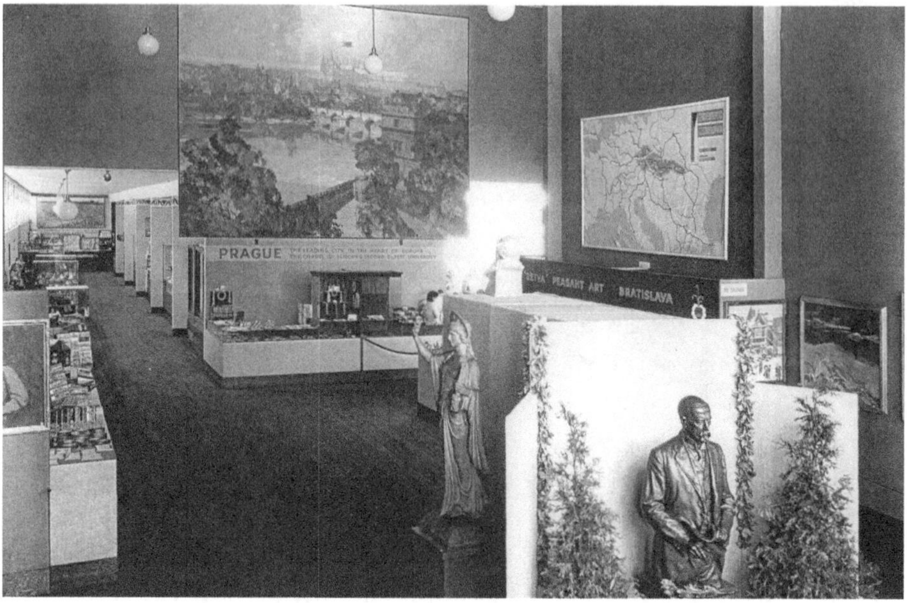

Fig.12. Interior of the Czechoslovak pavilion, Chicago, 1933.

world's fairs guidebooks generally did not devote much space to descriptions of "minor" national pavilions. The *Official Guide* to the 1933 fair, for instance, only noted that the Czechoslovak display contained a "gorgeous display of products of its varied industries, colorful and gay, the glassware and needlework of this industrious nation. Handicrafts, Bohemian glass, porcelain synthetic and precious stones, garnet jewelry, and official tourist displays are the main features among the exhibits."[76]

The guide also mentioned Czechoslovakia in the description of possible trips visitors could make in Europe, mentioning for example the glassware from Gablonz (Jablonec nad Nisou), china from Karlsbad (Karlovy Vary) or "peasant style" textiles as objects originated in Czechoslovakia which could be purchased in the "great bazaar" of the pavilion.[77] These cursory descriptions of the exhibits and Czechoslovakia as a destination nevertheless suggest what was deemed important by external reporters. In this case, it was the displayed objects like crafts and applied arts alongside the fact that many of these could be purchased in the booths and stalls of small traders. Sales of artistic and commercial goods be-

76 *Official Guide. Book of the fair 1933* (Chicago: Century of Progress, Administration Building, 1933), 95.
77 *Official Guide*, 195.

came a common addition to the exhibitions and were often purveyed by traders of émigré origin.

Because of the lack of presence of large industries in the Czechoslovak pavilion in 1933, the display was ultimately orientated towards the promotion of tourism to Czechoslovakia, and this included advertising not only Karlovy Vary as the home of glassmaking, china and the spa industry, but also destinations like the mountain ranges of the Slovak Tatras and the historic sites of Prague. The Czechoslovak pavilion in Chicago was therefore conceived to be a condensation of the country's most recognizable features with which foreign audiences as well as the Czech and Slovak diaspora in the USA could identify.

Colonial company: *Brussels 1935*

Marketing countries as tourist destinations at world's fairs developed gradually during the interwar period. The Universal and International Exposition (Exposition Universelle et Internationale) that took place in Brussels in 1935 combined the commercial aspects of tourism with the justification of colonialism [fig. 13]. Following the model of the French Colonial Exposition of 1931, the Belgian equivalent lasted between April 27 and November 6, 1935, and fo-

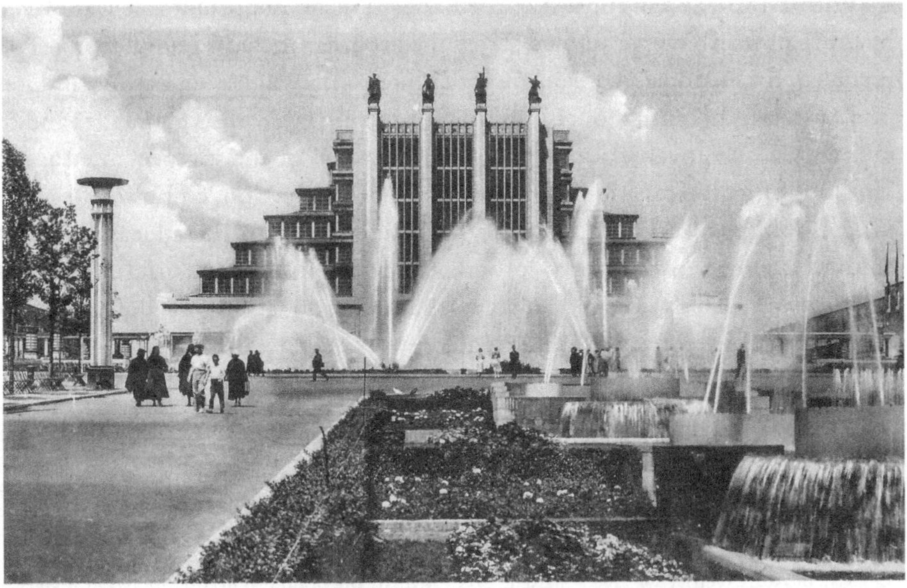

Fig.13. View of the The Universal and International Exposition in Brussels, 1935, postcard.

cused on transportation and colonization with the upbeat motto "Peace among the Races."[78] Though Czechoslovakia was free of direct colonial involvement, it nevertheless fostered close trade relationships with colonial powers as well as their colonies. Some of the contacts had been inherited from the Habsburg monarchy and some were established during the interwar period. Czechoslovakia therefore took part among the largest colonial powers of the time which included the host country, France, and the United Kingdom. Italy and France constructed several pavilions (sixteen in Italy's case), showcasing their colonies and the different cultural, social and administrative features of the respective countries. A total of twenty-five national pavilions comprised of Austria, Hungary and Poland, for example, that by this time were under dictatorial regimes.

Regarding the architecture of the world's fairs, many pavilions (Sweden, Austria, and Turkey included) embraced modernism, while others followed a more monumental expression with inspiration taken from classicism. This would be the case of the United Kingdom and some of the French pavilions.[79] The colonial section, dominated by the Belgian Congo, presented a mixture of regionalism and exoticism in its buildings. Many other structures came out as hybrid in their form, blending various historical references: for example, the Grand Palais resembled a ziggurat, while the massive Catholic Life pavilion with its three domes and six obelisks formed an intersection between a Byzantine church and a mosque.[80]

Within the colonial context of the Exposition, an important role was played by tourism which was promoted across the grounds, including advertising. This included, for example, African countries like Congo, Egypt or European states like Austria as potential tourist destinations using large photomontages and dioramas. Czechoslovakia also put emphasis on tourism and was presented as the "land full of historical and natural wonders, of production, cultural and social progress and a country that treats its minorities fairly."[81] These ideas would be epitomized in sculptures and photographs of various sites and vistas, images of folk festivals in Slovakia and Sokol assemblies, products of vocational schools, promotion of industries, as well as various works of art.

---

78 *Official Guide and Plan of the Brussels International Exhibition 1935* (Brussels: Fack-Roussel, 1935). The Brussels mayor, Adolfe Max, was one of the key men behind the organization, with Count Adrien van der Burch the General Commissioner and the architect Joseph Van Neck in charge of the site.

79 Johan Lagae, "Brussels 1935," in *Encyclopedia of World's Fairs and Expositions*, eds. Findling and Pelle, 278.

80 Lagae, "Brussels 1935," 279.

81 "Čsl. účast na světové výstavě v Bruselu 1935," Ministr obchodu pro předsednictvo ministerské rady, August 30, 1934, Výstavy v Belgii, no. 944, file 115, D6983/34, APRO.

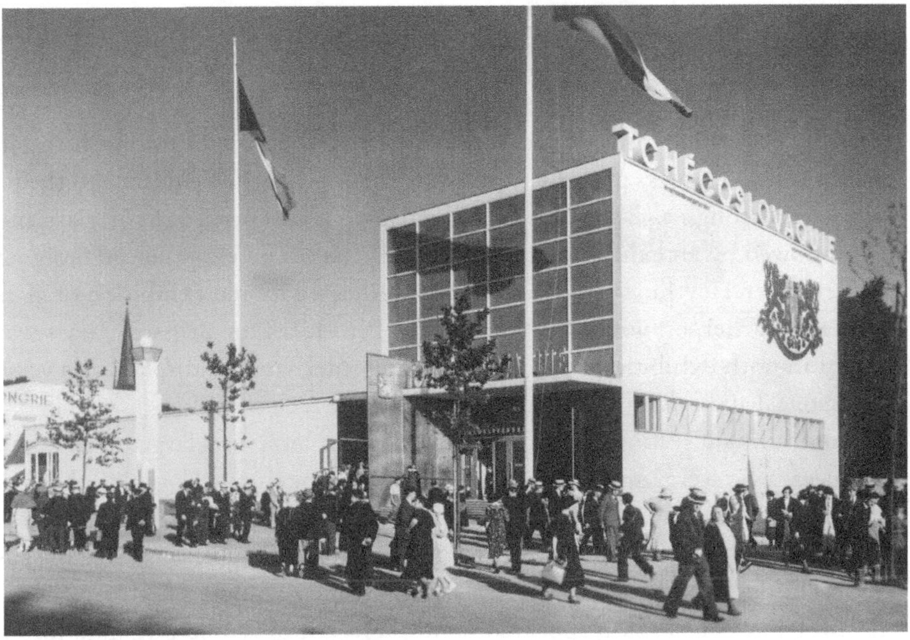

Fig.14. The Czechoslovak pavilion in Brussels, 1935.

The Czechoslovak pavilion, overseen—once again—by the exhibition commissioner Turnovský, was in the vicinity of Luxemburg, Italy, Poland, Hungary, and Sweden, opening on May 15, 1935. It was designed by the architect and designer Antonín Heythum (1901–1954) and built by the engineer Sirotek whose construction was made mostly of wood combined with glass [fig. 14]. The oblong-shaped building also included a courtyard with a fountain and a stream. The interior was lit by skylights and mirrors.[82] The large entrance hall contained presentations of various ministries, including the Ministry of Posts, Telegraphs and Telephones displaying stamps, and the Ministry of the Public Works showing radium mining, bridge models, and artworks.[83] Apart from Štursa's sculpture of Masaryk and a large panorama of the High Tatra mountains, an airplane suspended from the ceiling dominated the room. The subsequent spaces presented the visitors with Czechoslovak agriculture and industries which emphasized glass and ceramics production.[84]

82 Komora, *Hospodářské výstavy*, 224.
83 *Official Guide and Plan*, unpag.
84 Č., "Československo na světové výstavě v Bruselu 1935," *Sklářské rozhledy* 12, no. 1 (1935): 12–13.

In the Czechoslovak pavilions, glass played an important role as it could be turned into both an export commodity (in 1935 glass export occupied second place behind iron ore) and a representative artefact.[85] Its application was versatile, from decorations and practical objects at home to its use in architecture and construction. This was prominent in Brussels too where the entrance to the pavilion led through a glass doorway of transparent and black mirror glass and was followed by an exhibition of decorative glass. As an experienced designer and architect, Heythum had already put together a national exhibition of glass two years earlier, on which he built some of his ideas for Brussels. Glass was shown in both exhibitions and the following international displays as a versatile material and an important export article of Czechoslovakia.

The dominant section of glass was accompanied by several larger objects and producers who were limited to the Baťa company, the ironworks Poldi, the mining and metallurgy company Báňská a hutní and three car companies Škoda, Tatra, and Aero.[86] The industrial section also included a sizeable showcase of spas in Czechoslovakia, textiles, and musical instruments.[87] Just like the pavilion in Chicago, a sales section appeared here where visitors could purchase for example toys, glass or leather goods and Slovak cheese.[88] The pavilion as a whole was considered successful and many Czechoslovak exhibits received a total of 23 Grand Prixes and 26 golden medals from the organizers.[89]

Here too, the Czechoslovak participation was represented in more than one location. A separate modern art exhibition was shown in the General Hall of Arts under the direction of the art historian Antonín Matějček and the director of the press services of the Ministry of Foreign Affairs, Emil Purghart. The art section as a whole included artworks in different media and the Czechoslovak entry consisted of works by many authors who are today associated with classical modernism, like Emil Filla, Josef Čapek, Jan Zrzavý, sculptors Otto Gutfreund, Jan Štursa, architects Josef Gočár, Bohuslav Fuchs, Jaromír Krejcar, and

85 V. Svoboda, "Československý sklářský průmysl ve světě statistiky," *Sklářské rozhledy. Věstník sklářského ústavu v Hradci Králové* XII, no. 10 (1935): 153–154. Marta Filipová, "Czech Glass or Bohemian Crystal? The Nationality of Design in the Czech Context," in *Designing Worlds: National Design Histories in an Age of Globalization*, eds. Kjetil Fallan and Grace Lees-Maffei (New York: Berghahn Books, 2016), 141–155.

86 Styblík, "Československo v Bruselu," *Večerní Lidové noviny* July 20, 1935, Newspaper Excerpts archive, MFA, Oh8.

87 "Čsl. pavilon na mezinárodní výstavě v Bruselu," *Národní listy* May 29, 1935, Newspaper Excerpts archive, MFA, Oh8.

88 Styblík, "Československo."

89 "Čs. Úspěch na výstavě v Bruselu," *Lidové noviny* June 29, 1935. Newspaper Excerpts archive, MFA, Oh8.

Pavel Janák.[90] They were predominantly of Czech origin and male; the only female artist was the sculptress Mary Duras while the non-Czechs of Czechoslovakia included a handful of Jewish, German and Slovak artists.

Art exhibitions at world's fairs provide an important insight into the way the government, by means of the curators and commissioners, tried to shape the presentation of the country as modern. The questions of modernity and its interpretation were another ongoing concern of the interwar period and its exhibitions. What kind of modernism should be included, who the authors should be and what subject matters best represented Czechoslovakia were issues that had a direct impact on the overall image of the state both externally and internally.

Flow of display: *Paris 1937*

The modernity of world's fairs can be often related to the canonical buildings or structures built on these occasions, whether they survived the exhibition or not. The Crystal Palace of the 1851 Great Exhibition was an impressive, half a kilometer long construction of glass and cast iron which was eventually destroyed by fire in 1936. The Eiffel tower became the "clou," in other words, the chief attraction, of the Parisian Exposition in 1889, while the modernist German pavilion by Mies van der Rohe is what is left of the Barcelona 1929 world's fair. The iconic juxtaposition between the Soviet and German pavilions was a key feature of the Exposition Internationale des Arts et Techniques dans la Vie Moderne (The International Exposition of Art and Technology in Modern Life) in Paris in 1937, indicative of the political and ideological divisions in Europe [fig. 15].

Designed by Boris Iofan and Albert Speer respectively, the Soviet and German structures were built as responses to the modernity proclaimed by the exposition's title and arrived at a similar combination of historicism and modernism.[91] Both structures symbolized the radicalizing atmosphere of the late 1930s visible in many other national pavilions.[92] These divisions were also not the only ones

90 "Tchecoslovaquie," in *Exposition internationale d'Art moderne catalogue* (Dietrich and co, 1935), 244–254.

91 Danilo Udovički-Selb, "L'Exposition de 1937 n'aura pas lieu: the invention of the Paris International Expo and the Soviet and German pavilions," in *Architecture of Great Expositions 1937–1959: Messages of Peace, Images of War*, eds. Rika Devos, Alexander Ortenberg, and Vladimir Paperny (Farnham : Ashgate Publishing, 2015), 32; Danilo Udovički-Selb, "Facing Hitler's Pavilion: The Use of Modernity in the Soviet Pavilion at the 1937 Paris International Exhibition," *Journal of Contemporary History* 47, no. 1 (January 2012): 13–47; Karl Schlögel, *Moscow 1937* (Cambridge: Polity, 2013).

92 Apart from Udovički-Selb, e.g., Karen Fiss, *Grand Illusion: Third Reich, the Paris Exposition, and the Cultural Seduction of France* (Chicago: University of Chicago Press, 2009) and Karen Fiss, "In Hit-

that shaped the exhibition; the dualism between the capital of Paris and the French regions, between France and its colonies, and between socialism and capitalism were prominent around the grounds.[93]

The Exposition's official theme of arts and technology in modern life attracted the participation of forty-five nations.[94] Opened on May 24, 1937, by the French president Albert Le Brun, the fair occupied both sides of the Seine and the sites of the previous international exhibitions. It closed on 25 November having been visited by some 31 million people.[95] Ephemeral structures were built here too, either as national pavilions or specialized halls. This included Alvar Aalto's Finnish pavilion, or Jose-Luis Sert and Luis Lacasa's Spanish pavilion which featured Picasso's *Guernica* and Alexander Calder's kinetic fountain. Art therefore had a crucial place at the Exposition and artworks appeared in various build-

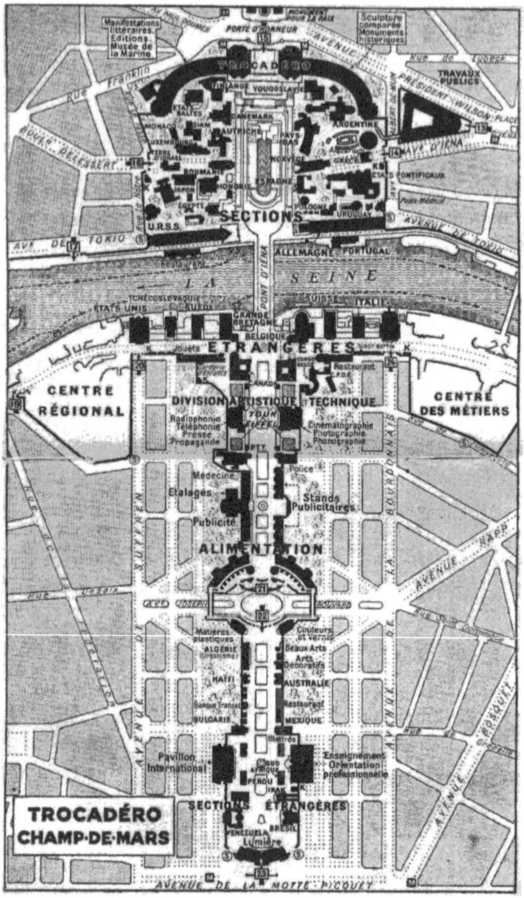

Fig.15. Site map of the Paris World's Fair 1937.

ler's Salon. The German Pavilion at 1937 Paris Exposition Internationale," in *Art, Culture and Media under the Third Reich*, ed. Richard A. Etlin (Chicago: University of Chicago Press, 2002), 316–43.

93 Arthur Chandler and Philip Whalen, "Paris 1937," in *Encyclopedia of World's Fairs and Expositions*, eds. Findling and Pelle, 290.

94 James D. Herbert, *Paris 1937. Worlds on Exhibition* (Cornell University Press, 2018); Julia Kostova, *Spectacles of Modernity. Anxiety and Contradiction at the Interwar Paris Fair of 1925, 1931 and 1937*. A dissertation at the State University of New Jersey, Rutgers, May 2011; "Paris" in Brigitte Schroeder-Gudehus and Anne Rasmussen, *Les Fastes du Progrès. Le guide des Expositions universelles 1851–1992* (Paris: Flammarion, 1992), 192–200; Robert H. Kargon et al., "Modernity a la française. The 1937 Paris Exposition," in Robert H. Kargon et al., *World's Fairs on the Eve of War: Science, Technology and Modernity, 1937–1942* (University of Pittsburgh Press, 2015), 7–29.

95 Brigitte Schroeder-Gudehus, "Paris," *Les Fastes du Progrès. Le guide des Expositions universelles 1851–1992* (Paris: Flammarion, 1992), 192.

ings, including, for instance, Sonia and Robert Delaunay's plane in the Aviation pavilion or Fernand Léger's painting of Transport des Forces (Power Transmission) in the Palais de la Découverte (the Discovery Palace). The visual arts exhibitions were also located in the pavilions of fine arts, decorative arts and the purpose-built Museum of Modern Art.[96]

The Exposition's theme of modern life included technology, which had a prominent place here. Apart from the Discovery Palace and the Aviation pavilion, other technological attractions included the extensive use of electricity and lighting across the grounds, illuminating buildings and the many fountains at night. The Czechoslovak pavilion also made extensive use of light on the glass façade and embraced the focus on technology in several industrial exhibits of various sizes. This included the 10t heavy Pelton turbine by the company Škoda and several large, heavy products by Vítkovice Steel Works, which were placed outside of the national pavilion.[97] The reasons for their placement in the open air were the fact that they were visually attractive, sizable exhibits that benefited from the display in the exterior. On a more practical level, they could not be incorporated indoors for their weight would likely damage the structure.

The Czechoslovak participation in Paris was the result of another extended discussion about the look, content and aims of the national pavilion. The debate revolved again around the ideological focus of the exhibition.[98] The Commissioner General of the pavilion was Jan Krčmář from the Ministry of Education and this time, Turnovský was appointed his deputy. Apart from the main national pavilion, Czechoslovakia was represented in several specialized halls, which included an exhibition of local print culture in the Press pavilion, Czechoslovak agriculture at the Annex Maillot, and industries in the International Pavilion. Czechoslovakia also had an exhibit in the Theatre Pavilion and contemporary fine arts located in the Charpentier Palace outside the main grounds. A guidebook to the exhibition also listed a pavilion of the Czechoslovak industrial arts with drawings of folk costumes, glass and other domestic artifacts, where tobacco products were sold and refreshments like beer and ham served.[99]

96 Chandler and Whalen, "Paris 1937," 292.

97 The Vítkovice company also provided material for the construction of the pavilion. "My na světové výstavě v Paříži," *Venkov*, August 22, 1937, 5.

98 "Zpráva odborné komise pro výtvarnou práci z prvé schůze dne 11.1.1937," Výstavy v cizině. Francie, Paříž 1937, sec. III, part V, no. 6328, MFA Archive.

99 V. Darras, "Praktický průvodce po pařížské výstavě umění a techniky v moderním životě," *L'independence Tchecoslovaque*, no. 11, 4.

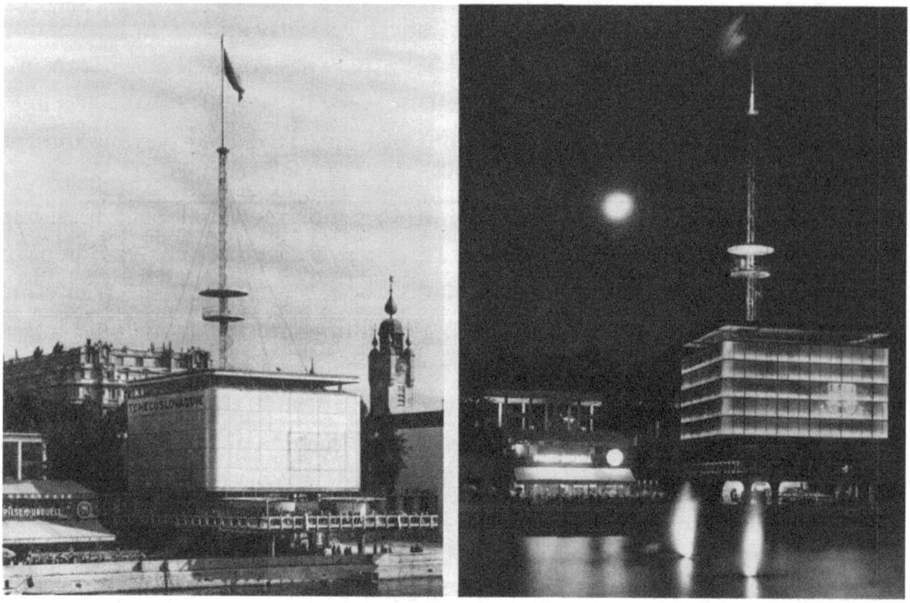

Fig.16. View of the Czechoslovak pavilion, Paris, 1937.

The design of the Czechoslovak pavilion came out of a competition which attracted many entries.[100] The winning structure by the architects and designers Jaromír Krejcar, Zdeněk Kejř, Bohuslav Soumar and Ladislav Sutnar was made of steel and glass and had to negotiate location on a slope by the river [fig. 16]. The pavilion, which an anonymous commentator poetically called a "fairy tale of steel and glass," had a cubic shape on steel pylons and a slim and tall tower, supported by iron rods.[101] A prominent feature was a terrace, which extended over the Seine, serving as an extension of the indoor exhibition space. The visit to the pavilion was planned as a predetermined route that started on the ground floor. This space contained the official Hall of Honor and led to a section of specialized (vocational) schools. Showcases of glass and porcelain followed, finished by textiles and musical instruments before the stairs took visitors to the first floor with the Industries Hall. Back on the ground floor, the display continued with agriculture and exited to the terrace, into the tourism section. Here a panorama of Prague was shown, this time by the painters Jan Slavíček and Richard

100 "Veřejná soutěž na ideové náčrty na čsl. státní pavilon pro mezinárodní výstavu Paříži 1937," "Umění a technika v moderním životě" v Paříži 1937, 20041215/07 f.140 Kejř, AACE NTM.

101 "Jako skvělá pohádka z oceli a sklad bude zářit skleněný čs. pavilon na světové výstavě v Paříži," *Venkov* 32, February 7, 1937, 8.

Wiesner, alongside folk artifacts and a kinetic fountain by the sculptor Zdeněk Pešánek which advertised Czechoslovak spas.[102]

In many respects, the pavilion was a continuation of exhibition techniques applied by Heythum in Brussels. His emphasis on glass as a key material of Czechoslovak representation and the planned route through the building were developed further in Paris. Sutnar and his colleagues distributed the space and objects as a controlled flow, a topic returned to in the following chapter. This strategy was planned to be applied at the subsequent, and final, Czechoslovak national pavilion of the interwar period that was planned for New York in 1939–40 and that was affected by the events of the Second World War

End of an era: *New York and San Francisco, 1939–1940*

The World of Tomorrow in New York City was the last world's fair at which interwar Czechoslovakia had its own national pavilion. Or at least, it planned and built one in the name of Czechoslovakia. Events leading to the Second World War then took their course; first, the border regions with Germany and Austria with large German populations were ceded to Germany in the Munich Agreement of September 1938. In October 1938, Slovak autonomy was declared and the new name of the state, separating the Czech and Slovak by a hyphen confirmed in the constitutional law.[103] It signaled growing Slovak emancipation, which escalated in the declaration of independence on March 14, 1939, leading to the creation of the Slovak Republic, a puppet state of the German Reich, on the one hand and the Protectorate of Bohemia, Moravia and Silesia on the other. The national pavilion in a faraway USA, the construction of which had already been underway, turned into one of the symbols of the disappeared interwar Czechoslovakia. A similar fate was met by another participation in America that took place simultaneously with New York; the Czechoslovak section at the Golden Gate International Exposition in San Francisco.

Both fairs had been prepared for years by the respective cities and competed for exclusivity and recognition by the BIE. Even though the World of Tomorrow in New York was officially granted a world expo status, in the end, both events went ahead because Golden Gate received support from the American

102 "Préambule," *Tchecoslovaquie à l'Exposition Internationale des Artes et Techniques dans la Vie Moderne Paris 1937*, 5–6.

103 Ústavný zákon zo dňa 22. novembra 1938 o autonomii Slovenskej krajiny, no. 299, *Sbírka zákonů a nařízení státu československého* 1938, 1161–1164.

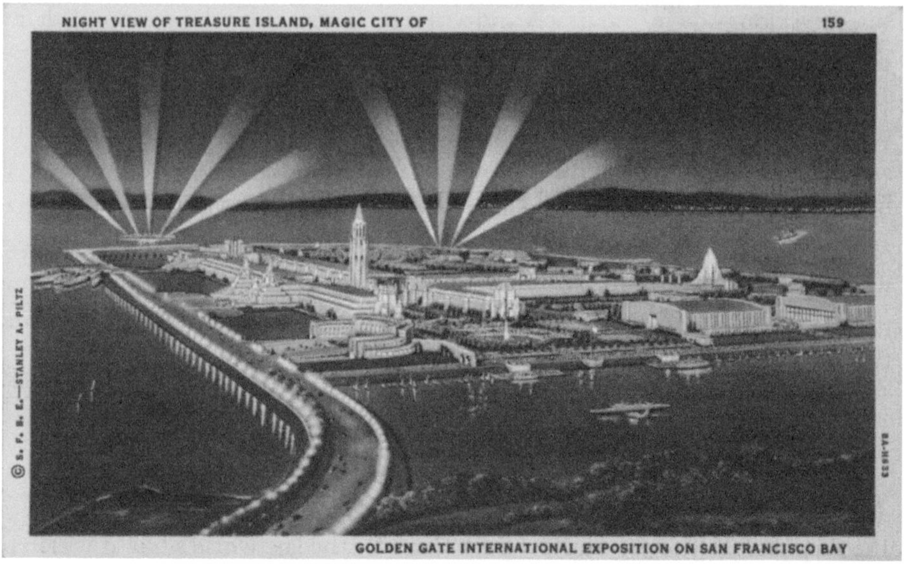

Fig.17. Treasure Island, Magic City of Golden Gate International Fxposition, 1939.

President Franklin D. Roosevelt. Each exhibition could therefore serve their respective coasts. The San Francisco fair ran between February 18, 1939, and September 29, 1940, with a winter break. Its title was "The Pageant of the Pacific," consciously built on the legacy of the 1915 Panama Pacific International Exposition in the same city. The 1939 fair celebrated achievements in transportation, and world trade and set out to be "an important world goodwill center."[104] It was organized by the San Francisco Bay Exposition Corporation, which was established for this purpose in 1934 and consisted of the Bay area's financial, industrial and commercial leaders[105] [fig. 17].

Most of the 1939 buildings were ephemeral and located on Treasure Island. The artificial island was purpose-built for the exhibition in the San Francisco Bay, as there was no more available space in the city. Its Art Deco architecture was designed to impress by its size and references to historical styles and distant countries. Pyramids, Native American and Asian structures tried to revive the

104 *Official Guidebook. Golden Gate International Exposition* (San Francisco, 1939), 12.

105 *Official Guidebook*, 33; Jack James and Earle Weller, *Treasure Island. "The Magic City" 1939–1940. The Story of the Golden Gate International Exposition* (San Francisco, 1941); Patricia F. Carpenter and Paul Totah, eds., *The San Francisco Fair. Treasure Island 1939–1940* (San Francisco: Scottwall Associates, 1989); *Golden Gate International Exposition. A Pageant of the Pacific 1939* (1936), Andrew M. Shanken, *Into the Void Pacific. Building the 1939 San Francisco World's Fair* (Berkeley: UC Press, 2015). The exhibition president was Leland W. Cutler, an insurance businessman, followed by Marshall Dill of the Housing Authority Donald G. Larson, "San Francisco," *Encyclopaedia*, 309.

success of the replica of the famous Angkor Wat temple that was built at the 1931 Colonial Exposition in Paris.[106] Two large towers of the Portals of the Pacific served as an entrance to the site while the Court of Honor with another tall Tower of the Sun was the central landmark. The designs of the Court of the Seven Seas, the circular Court of Pacifica, the Court of the Moon, the Federal Building with murals and forty-eight columns representing the American states, were an attempt at a hybrid combination of architectural elements from the eastern and western coasts of the USA.[107]

The site included many public works of art like sculptures and murals, one of the most prominent ones by Diego Rivera. In 1940, he created a mural of the Pan American Unity, as it is known today, for the Art in Action pavilion. The mural was both a political and artistic commentary on the state of the world with its many cultural and political references.[108] The grounds also contained fountains and pools, as well as the entertainment section of the Gayway. There was also the interactive dinosaurs exhibit, carousels and Ferris wheels, a Scottish village and a series of "exotic" performances, including those by the nude dancer Sally Rand who had already been making her widely publicized mark in the Chicago fair in 1933.

Various colonial and indigenous exhibits were also included either in the entertainment zone or in various exhibition halls. Official foreign participation at Golden Gate was small and limited to the national pavilions of France, Argentina, Brazil, Italy and Norway because most countries concentrated their efforts and finances on New York. An international section was located around the Lake of Nations where the International Hall included the Czecho-Slovak section designed by Antonín Heythum, known for his work on the Czechoslovak pavilion in Brussels [fig. 18].

The Czechoslovak section in the Golden Gate Exhibition officially opened on April 3, 1939, three weeks after the German takeover, and renewed on May 25, 1940. By the time of the planned national day on July 8,1939, Czechoslovakia was no more. At the world's fairs in San Francisco, as well as in New York,

---

106 Anna Burrows, "The San Francisco Golden Gate Exhibition 1939–1940," The Digital Collection, University of Maryland libraries, December 5, 2023, https://digital.lib.umd.edu/worldsfairs/result/id/umd:1010.

107 "A New Style of Architecture Designed by Californians," *Official Guidebook,* 75–77.

108 Consisting of ten panels, Rivera's work represented a blend of indigenous and modern features on Northern America, with portraits of his contemporaries and historic figures (including Frida Kahlo, the actress Paulette Goddard, the American diver of Croatian descent Helen Crlenkovich, as well as politicians like Stalin, Hitler, and Mussolini. Will Maynez, "Art and Danger at the Fair: Diego Rivera Escapes to the GGIE," a talk at the Treasure Island Museum, November 16, 2019. Diego Rivera Mural Project at riveramural.org.

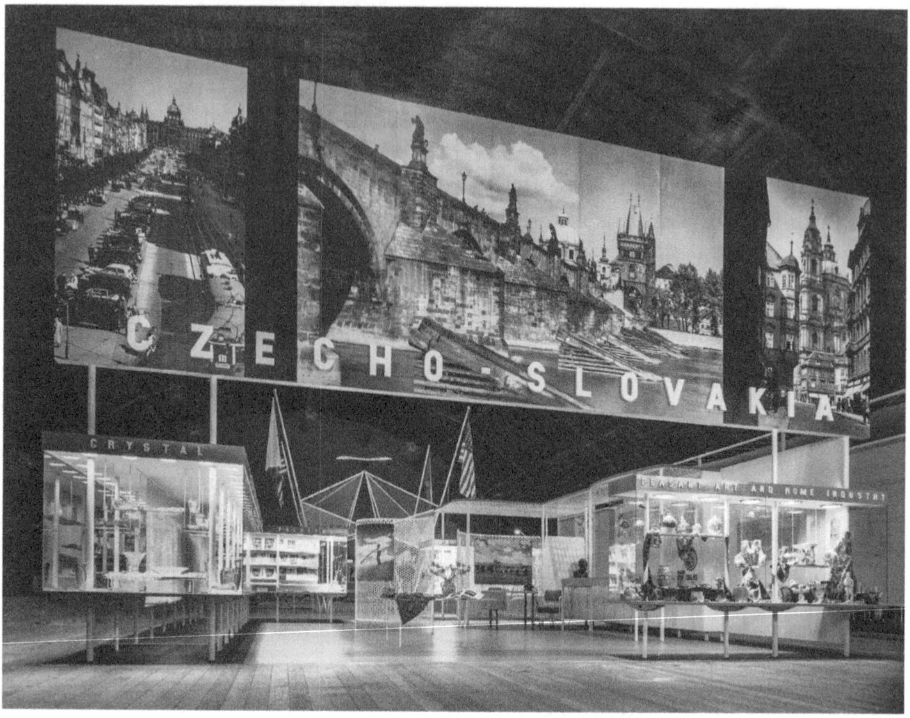

Fig.18. Czechoslovak exhibit at the Golden Gate Fair, San Francisco.

Czechoslovakia became Czecho-Slovakia in response to the political developments in Central Europe. The country's name was hyphenated to reflect the political and ethnic separation but also to indicate that American organizers as well as politicians did not accept the German takeover as legal and continued to recognize the country in its interwar form [fig. 19].

The Czecho-Slovak section in San Francisco was affected by these events and reduced in size and content. It consisted of large-scale photographs of the country's tourist sites and landscapes, accompanied by smaller exhibits, such as glass, porcelain, ceramics, toys, folk dresses and embroideries, which were all easy to transport and display here.[109] The photographs were placed on metal mash constructions and a bust of Masaryk by the Czech sculptor Josef Mařatka was located in the center.[110] Masaryk, who passed in 1937, continued his symbolic presence in exhibitions, which became especially important in the times when the country's political sovereignty was endangered.

109 *Official Guidebook*, 55.
110 The bust is today located in the Rose Garden of the Golden Gate Park.

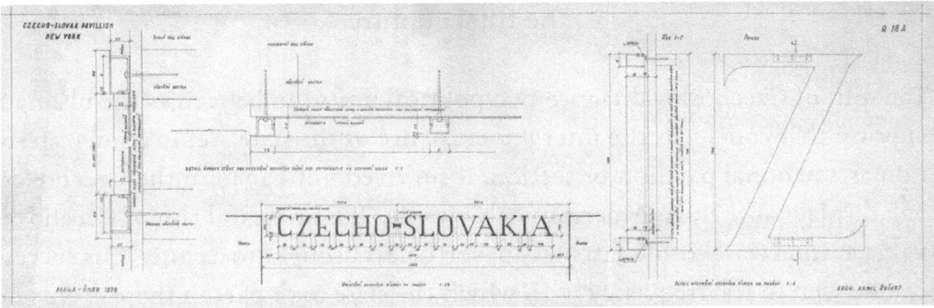

Fig.19. Inscription Czecho-Slovakia, design for the pavilion in New York, 1939.

After several exhibits failed to be dispatched from Europe or arrive at their destination, they were replaced with substitutes on the spot. In San Francisco, newly added objects included, for instance, exhibits and promotional literature focused on resistance or a collection pot for donations towards its upkeep. Financial and legal issues also arose in this relation. The contract of participation at both world's fairs became invalid because the legal entity entered, i.e., Czechoslovakia, ceased to exist. Similarly, the ownership of the exhibits became unclear which led to questions about who should pay the customs fees, shipping and storage of the objects and who should cover their insurance. However, Czechoslovakia was not the only victim of the war events.

Poland, Holland, Belgium, Norway and France were also occupied by Germany which impacted both their political sovereignty in Europe and the national pavilions in the American world's fairs. In any case, this was not a novel situation. The Panama-Pacific International Exposition of 1915 that took place in San Francisco was similarly affected by the First World War and the political changes in Europe. Many exhibits from for example Austria, Germany or Finland had been withheld in the USA well into the 1920s in disputes about where they should be returned.[111] In 1940, the war in Europe had started and the return of exhibits and many people from the USA to their home countries was made impossible.

111 "List of Austrian, Hungarian, Finnish and Norwegian Exhibits Consisting of Paintings, Graphics, Bronzes, Porcelains, Sculpture, Embroideries and Rugs Originating Abroad Through J. Nilsen Laurvik and on Exhibition at the Palace of Fine Arts by the San Francisco Art Association," Panama Pacific International Exposition, Records 1911-1929, carton 167, Manuscript Collection, BANC MSS C-A 190, U.C. Berkley Bancroft Library.

The final rupture

The split of Czechoslovakia into two political units can be seen as the ultimate rupture that concludes the interwar existence of the state. Yet in its manifestation as a national pavilion or section, it survived until 1940 in the Czecho-Slovak displays at the two American exhibitions. The successful run of Czechoslovakia at interwar world's fairs was nevertheless brought to a close. This success can be seen in the frequency with which the state took part in these events despite the ongoing and frequent criticism of the cost and various problems, including economic downturns, arguments about the content, late installations and last-minute changes in the exhibits. It is almost impossible to measure the achievements in terms of what and how much trade the participation generated. At the same time, Czechoslovak pavilions were—often in retrospect—seen as successful in their representative function and for the number of prizes and awards received for various exhibits and projects. While at the start of the interwar period, Czechoslovakia was a new, little-known state announcing itself to the world, over the course of only twenty years it developed a recognizable brand at least in the space of the world's fairs.

The last national pavilion was built in New York and this participation had been planned as twofold: a national display in a pavilion and an exhibit in the Hall of Nations, to which eight different ministries would contribute.[112] Following the Munich Agreement, however, the extent of the involvement was reduced by a third.[113] In March 1939, very shortly after the German takeover, the new government in Prague decided to liquidate the exhibit in the Hall of Nations completely, sell or lease the half-built national pavilion and put a stop to all transport of exhibits. Yet as the German interventions in Central Europe was not recognized, the organizers placed the Czechoslovak pavilion into the custody of the Czechoslovak ambassador Hurban, the mayor of New York City, Fiorello La Guardia, the president of the fair Grover A. Whalen, and the Czechoslovak commissioner general George Janeček who continued with the exhibit.[114] The unfortunate situation of the pavilion (and the state) was publicized in American press that referred to the building as a "silent pavilion," "an orphan" and

112 Ministerstvo obchodu, průmyslu a živností, June 9, 1939, *Ministerstvo obchodu* SV-NY, 1939–1940. NACR, 2.
113 Ministerstvo obchodu, 2.
114 "Czech Fair Centre is Now an Orphan," *New York Times*, March 17, 1939, clippings in Series II. Ladislav Sutnar papers (Accession no. 2013.M.6), 1897–1976, GRI.

"a monument to the tragedy of the Czecho-Slovak Republic."[115]

The entire organization and course of the New York world's fair was tinged by the increasing political tensions and the eventual outbreak of the war. The fair ran between April 30 and October 31 in the first year and between May 11 and October 27 in the second year.[116] Located in Flushing Meadows (at the time the city's dump and swamp) in Queens, the optimistic theme of Building the World of Tomorrow looked into the future of science, technology, consumerism, and capitalism [fig. 20]. The fair also commemorated 150 years since the inauguration of George Washington as president.

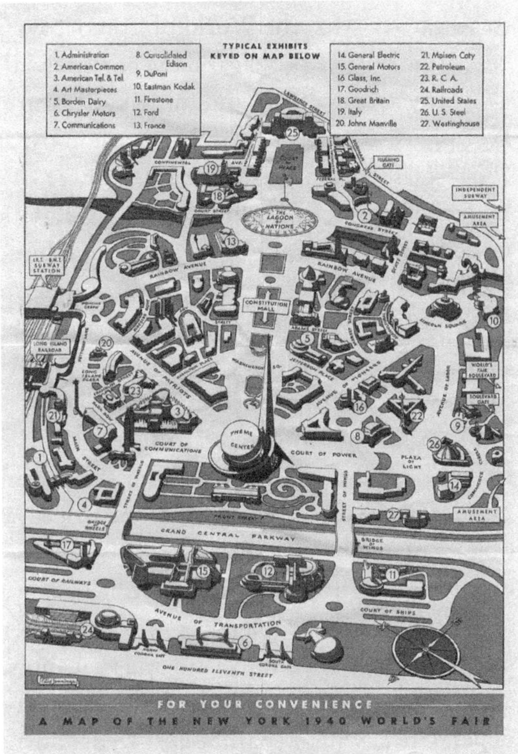

Fig. 20. Foldout from the General Motors brochure, "Things to see at Highways and Horizons," 1939.

The event saw the participation of 58 nations, 22 national pavilions, 26 United States pavilions, and many other commercial and entertainment halls built on the grounds. Foreign participation included the Soviet Union, Great Britain, Japan, Italy and Poland. Imperial powers also showcased their colonies; there was the Belgian Congo "depicting life in Belgian Africa," and the British Colonial Empire "showing colonial life and products from four corners of the world."[117] Similar

115 "Czech Fair Centre," "The Silent Pavilion," *New York Times*, May 3, 1939, "Czech Pavilion is Now an Orphan," *New York Times* March 17, 1939, clippings in Ladislav Sutnar Papers, GRI.

116 Rydell and Schiavo, eds., *Designing Tomorrow*; Andrew F. Wood, *New York's 1939–1940 World's Fair* (Charleston, SC: Arcadia, 2004); Richard Wurts and Stanley Applebaum, et.al., *The New York World's Fair 1939/1940 in 155 photographs* (New York: Dover Publications, Inc., 1977); *Official Guide Book of the New York World's Fair 1939* (New York: Exposition Publications, Inc. 1939); Roland Baker, Official *Guidebook: New York World's Fair 1939 "Building the World of Tomorrow"* (New York: Exposition Publications, 1939); *Official Souvenir Book, New York World's Fair Bulletin 1939* (New York: Exposition Publications, 1939).

117 "Fair Facts," *Officially Licenced New York World's Fair Comics 1940* (New York: Detective Comics, 1940), 46.

to all earlier fairs, the World of Tomorrow was a medley of entertainment and education, visible propaganda of commercial interests and political declarations.

Many architectural and futuristic features defined the image of the fair, the Perisphere and the Trylon being the most spectacular and visible.[118] Wallis K. Harrison's Trylon was a 212 high tower of a triangular shape, the Perisphere by Jacques-André Fouilhoux a massive globe of 65 meters in diameter. It contained a diorama envisaging a city one hundred years in the future, designed by the designer Henry Dreyfuss. Another diorama of a futuristic 1960 city was the highly popular Futurama of Norman Bel Geddes in the General Motors section.[119] Yet despite its 45 million visitors over the two years, the World of Tomorrow eventually ended in a financial loss.[120]

In this future-oriented setting, the Czechoslovak contribution was of a country without a certain future. The exhibit in the Hall of Nations became a reminder of the disappeared state. The space was left deliberately empty.[121] The few objects placed here in the end consisted of a few works of art, including a sculpture of the Fighting Republic by the Czech artist Jaroslav Brůha.

Anti-German propaganda was prominent in the national pavilion where one part was turned into a cinema that showed newsreels and short films about Czechoslovakia.[122] Mottos and quotations reminiscent of the country's lost independence, which shall be restored, adorned the walls in the interior. Karel Čapek's reminder of the peaceful nature of the nation, taken from his text on Czechoslovakia as the heart of Europe, was included here alongside a lengthy inscription explaining the lack of exhibits. They recalled that the "young republic became a victim of a whole series of betrayals and broken pledges" and ultimate ruthless invasion.[123] The motto "After the tempest of wrath has passed the rule of thy country will return to thee, O Czech people" appeared prominently above the main entrance. Originating from the 17th century Bohemian Protestant thinker and reformer Jan Amos Komenský (1592–1670), it suggested that the hardship Czechoslovakia was going through would be eventually overcome and the people would be free again.

---

118 Paul T. Sayers, "New York 1939–1940," in *Encyclopedia of World's Fairs and Expositions*, 303.

119 Robert H. Kargon, "Whose Modernity? Utopia and Commerce at the 1939 New York World's Fair," in *World's Fairs on the Eve of War*, 72–75.

120 Sayers, "New York," 303.

121 Sutnar's notes for *The Czechoslovak Pavilion*, 1897–1976 (Accession no. 2013.M.6), 1897–1976, GRI Los Angeles.

122 Sutnar's notes, GRI; Film 1938–1939, NY-SV, NACR.

123 Ladislav Sutnar, *The Czechoslovak Pavilion*, pamphlet, Ladislav Sutnar Papers, (Accession no. 2013.M.6), 1897–1976, GRI Los Angeles.

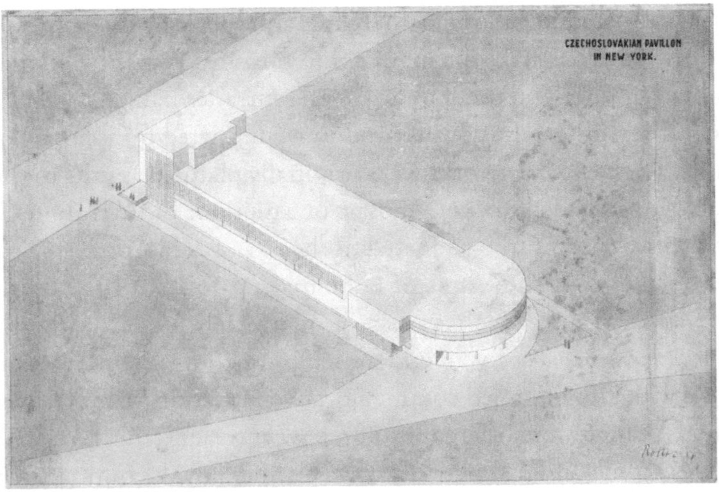

Fig. 21. Kamil Roškot, Design of the Czechoslovak pavilion, 1939.

Disruptions occurred also in more material aspects of the Czechoslovak participation. The building was designed by the architect Kamil Roškot who had previously worked on the national pavilion in Chicago. It was an elongated structure with a tall front of a solid wall, behind which the main exhibition part benefited from large glass areas finished by a curved end [fig. 21]. The construction was overseen by the engineer Jaroslav Polívka and consisted of steel delivered from the ironworks of Vítkovice in Czechoslovakia. The interior and its sections were originally co-designed by other architects, although serious on-the-spot adjustments had to be made considering the changed political situation. Just like in San Francisco, many exhibits failed to arrive in New York from Czechoslovakia and the display had to be profoundly modified using available resources. The exhibition committee appealed to the Czech and Slovak diaspora for objects of interest to be lent to the exhibition and to the public for monetary donations towards the pavilion. The pavilion became a symbolic site of resistance for the diaspora and politicians alike and eventually opened on May 31, 1939, a month later than the actual fair with flags in front of it at half-staff. Exiled President Edvard Beneš, who lived in the USA from February to June 1939, delivered one of the opening speeches, reassuring the crowds that Czechoslovakia would continue to live, evidenced by the existence of the national pavilion.[124]

124 Černovský, "Cestovní zpráva. Referát amerických novin o otevření čsl. pavilonu 31.V.1939," June 10, 1939, Korespondence ředitelny fy Baťa s USA, I/4 Ředitelna, inv.no. 412 fond Baťa 1939, MRA in Brno, branch Zlín.

The interior was also a mixture of original ideas and subsequent changes. It was divided into the entrance hall, the main exhibition area across two floors and a restaurant with indoor and outdoor areas. The entrance hall provided an overview of the history of Czechoslovakia and was dominated by a mosaic of medieval Prague by Roškot. Many works were originally planned as gifts to the USA after the fair, this included Roškot's mosaic or a diorama of Washington's inauguration which was meant to be donated to the city of New York.[125]

There was also a mural of the Kings of Bohemia, a painting of the tenth-century Prince Wenceslas, and a replica of the crown jewels. Underneath was a sculpture of *The Sleeping Republic* by an émigré sculptor from Bohemia, Mario Korbel (1882-1954) who also designed commemorative medals in bronze, silver and golden versions, which were offered for purchase at the entrance to the pavilion to generate income and distributed to Czech and Slovak communities across the USA.[126] The main hall contained objects more typical of trade exhibitions, like examples of light and heavy industries. The Kaplan turbine and a coil spring were suspended from the ceiling as impressive symbols of modernity, while smaller items like linen or glass were to demonstrate the quality of Czechoslovak export industries.

The pavilion's upper floor was dedicated to "the working hands of Czechoslovakia," and part of it had been reserved for Baťa. The company's exhibits included a large, curved window "Hymn to Work" by the painter Cyril Bouda depicting the evolution of the company from a humble shoemaking by a cobbler to a modernized shoe production by a global giant. This section was also equipped with a carpet sponsored by Baťa, hand woven by patients of an asylum in Prague, and with cushion chairs for visitors. The other parts of the upper floor featured folk artifacts including a maypole and small handmade articles available for purchase. Folk art again played an important role as a constant and safe referent to the cultural uniqueness of a nation that may have temporarily lost its statehood but would survive otherwise.

125 Markéta Formanová and Zdeněk Kuchyňka, *Ztracené diorámа: příběh unikátního československého exponátu na Světové výstavě v New Yorku v roce 1939 = The Lost Diorama: The Story of a Unique Czechoslovak Exhibit at the 1939 World's Fair in New York City* (Plzeň: Západočeské muzeum, 2019).

126 E.g., "Československo bude zase svobodno!," *Věstník Česko-Americké jednoty* XXVI (July 1939): 4.

Conclusion

The repeated appearance of specific objects in the national pavilions established familiarity with the main features of the Czechoslovak exhibit, so important for the recognition of the country as a self-sufficient and visually recognizable entity. The sculpture of President Masaryk, for example, evoked the permanent quality of the state's democracy, while folk art would point to what was presented as authentic and historical in Czechoslovak culture. Touristic sights like Prague or the Tatra mountains that also appeared regularly in the pavilions further contributed to the uniqueness of Czechoslovakia. It was not only objects but also people that ensured, however consciously, the persistence of these visual tropes. The exhibition commissioner Turnovský, the ministerial aid Štech, the designer Sutnar are examples of those who recurrently appeared in key roles at different stages of building the national presentation and ensured the continuity of exhibits and exhibition ideas.

With such objects, people and ideas, Czechoslovak pavilions created their own historical time not because they would stand outside political and economic events; on the contrary, they stood in the middle of them. A correct sense of historical time in national pavilions was not so crucial and the past played a role only to the extent to which it informed the displayed present and future. During the twenty interwar years, the past—as presented by Czechoslovakia—remained relatively constant and represented by the same artifacts. The main emphasis was on the modern present and the future that could be shaped and selected. Artifacts like paintings of attractive places or products of folk culture therefore played a subsidiary role in objects representing modernism and modernity of the state and its people. Technologically innovative products of companies, the latest architecture of the pavilions and their materials, alongside novel design ideas, linked Czechoslovakia to the universal progress and future orientation of the world's fairs.

Chapter Two

EXHIBITION SPACES

Explaining their project design of the Czechoslovak pavilion for The International Exposition of Art and Technology in Modern Life in Paris, 1937, the winning team noted:

> We may say that 90% of people [visiting exhibitions] are not exhibition experts and do not share the cultural demands of the "upper ten thousands," and as such they spend very limited time looking around exhibition pavilions. ... The content of the pavilion should therefore be such that it captures masses of these ordinary visitors while satisfying the demands of the culturally mature visitor.[1]

Despite their somewhat patronizing tone towards the cultural level of exhibition visitors, the designers expressed their clear awareness of the way people behave in large exhibitions. They observed that visitors only skimmed some of the displays while focusing on entertainment and spectating. Architects and designers therefore needed to carefully consider not only what is displayed but also how.[2] The team of four Czech designers and architects, Jaromír Krejcar, Zdeněk Kejř, Bohuslav Soumar and Ladislav Sutnar, continued their explanatory notes for their project and proposed that "when designing the façade and all interior spaces, the psychology of the visitors should be affected. The psychology of a visitor of world's fairs is very different from the psychology of a visitor of smaller and specialized exhibitions."[3]

---

1 "Vysvětlivky k průvodní zprávě projektu československého pavilonu na Výstavě umění a techniky v moderním životě v Paříži v roce 1937 k soukromému použití poroty," f.140, Kejř, 5, AACE NTM.

2 "Vysvětlivky," 5.

3 "Vysvětlivky," 5.

The Czechoslovak pavilion at the Parisian exposition in 1937 can be considered one of the most successful examples of exhibition architecture for reasons that I will discuss in this chapter alongside the designs that led to it. With the last national pavilion of the interwar period at the 1939 New York world's fair ending as a curtailed version of the original plans, Paris represents the peak of the joint exhibition efforts of the Czechoslovak government, the creators and the exhibitors who had been perfecting their exhibition approaches and techniques for almost two decades.

At The International Exposition of Art and Technology in Modern Life, Czechoslovakia's main official representation was concentrated in the national pavilion. The Czechoslovak pavilion was built out of a public competition with quite a strict brief; the design was to consider a potentially high volume of visitors who would be flowing into the pavilion with an estimated 50,000 people a day. The competing designers were asked to address this issue, which the winning team did by employing techniques and strategies that I explore in this chapter.

Designs of interiors and exteriors of exhibition pavilions habitually had to take into consideration a wide range of factors, including visitors' numbers, their attention span and physical and mental fatigue, and therefore were elaborate exercises in information design. During the interwar period, designers started paying more and more attention not just to the physical and technological features of the exhibitions but also the psychology of display. Designs and projects, however, often clashed with the reality of various demands from the sponsors, exhibitors and the state.

While the commercial involvement and business interests in national displays are the subject of the next chapter, here I concentrate on the interplay between the practical questions of designing a physical space to house a set of objects. I also examine the creative approach to shaping the experience of a visit to such a space. In this process, the state (often reduced to an abstracted entity) can be seen as the pavilion initiator, whereas the execution and content creation were in the hands of other agents. Who and what ultimately shaped the display, and its meaning is therefore a question I explore here in relation to the theories from various design disciplines that were used in exhibition design. What is behind the façade of the Czechoslovak pavilion in the Paris Exposition of 1937 serves as a means of looking at other national pavilions, their customary content and ideas and the way they were laid out. Ultimately, I identify discrepancies between the theory and practice of exhibition design; the discrepancies between the original, theoretical ideas of architects, designers and theorists on the one hand and the final execution of the national pavilion on the other hand.

Ephemerality of design

The French organizers of The International Exposition of Art and Technology in Modern Life granted the Czechoslovak pavilion a total area of 2,440 m^2, out of which 1,500m^2 was meant for the pavilion.[4] It was planned on the Quai d'Orsay with Sweden and the USA as neighbors [fig. 22]. Several requirements were laid out, asking for a single-floor building, which would consider the train line under the ground and include a sheltered portico that the neighboring pavilions would share too. The artistic and technical details of the project were left to the Czechoslovak organizers. The competition rules published in Czechoslovakia further stipulated that participants had to be of Czechoslovak nationality and that the height of the main part of the building should not exceed 18m. The plot was understood as challenging; it was located on a slope above the underground, and this meant that weight had to be distributed carefully.

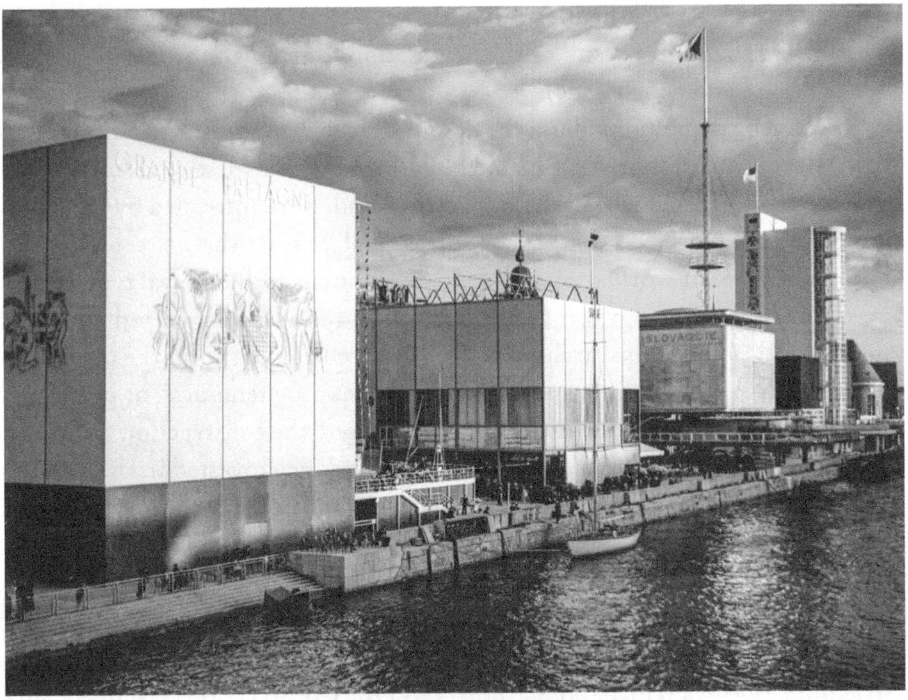

Fig.22. Views of the Paris Exhibition, 1937.

4 "Article 9," Podmínky soutěže (Tender terms and conditions) 20041215/07 f.140 Kejř (mezinárodní výstavy), AACE NTM, unpag.

The competition had to take place in two rounds because no winner was selected in the first round. When judging the entries, the committee consisting of governmental representatives and architects, expressed several objections to the Krejcar-Kejř-Sutnar-Soumar project.[5] Under the name "Folio 1937," they submitted a structure in steel and glass which the selection committee appreciated for its aesthetic quality but deemed impractical as an exhibition space. The committee questioned the main rectangular area as not sufficiently spacious for all the exhibits and did not provide enough light. The designers took the objections into account and resubmitted the project with modifications for the second round. Their winning design was a structure with four main load bearing points of steel with an entrance suspended above the underground and emphasized by a tall and slim lookout tower which supported a terrace by iron rods, extending over the Seine. The façade of the building was covered by Thermolux glass, which gave the pavilion transparency and translucency. The pavilion proved an enormous success, with over 2,200 prizes given to its different aspects and people responsible for them. The awards went to things material and tangible which were the visible result of the negotiations during all the stages of the pavilion design and construction.

The practical questions about the design were driven by considerations of the spatial layout, exhibition content and building material. These were, indeed, also practical concerns which, in the interwar period, were informed by decades of large exhibition projects that had been taking place since the 19th century. Exhibition pavilions at world's fairs were an exceptional architectural type, the primary function of which was to house exhibits for a limited period. In designing exhibition buildings, architects would therefore start from the decision of whether pavilions were meant as ephemeral, meaning temporal, or permanent structures. This determined the material to be used, the construction techniques and technologies, and related to the amount of money invested in the building.

Architects and designers created increasingly elaborate exteriors and interiors for the purposes of showcasing Czechoslovakia. How to display the set objects had been the subject of various theoretical reflections by both the creators and the critics. In Czechoslovakia, articles on exhibition architecture were published in journals specialized in architecture and construction like *Stavba* (Construction), *Horizont* (Horizon), *Stavitelské listy* (The Construction Gazette) but also

5 The committee consisted of Evžen Syrovátka, V. V. Štech and the construction councillor Vladimír Pacold from the ministry of public works, the art historian Zdeněk Wirth from the ministries of education, Ladislav Turnovský from the ministry of trade, Emil Purkhart of foreign affairs, and architects from Prague, Brno and Bratislava.

in the daily press, which indicated the level of interest and importance of the topic of state representation abroad. Newspapers frequently reported on the numerous prizes and awards the Czechoslovak pavilions, exhibitors and exhibits collected, by which they sanctioned the approach to exhibiting. Critical voices in architectural and artistic journals, on the other hand, pointed to faults in execution, changes of designs, insufficiencies of displays or irregularities in the competitions, which they used as arguments for improvement of future displays or their abandonment.

Following these debates also suggests how thinking about exhibition architecture and the general topic of exhibition design had developed and what informed the decision-making of the designers. Increasingly, many of them, including Sutnar, Krejcar or Jiří Kroha, whom I will discuss in detail later, entered the field of exhibition design with a background in a variety of disciplines. In several cases, they worked across many areas, including information and graphic design, interior design, and stage design. They brought experience from such commissions into exhibition design and enhanced the visual as well as informative effects of the exposition. At the same time, they turned their attention to visitors and their potential experience of the pavilions and the fair ground. Such concerns became important, albeit often theoretical.

Before the theoretical debates about exhibitions turned towards contemplating visitors experience, architects discussed more practical questions. They included topics of how to build exhibitions, how to display objects and represent the state or the various companies in the most precise and effective way, focusing primarily on the content and layout of the exhibitions. An article published in the Czech journal *Styl* as early as 1909 outlined the future needs of artistic and industrial exhibitions and identified specific considerations and objectives when organizing a large exhibition.[6] The author, Alois Koch, who was an architect in Vienna, claimed that exhibitions should represent actual needs, have an educational aim, and practical outcome, they should encourage sales and support local industry.

These were, indeed, the primary goals of such enterprises that were oriented towards the benefits of the exhibitors and organizers. When reception was considered, it was more related to the consumption of goods and knowledge. Apart from the main interest in the practicalities of displays, Koch nevertheless also paid some attention to how to tempt people to visit exhibitions. He suggested that "... everyone who is interested in the cultural tasks of our time must feel the

6 Alois Koch, "Budoucnost umělecko průmyslových výstav," *Styl* I (1909): 174–80.

need to see the exhibition and purchase something there."[7] Going to exhibitions, whether large or small, was considered a common pastime, as also fiction attests.[8]

The emphasis on the practical questions in exhibition architecture continued into the interwar period. Emil Edgar (1884–1963), an architect, theorist and editor-in-chief of *Stavitelské listy*, criticized the trend of building historicizing pavilions at exhibitions.[9] In 1921 he argued against recreating romantic castles and idealized architecture, which one could see for instance in the Pavilion of the Paper Industries built as an Egyptian temple at the Jubilee exhibition in 1891 in Prague. Instead, exhibition structures should acknowledge their function, he recommended, including their "ephemerality," to use Paul Greenhalgh's term. As many of the temporary pavilions were built from cheaper materials like wood and plaster, Edgar called for exhibition architecture that could use materials easy to assemble and that would have an overall architectural awareness of the exhibition ground. In Czechoslovakia, these ideas were applied at many national and regional exhibitions. Edgar named for instance the Prague Sample Fair organized from 1920 at the extensive grounds in the capital twice a year, where buildings by Bohuslav Fuchs appeared next to historicist architecture of the late 19th century. Various regional and national fairs outside of Prague in, for instance, Mladá Boleslav in central Bohemia (1927) or in Uherské Hradiště in eastern Moravia (1937) featured many ephemeral pavilions that were destroyed immediately or shortly after the events ended.

For their short lifespan, ephemeral exhibition structures could be considered as practicing grounds for the designs and concepts of national pavilions at world's fairs. Architects and designers could similarly consider interiors. As a very specific building type, the pavilion and its internal disposition could be subjected to creative approaches which would look for various solutions to the arrangement of spaces and objects. Temporality and ephemerality therefore brought certain positives.[10] If these buildings were not meant to last, these structures could be exceptional in their design and construction, and they could afford being more experimental and therefore more attractive to visitors. This was certainly implemented in the designs of Czechoslovak pavilions that used various experiments to appeal to audiences and their experience.

7 Koch, "Budoucnost," 176.
8 For example, Svatopluk Čech, *Matěj Brouček na výstavě* (1892), E. L. Doctorow, *World's Fair: A Novel* (New York: Plume, 1985), Eric Larson, *The Devil in the White City* (New York: Crown Publishers, 2003), Rob Levandoski, *Going to Chicago* (Sag Harbor, New York: Permanent Press, 1997).
9 Emil Edgar, "Výstavní architektura," *Život a umění. Soudobá architektura* (Prague, 1921), 88.
10 Jiří Kroha, "Několik poznámek k výstavní architektuře," *Horizont* (1927): 99.

Various commentators had paid attention to visitors and their experience at exhibitions already at the end of the nineteenth century. They often pointed to the ever-increasing size of these events which prevented visitors from meaningfully inspecting the exhibition grounds.[11] As early as 1880, the German professor of mechanical technology Egbert Hoyer introduced the term "exhibition fatigue" [Ausstellungsmüdigkeit], which he used to describe a feeling from the abundance of great exhibitions and from a visit to each of these overwhelming events with thousands of exhibits.[12] Later exhibition critics agreed that such a state caused by prolonged exertion and exposure to displayed objects can lead to information overload, distraction, limited cognitive capacity and higher selectiveness of what to view or visit.[13] By the First World War, the end of the exhibitions era was discussed and a move away from these large enterprises suggested.[14] Yet the war changed that and brought a renewed need and eagerness for large exhibitions which, as was believed, could help political, cultural and social recovery, while also engaging the visitor.

Contemplating exhibition spaces

After the First World War, exhibition design became a more self-aware activity for many architects across Europe. Designers, architects and visual artists like Walter Gropius, Mies van der Rohe, László Moholy-Nagy, Herbert Bayer and El Lissitzky, linked to the Bauhaus, were engaged in rethinking exhibition spaces.[15] Their work was reflected in Czechoslovakia, where individual designers, organizations and schools established exchanges with the Bauhaus.[16] Some of

11 Alexander C. T. Geppert, *Fleeting Cities. Imperial Expositions in Fin-de-Siècle Europe* (Palgrave Macmillan, 2010), 211.

12 Egbert Hoyer, "Über die heutige Praxis der Ausstellungen," *Zeitschrift für deutsche Volkswirtschaft* 1 (1880): 16–23. Quoted in Geppert, *Fleeting Cities*, 210–11.

13 Cf. for example Stephen Bitgood, "Museum Fatigue: A Critical Review," *Visitor Studies* 12, no. 2 (2009): 93–111; Stephen Bitgood, "When is 'Museum Fatigue' not Fatigue?" *Curator: The Museum Journal* 52 (2009): 193–202; Gareth Davey, "What is Museum Fatigue?" *Visitor Studies Today* 8 no. 3 (2005): 17–21; Benjamin Gilman, "Museum Fatigue," *Scientific Monthly* 12 (1916): 67–74.

14 A. Anderson, "The Paris Exhibition and Some of its Buildings," *Architectural Review* 7 (1900): 29–37; George Collins Levey, "Exhibition," in *Encyclopedia Britannica*, vol. 9, 13th ed. (London: Encyclopedia Britannica, 1926), 67–71; cf. Geppert, 215.

15 Olivier Lugon, "Dynamic Paths of Though: Exhibition Design, Photography and Circulation in the Work of Herbert Bayer," in *Cinema beyond Film: Media Epistemology in the Modern Era*, eds. François Albera and Maria Tortajada (Amsterdam University Press, 2021), 118.

16 Markéta Svobodová, *Bauhaus a Československo 1919–1938: studenti, koncepty, kontakty. The Bauhaus and Czechoslovakia 1919–1938: Students, Concepts, Contacts* (Prague: Kant, 2016); Nicholas Sawicki, "Czechoslovakia: Bauhaus Students and Associates," *Centropa* 3 no.1 (2003): 27–40; Leo Kohut, "Bauhaus. Un-

the ideas also made their way into the Czechoslovak pavilion in 1937 and therefore merit a short overview.

El Lissitzky was particularly active in designing exhibition spaces, especially in Germany, and started considering the effects that exhibition space had on the visitors. The static, inactive onlooker was to be surpassed by an active viewer, whose activity, however, was not left to chance and was regulated by interior design.[17] El Lissitzky developed these ideas in Proun rooms [Prounenraum], or demonstration rooms, which were installations of various materials that included for instance paint, wood, or foil, developed between 1919 and 1927. They had to be experienced spatially, El Lissitzky believed, giving the visitor an active role.[18] Alan C. Birnholz has placed these attempts at inciting active participation of the visitor to the Russian devotional traditions of physical and emotional interactions with icons as well as the performativity of Jewish rituals.[19] Active participation in festivals, public celebrations and pageants, which brought art into the streets, was also an important input for the Russian avant-garde. What these events share with exhibition design is the need to direct large masses of people, make them visually engaging and attract their attention.

El Lissitzky developed his innovative approach further, for example, in the Great Berlin Art Exhibition in 1923 and the International Art Exhibition in Dresden in 1926. In the latter, he tried to conceive of the space in a way that did not overwhelm the visitor at once nor brought them "into a numb state of passivity."[20] He intended to make the visitor active with his design and tried to organize the display in such a way that would automatically invite everyone to perambulate in it.[21] "Perambulation," or the way people walk around a space,

---

garn-Tschechoslowakei. Zur Bauhaus-Rezeption in Ost-europa," in *Bauhaus-Archiv, Museum. Sammlungs-Katalog, (Auswahl), Architektur, Design, Malerei, Graphik, Kunstpädagogik*, eds. Hans Maria Wingler et al. (Berlin: Gebr. Mann, 1981), 283–287.

17 Alan C. Birnholz, "El Lissitzky and the Spectator: From Passivity to Participation," in *The Avant-garde in Russia 1910–1930. New Perspectives*, eds. Jeanne D'Andrea and Stephen West (Cambridge, Mass.: MIT Press, 1980), 98.

18 Olivia Crough, "El Lissitzky's Screening Rooms," in *Screens. From Materiality to Spectatorship – A Historical and Theoretical Reassessment*, eds. Dominique Chateau and José Moure (Amsterdam University Press, 2016), 224; Éva Forgács, "Definitive Space: The Many Utopias of El Lissitzky's *Proun Room*," in *Situating El Lissitzky: Vitebsk, Berlin, Moscow*, eds. Nancy Perloff and Brian Reed (Getty Publications, 2003), 47–75.

19 Birnholz, "El Lissitzky," 98.

20 El Lissitzky, "Exhibition Rooms," *Russia: An Architecture for World Revolution*, transl. Eric Dluhosch (Cambridge, Mass.: MIT Press, 1970), 149.

21 El Lissitzky, "Proun Space, The Great Berlin Art Exhibition of 1923," in El Lissitzky, *Russia: An Architecture for World Revolution*, 139.

was meant to be a pleasurable activity, which El Lissitzky also discussed in his theoretical texts, most importantly in *Exhibition Rooms*.[22]

The ephemerality of El Lissitzky's exhibition projects provided an opportunity for experimenting. One of the most discussed designs was for the Soviet section at the International Press Exhibition (Pressa) in 1928 in Cologne with a large photomontage, the so-called Photo-Fries, depicting "The Task of the Press is the Education of the Masses."[23] Lissitzky created the photomontage with Gustav Klucis in Moscow and made use of a dynamic and asymmetrical space design which included unconventional material like cellophane.[24] Even though this approach to space, complicated with movement, letterings and pictures of various sizes, was not adopted in Czechoslovakia completely, the ideas of the unity of space and spectator's interaction with the exhibits were. Lissitzky's theoretical and practical ideas about the exhibition space in general resonated in the work of several Czech designers in the latter half of the 1920s and in the 1930s if not in their realizations, then in many planned designs.[25]

Art and design periodicals in Czechoslovakia regularly published articles by international designers and about international and national exhibitions.[26] Czech authors would also write texts about exhibition design which theorized and reflected on exhibition spaces, looking for possible new directions and approaches. The architect and designer Jiří Kroha (1893–1974) was one of the first in Czechoslovakia to engage with exhibition design. His interest in the display techniques, unity of space, and the spectator overlap with El Lissitzky's attempts which he combined with awareness of the buildings ephemerality. In his article on exhibition architecture published in 1927, Kroha described ephemeral pavilions as "a manifestation of modern architecture," and as structures that, at least, aspired to be such manifestations.[27] Their modernity could be detected in the latest technologies and construction techniques used, and in what Kroha termed "psychology of building."[28]

In Czechoslovakia, debates that considered not only the physical but also the psychological side of exhibition architecture appeared in various articles on

22 El Lissitzky, "Proun Space," 139.
23 Jeremy Aynsley, "Pressa Cologne, 1928: Exhibitions and Publication Design in the Weimar Period," *Design Issues* 10, 3 (1994): 71.
24 Herbert Bayer, "Aspects of Design of Exhibitions and Museums," *Curator* 4, no. 3 (1961): 267.
25 I am deliberately using the adjective Czech, rather than Czechoslovak here, in relation to designers and architects of Czechoslovak expositions because Czechs constituted overwhelming majority in these roles. The reasons for this are explained throughout the book.
26 Articles were published in journals *Výtvarné snahy*, *Typografia*, *Index*, *Stavba* etc.
27 Jiří Kroha, "Několik poznámek k výstavní architektuře," *Horizont* (1927): 99 .
28 Kroha, "Několik poznámek," 99.

exhibitions from this period that indicate designers' growing awareness of the role of such architecture. Pavilions were no longer considered as mere shells for the exhibits; they were means of promotion and propaganda aimed at the "customer." This was a term Kroha used to describe the recipients of the visual presentations put on display either at exhibitions or in more commercial spaces. Kroha was also a renowned designer of shop windows and department stores interiors, which informed his work for exhibitions. Careful thinking about the distribution of information via the objects was therefore essential and cognizant of disciplines which applied the established means of visual communication, like advertising and typography but also stage design.

This translated into looking for new approaches to the organization of space and distribution of objects. Kroha himself pointed out that designers treated the relationship between the exhibited objects and the exhibition space as novel and different from how the object would appear in real life. In doing so, designers should rouse new interest of the visitors/customers in the displays. A chair, a table and a vase thus did not have to be shown in the utilitarian function they have at home; in exhibitions, they can be shown inventively against the conventions and traditions.

Many of Kroha's ideas were reflected in the Czechoslovak exhibition architecture of the 1930s. Moreover, his work for the 1927 and 1928 National Exhibitions in Mladá Boleslav and Brno was an important step in the development of pavilion architecture and design. Yet he himself never designed a national pavilion for an international exhibition; the reason most probably being the fact he stood outside the right circles of influence. Although Kroha was one of the leftist architects who were ardent promoters of socialist values in architecture, he was not considered a member of the avant-garde. For example, he was not one of the artists of Devětsil, a short-lived but significant and vocal association that had an impact on how modern art and architecture were framed in interwar Czechoslovakia.

Kroha's exhibition architecture also received harsh criticism from Karel Teige, the art theorist, art critic and one of the leading members of Devětsil, an important voice of the interwar art world and the Czechoslovak avant-garde.[29] Teige called Kroha's architecture a misunderstood constructivism with its arbitrary forms, ostensibly conceived in the modern spirit and yet alien to rational archi-

29 Mariana Holá, "Teige, Karel (1900–1951)," in *Routledge Encyclopedia of Modernism* (London: Routledge, 2016), doi:10.4324/9781135000356-rem1653-1; Eric Dluhosch and Rostislav Šváchá, eds., *Karel Teige / 1900-1951: L'enfant terrible of the Czech modernist avant-garde* (Cambridge, MA: MIT Press. 1999); Rea Michalová, *Karel Teige: captain of the Avant-Garde* (Prague: KANT, 2018).

tectural conception.[30] Constructivism was, in Teige's view, not a simple denial of ornament, adoption of flat roofs, and corner windows which lacked revolutionary perspectives and elasticity.

Teige was one of the most vocal critics of Czechoslovak exhibitions, both national and international, and I return to his critical comments on the content of the Czechoslovak pavilions at both the 1925 and 1937 Parisian Expositions in Chapter Three. He also offered his views on what a successful modern exhibition should look like when he criticized the local exhibition efforts which, in his view, failed to comprehend the very substance of this enterprise. He found a great example of a modern exhibit in the Soviet Constructivist section at Pressa and its "beautifully laid out catalogue" which displayed modern and effective exhibition techniques.[31]

Teige also expressed his disapproval of the national Exhibition of Contemporary Culture that took place in Brno in 1928. It was conceived as a summary of the best achievements of the new state during the first ten years of its existence, for which an entirely new exhibition ground was built with pavilions by contemporary architects like Bohuslav Fuchs, Pavel Janák or Josef Gočár, featuring industries, agriculture, businesses and art schools. This event allowed Teige to contemplate "the art of exhibiting" more generally and conclude that in the Czech context it was flawed. Organizers and exhibitors often forget "that the art of exhibiting is primarily the art of organization, the art of grand plans and clear dispositions, not only the art of window dressing and decorations."[32]

The exhibition overall showed, Teige thought, how the Czechs lagged not only in the art of exhibiting but also in the art of propaganda and advertising.[33] He argued that the unimaginative, unattractive and undistinguished exhibits and installations were the result of the petit bourgeois mentality of the organizers and exhibitors. These would be the most visible and influential individuals and organizations behind Czechoslovak exhibits displayed abroad, linked to the government, key institutions and businesses.

Teige's criticism targeted the Czech regional and national exhibitions for lacking an overall exhibition concept and originality in the display. Although Teige directed his remarks primarily against the Brno exhibition, his recommendations for overall plans and disposition of the exhibits can also be applied to national pavilions at international exhibitions. However, El Lissitzky's ideas,

30 Karel Teige, "Výtvarné umění na brněnské Výstavě soudobé kultury," *Stavba* 7, no. 3 (1928/29): 45.
31 Teige, "Výtvarné umění," 44.
32 Teige, "Výtvarné umění," 44.
33 Teige, "Výtvarné umění," 47.

which Teige saw as a suitable model, could not be fully translated into the design of regional or national pavilions in Czechoslovakia. Here, the educational and commercial motivations of exhibitions took priority over visually impressive installations which would allow a degree of experimentation with design and display, the goal being to lure in random visitors.

On the national level, the messages needed to be as unambiguous as possible, the opposite of open-ended or abstract. While this partly applied to international exhibitions too—the products being shown needed a clear and explicit display—the extent of experimentation with the presentation and setting of exhibits was greater in these contexts. The visual display and its relationship to clear communication of information therefore became a topic debated by Czech designers, theorists, and the bureaucrats involved in the Czechoslovak display.

Three-dimensional graphic design: Ladislav Sutnar

Before the escalation in the debates about the concept of national displays at the Paris Exposition in 1937, new display techniques had been gradually introduced into Czechoslovak presentations. They developed significantly in the way they approached the exhibition space and its relation to exhibits and visitors. The first large pavilions that represented Czechoslovakia as an independent country in the interwar period in Rio de Janeiro in 1922 and Paris in 1925 were experimenting with the exterior of the architecture, but inside they were very much traditionally conceived exhibition spaces with display walls and glass cabinets full of objects [fig.23].

The Czechoslovak pavilion aside, the Parisian Exposition of 1925 could be considered a crucial moment in the history of exhibition design. Some pavilions built here adhered to a more conventional display in cabinets and exhibition rooms and this applied to the Czechoslovak, Austrian or Polish pavilions. Other structures, however, like Konstantin Melnikov's Soviet pavilion or Le Corbusier's L'Esprit Nouveau, were much more innovative in their handling of the physical form, minimalist approach to architecture and ideological content.

It was also around and after 1925 that a new generation of designers and architects in Czechoslovakia started approaching exhibition design as a self-sufficient field of work. One of the most prominent figures was Ladislav Sutnar, a graphic and product designer, designer of textiles, stage sets and toys, who also created marionettes as well as paintings. He was the director of the State Graphic School in Prague and worked as a book designer and typographer for

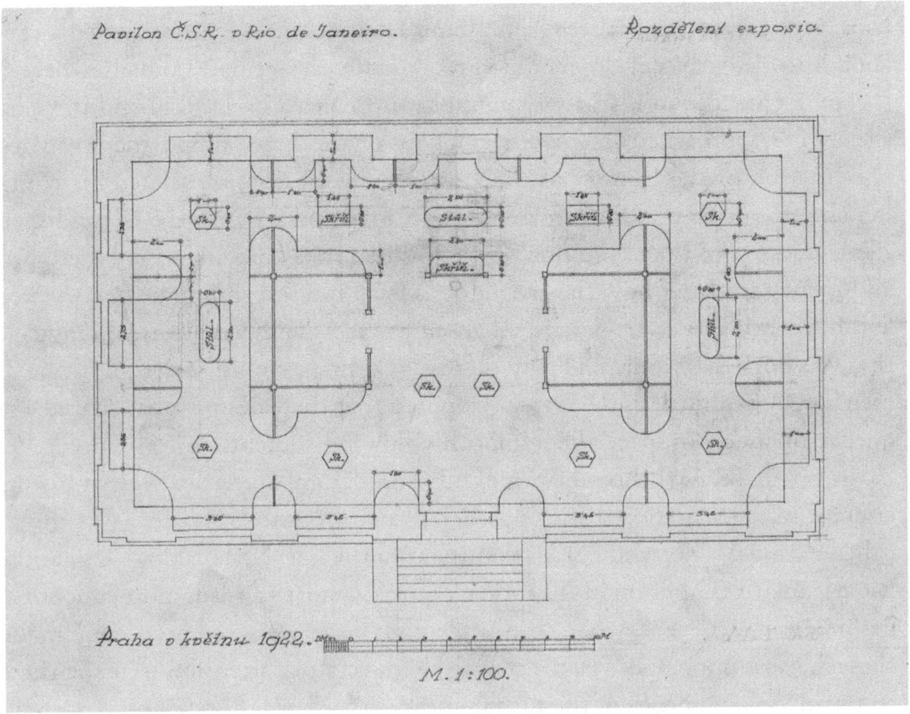

Fig. 23. Pavel Janák, Floor plan of the Czechoslovak pavilion in Rio de Janeiro, 1922.

the organization Družstevní práce (The Cooperative Work) and many publishing houses in Prague. His career in Czechoslovakia ended in 1939 when he left to dismantle part of the Czechoslovak exhibit in New York never to permanently return.

Sutnar collaborated on the Czechoslovak pavilion in Paris in 1937 and his designs were based on his earlier theoretical and practical work with exhibition design, informed by his experience from other disciplines. Through the Association of the Czechoslovak Werkbund, the Czech response to the Deutsche Werkbund, and a good network of contacts with artists, publishers, companies and governmental officials, Sutnar also participated in more than twenty regional, specialized, national and international exhibitions.[34] Here, he introduced many innovative ideas into exhibition design. He had personal contact with the Bau-

34 These included the Leipzig International Book Exhibition in 1927, The 3rd International Book Fair in Florence in 1928, The Bucharest Exhibition of Czechoslovak architecture in 1930, Poznań International Hygiene Exhibition of 1933, The 6th Triennale di Milano in 1936. Ladislav Sutnar Papers, Box 18, f/ 3, GRI.

haus, especially with Walter Gropius and Hannes Mayer, and possessed a very good knowledge of developments in typography in Germany and elsewhere.[35]

Jan Tschichold in his new typography, influenced by De Stijl and the Constructivists, approached design with emphasis on the direct, simple and organized presentation of the content to the viewer.[36] Sutnar, too, translated his thinking about how to present information on a flat surface into the three-dimensional context of the exhibition space. He identified that after 1925 there was an urgency to think how the spectators' visual interest should be best evoked and how they should be persuaded about the facts with a new visual language that was both functional and convincing.[37] Calling exhibition design "three-dimensional graphic design," Sutnar proposed that these design areas shared a sequence of effects in space and a smoothly flowing content.

In his theoretical thoughts on exhibition design, Sutnar expressed his discontent with the nationalistic approach drawn from folk art. This was the so-called national style, which was prominent in the Czechoslovak pavilion at the Rio de Janeiro exhibition in 1922. He identified a move towards more functional design after this event, when thinking about visitors started preoccupying designers. Most important was attracting the spectator's attention to make them stop and look by means of the distribution of space or other devices.[38] Sutnar's use of such techniques were for example the colored abstract design of the ceiling at the Book Fair in Leipzig in 1927. The "bright colors, simple and witty composition of the ceiling painting, [and] functional lighting of the cabinets" and show windows that were built into the walls provided visual excitement for the visitors.[39] In the Ausstellung Europanisches Kunstgewerbe in the same year, which showed the decorative arts of ten European countries,[40] Sutnar's wall decoration of the display for the Association of the Czechoslovak Werk-

35 Ladislav Sutnar, "The Bauhaus, as seen by its neighbors South of the Border," Manuscript, 1971, Ladislav Sutnar Papers, Box 22, Biographical Information, Miscellaneous Correspondence, Exhibitions, Cooper Hewitt Museum archive, 2.

36 Ladislav Sutnar, "Early modern design concepts," in *Visual Design in Action. Principles, Purposes* (Zurich: Lars Müller Publishers, 2015); Paul Stirton, *Jan Tschichold and the New Typography: Graphic Design between the World Wars* (New York City: Bard Graduate Center - New Haven: Yale University Press, 2019).

37 Sutnar, "Early modern design concepts," units c/7-c/10.

38 Sutnar, "Early modern design concepts," c/11.

39 Oldřich Poskočil, "Mezinárodní knižní umění. Poznámky k lipské výstavě," *Typografia* 34, no. 10 (1927): 243.

40 Germany, Belgium, Denmark, France, Great Britain, Italy, the Netherlands, Austria, Switzerland, and Czechoslovakia. See Richard Graul, *Ausstellung europäisches Kunstgewerbe 1927* (Leipzig: Rudolf Schick, 1927).

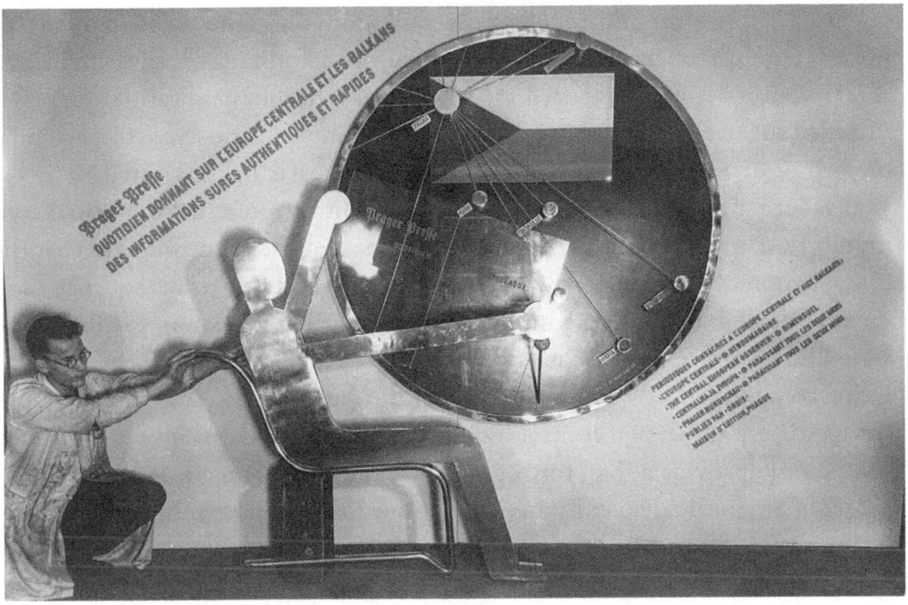

Fig.24. Ladislav Sutnar, Prager Presse exhibit, The Press pavilion, Paris, 1937.

bund combined striking red, white, yellow and grey colors which showed clear influence of El Lissitzky's Prouns.

His later exhibition designs, at for example the Florence Book Fair in 1928, the Barcelona international exhibition in 1929, or the Exhibition of Czechoslovak Architecture in Bucharest in 1930, also creatively applied the use of shapes, combination of images and text as well as light which were highly attractive to visitors according to reports and Sutnar himself.[41] In Paris 1937, for example, his newspaper exhibit in the Press Pavilion included an impressive life-sized, two-dimensional man holding a glass newspaper. The figure was depicted sitting underneath a map, which showed the locations of the paper's correspondents [fig. 24]. Such exhibits were soliciting what Sutnar termed *shock reactions* and often came as a climax at the rear of the pavilion.[42] Ideally, the spectators would follow the planned circulation route and take prescribed steps through the exhibition to reach the intended exhibition highlight.

---

41 "Československé oddělení na mezinárodním knižním veletrhu ve Florencii 1928. Podle zprávy výstavního komisaře Prause," *Zprávy Svazu československého díla* 9, no. 8 (1927–28): 60; Ladislav Sutnar, "Curriculum vitae draft," Ladislav Sutnar Papers, box 17 f. 6, GRI, 13-15.

42 Sutnar, "Early modern design concepts," c/11 and Iva Janáková, "Exhibition Design," *Ladislav Sutnar. Praha – New York, Design in Action* (Prague: UPM – Argo, 2003), 324–5.

Sutnar's concern—and not just with exhibitions—was the question of how to present complex scientific and educational information in "a lively, simple and direct manner."[43] His advanced visualization techniques and visual surprises included the use of giant photographs, photomontages, and three-dimensional designs of diagrams, graphs and maps, not dissimilar to those used by El Lissitzky. Sutnar also worked with another concept introduced from visual communication, namely the flow of visual information. He acknowledged that the introduction of a flow pattern which guided the spectator from sequence to sequence through the entire exhibition was a major innovation in exhibition design. The *controlled visual continuity*, as he called the flow, required that all exhibition components flowed into each other in a "relationship that provides both design variety and design continuity."[44] The national pavilion, Sutnar continued, should only provide a shell for the exhibits and should not interfere with any aspect of any individual exhibit. The exterior and interior architecture should be interconnected and work together, not against each other.

Sutnar's continuous space idea has parallels in similar thinking found not only in El Lissitzky's unity of space but also in Mies van der Rohe, Moholy-Nagy and Herbert Bayer. Van der Rohe worked with the *space-time continuum*, seen for instance in the German pavilion at Barcelona, 1929, while Moholy-Nagy employed fluidity and curved walls, for instance at the Werkbund exhibition in Berlin in 1930.[45] Bayer paid a lot of attention to the exhibition space which in his view should consist of the new vision techniques of El Lissitzky, which would be combined with color, scale, elevation and typography.[46] Similarly to Sutnar, Bayer also saw exhibitions as a combination of two-dimensional design like graphic arts and photography and three-dimensional design, i.e., architecture. And although he published his theories as late as in the 1950s, he put them into practice already in the interwar period.[47]

One of the best examples of Sutnar's controlled flow was his collaborations on the designs for the Czechoslovak pavilion in Paris in 1937 [fig. 25]. Many of the proposed innovations and experimentations in exhibition design were brought together here. Sutnar took part not in one but in two projects submitted for the competition in 1936; one submission was with the architect Bohuslav Fuchs, the other with the architects and designers Krejcar, Kejř and Soumar, the

43 Sutnar, "Early modern design concepts," c/12.
44 Sutnar, "Early modern design concepts," c/13.
45 Herbert Bayer, "Aspects of Design of Exhibitions and Museums," *Curator* 4 no.3 (1961): 260.
46 Christopher Phillips, "The Judgement Seat of Photography," *October* 22 (1982): 41–42.
47 Bayer quoted in Lugon, "Dynamic Paths," 121.

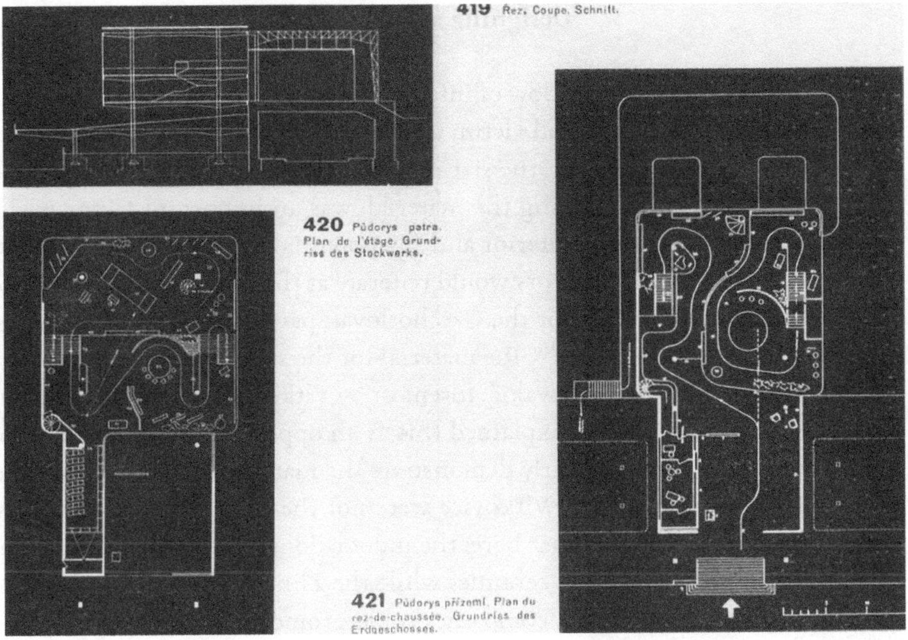

Fig. 25. Ground plan of the Czechoslovak pavilion, Paris.

latter being the winning one. Both concepts, nevertheless, show Sutnar's concern with the controlled visual flow and the plans include arrows indicating the direction of the visitors' flow. Sutnar conceived of the exhibition space where the three levels of the pavilion were connected by a route leading from the entrance at the river level to the upper floors.[48]

The winning project adopted some ideas from the project with Fuchs, called "Czechoslovakia: The Heart of Europe." It opened with a shopping street of Prague in which "typical" export products would be displayed in shopping windows. This idea was ultimately included as a way out of the Czechoslovak pavilion. The Fuchs project also included the construction of a hotel terrace in the Tatra mountains represented by a large-scale photograph. This design was also reused in the plans for the successful competition entry with Krejcar et al., where it created an effect close to a shock reaction of an unexpected exhibit.

48 Sutnar, "Early modern design concepts," c/13.

Designing a fairy tale

Sutnar's elaborate ideas of the flow of information, space and people represented the ideal setup which partly fell victim to subsequent decisions about the material to be used in the pavilion, the size of individual sections, and their actual content. The exterior, including the material, was an important factor in determining the layout of the interior and any potential design aspects of the exhibition. As many commentators would reiterate at the time of the Paris Exposition and later, the exterior of the Czechoslovak pavilion was constructed as a "fairy tale of steel and glass."[49] The material for the structure was supplied by companies based in Czechoslovakia, foremost by Vítkovické železárny (the Vítkovice Steel Works). Krejcar explained this as an opportunity for the Czechoslovak export industry to clearly demonstrate the materials that can be sold at international markets.[50] And Vítkovice were not the only company that was part exhibitor and part supplier here; the indoor floors, for example, were finished with Slovak marble and ceramics while the Zlinolit rubber flooring was provided by Baťa. In this way, companies could promote their products, while also contributing to the cost of the pavilion.

During the project assessment stage, the selection committee pointed to insufficient light in the pavilion. To tackle the issue, the main feature of the pavilion exterior became glass. This material was gaining popularity throughout the 1930s for its physical and metaphorical properties. The Czech designers described using glass as the new trend in pavilion architecture and Krejcar pointed out how prominent it was already at the Brussels Exposition in 1935. Here the most successful Norwegian, Swedish, Danish, Swiss, Austrian, Finnish and Lithuanian pavilions were in his view from large part built in glass.[51] Building materials had specific values, Krejcar believed, and while glass provided transparency; solid, opaque walls of for example the French and Italian pavilions indicated to him the monumentality and historicity of the states. The choice of material was a marker of national and international aspirations, because "only oriental and southern nations still build pavilions with solid walls [meant] to illustrate their ancient, historic, southern culture."[52]

---

49 "Jako skvělá pohádka z oceli a sklad bude zářit skleněný čs. pavilon na světové výstavě v Paříži," *Venkov* 32, February 7, 1937, 8.

50 "Průvodní zpráva k projektu arch. Jaromíra Krejcara a spolupracovníků," *Stavba* XIII (1936–37): 176.

51 Krejcar et al., "Československý pavilon," 175.

52 Jaromír Krejcar et al., "Československý pavilon na Mezinárodní výstavě umění a techniky v moderním životě," *Stavba* XIII (1936–37): 175.

Two years later in Paris, it was for example the British pavilion with its Canadian annex that conformed to Krejcar's view that only southern and colonial nations insisted on solid, monumental structures. The pavilion was criticized by the American architectural historian Henry Russell Hitchcock (1903–1987), who wrote an extensive report on the exposition for *The Architectural Forum* and identified the worst pavilions as those of the British Empire, Belgium and the USA for their insistence on what he saw as conventional modernism.[53]

In all fairness, the entire 1937 exposition consisted of pavilions subscribing to a whole range of modernisms. The most cited examples of monumental architecture have indeed been the Soviet and German pavilions discussed in Chapter One. Restrained, or tame modernist designs were visible in many other national pavilions. Hitchcock, for instance, noticed the conspicuous Italian pavilion in a classical tradition updated for the 20th century.[54] Its more inventive modernist court of honor by Ettore Rossi featured a large Thermolux glass window which provided attractive illumination at night. As a new type of heat resistant façade glass, Thermolux was patented in Italy and used on the façade of the Czechoslovak pavilion.

The extensive application of glass in exhibition architecture goes back to at least the 1925 Exposition in Paris and the Soviet pavilion by Konstantin Melnikov, which was conceived as a house of glass. By 1937, large glass windows or whole glass walls became commonplace. It was a primary feature in many national pavilions. The large glass window of the Austrian pavilion that was built in Paris by Oswald Haerdtl allowed a much-praised view of the gigantic wall photograph of the Alps, while the tree in front of the pavilion created an interesting shadow and contrast.

The use of large glass windows in some cases led to a novel approach to the interior but Hitchcock identified an important discrepancy in some pavilions of the Parisian exhibition. The interior of the Austrian pavilion, despite its transparent front, featured "dull and cluttered" exhibits in the Viennese tradition of Josef Hoffmann and the Wiener Werkstätte. Such conflict between the exterior and interior could be found in the Czechoslovak pavilion too which, although designed as a striking glass building with clearly distributed interior, ended with a similarly cluttered display.

For the Czech designers, the extensive use of glass in Paris was an indicator of modernity in many respects: the pavilion was transparent, and its manufac-

53 Henry Russell Hitchcock, "Paris 1937," *Architectural Forum* (September 1937), 169.
54 Hitchcock, "Paris 1937," 164.

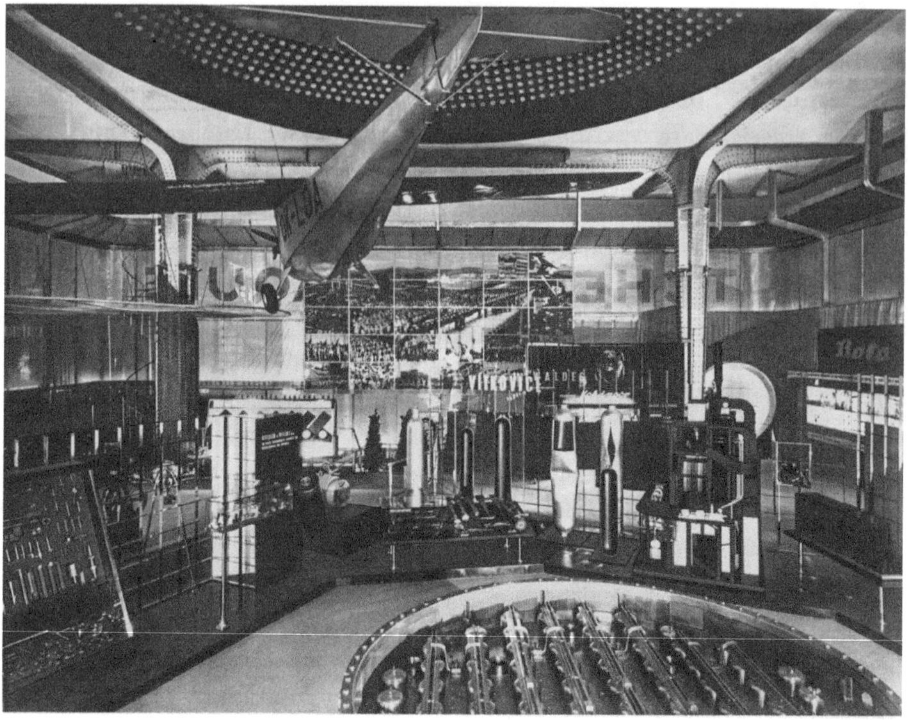

Fig.26. View of the Industries Hall, Czechoslovak pavilion, Paris, 1937.

turers were capable of using the latest technologies. And such qualities could be associated with the Czechoslovak state too. Using the same metaphor, the Czechoslovak state could be presented as transparent and therefore democratic, with its achievements in full view or visible through the glass.

In construction terms, the Thermolux in the walls provided "the exhibition space with such light that the visitor can concentrate well on the display..."[55] From the technical point of view, the glass let in enough natural light, which most designers preferred to artificial light. Effects of the material were applied in various places in the pavilion, most impressively at the ceiling of the Industries Hall, which was domed by a cupola of 12.5m in diameter from lenticular glass diffusing light from above[56] [fig. 26].

Light and lighting were indeed important features in exhibitions, breaking up spaces, emphasizing individual exhibits and providing a dynamic experi-

55 Krejcar et al., "Československý pavilon," 175.

56 María Pura Moreno, "Jaromir Krejcar, la apología de la técnica: el pabellón de Checoslovaquia en la Exposición Internacional de París, 1937," Rita: Revista Indexada de Textos Académicos, no.9 (2018): 142.

ence for the visitor.[57] In Paris, floodlights were used across the grounds, swirling lights were projected from the Eiffel Tower, lit fountains and the Seine. Fireworks and internal light sources illuminated the buildings from within. This included the German "cathedral of light," which provided impressive and colorful views at night.

Designers of the Czechoslovak pavilion also made good use of light; the strips of Thermolux glass came gradually alight at dusk and the Czechoslovak flag on the tall beam above the pavilion was lit by a floodlight. Thus, they applied the specific features of glass into the architecture of the pavilion and its interior. With glass, Krejcar could design the structure with rounded corners to achieve what he deemed as softness of the building.[58] It was, in his view, more in line with international tastes and a departure from the contemporary conventions of architecture in Czechoslovakia, by which he most probably meant the more functionalist and constructivist structures.[59]

Journey through the pavilion

The lack of sharpness of the exterior corresponded with the designed interior's flow and ideas of architects like Herbert Bayer who recommended rounding out the space and using curves in exhibition spaces. Bayer believed that "in the future, ground plans will be more readily conceived of in groups, lightened, dynamic, and less symmetrical and axial" to allow the traffic of visitors to flow in a continuous direction.[60] Even though Bayer's views were published in 1937, they were most probably well-known before that and especially Sutnar and Krejcar seem to employ similar ideas of the communication route and the flow of visitors. In 1930, Bayer devised the so-called "Diagram of Field of Vision." This space arrangement considered visitors' field of vision, which the designer tested for instance at his 1931 exhibition of the Building Workers' Unions in Berlin. The space, created together with Walter Gropius and László Moholy-Nagy, used raised, walking platforms from which the visitor could observe the ceiling, the floor and wall panels with photographs which were often mobile and could rotate.[61]

57 Bayer, "Aspects of Design," 278–79.
58 Krejcar et al., "Československý pavilon," 175.
59 Krejcar et al., "Československý pavilon," 175.
60 Bayer, "Fundamentals of Exhibition Design," *Production Manager* (1939/40): 19.
61 Vanessa Rocco, "Activist Photo Spaces: 'Situation Awareness' and the Exhibition of the Building Workers Unions," *The Journal of Curatorial Studies* 3, no. 1 (2014): 26–48.

Bayer also offered thoughts about the way exhibition material was gathered and grouped. If the presentation happened in "reasonable succession," the direction of visitors followed in a free manner, assisted by direction arrows and sound and by placing interesting items for observation in a direction from left to right, just like reading and writing. The habit of walking either left or right needed to be taken into consideration as a subconscious reaction of visitors towards the order of traffic.[62] Exhibitions which did not consider the direction of visitors' movement and divided the exhibition space geometrically into sectors were, in his view, failed designs.

While the arrangement of space in the Czechoslovak pavilion reflected some of these innovative propositions, not all architects and designers in Paris subscribed to them. The previously mentioned Soviet and German pavilions, for example, with their imposing exteriors did not experiment with the flow or movement of visitors inside, opting instead for conventionally arranged exhibition rooms. On the other hand, the Polish pavilion by Bohdan Pniewski and Stanisław Brukalski, or Alvar Aalto's Finnish pavilion and a number of French pavilions integrated the flow of people. They distributed the displays across varying levels and used color, lighting or mosaic lines on the floor, which was the case of the French Social Services pavilion, to direct visitors and distribute the exhibits.[63]

Apart from Sutnar's flow, another innovative solution to the issue of visitors' traffic suggested by the Czech designers was the inclusion of a "mirror" presentation. Displays on both sides of the walking route would show similar exhibits and individuals, who were not able to see both sides, could still observe enough. Similar items would therefore be shown left and right of the communication route.[64] It is clear from Krejcar's suggestions about visitors, mentioned at the start, that the architects and designers of the Czechoslovak pavilion considered the visitors experience and behavior too—they also recognized the "psychology of a world's fair visitor." An average visitor would spend a limited amount of time at national pavilions and therefore, as mentioned at the start, the content of a pavilion had to both *capture* and *satisfy* visitors of different cultural maturity.[65] To do so, Krejcar et al. put forward what they saw as a captivating archi-

62 Bayer, "Fundamentals," 19 and 23.
63 Talbot F. Hamlin, "Paris 1937. A Critique," *American Architect* (1937): 25–34.
64 "Projektová dokumentace," 8, Architecture archive, National Technical Museum, f. 140, Kejř Paříž.
65 "Československý pavilon," 176 and "Vysvětlivky k průvodní zprávě," 5.

tecture with an inviting entrance designed to bring the visitor into the exhibition hall from which they could glimpse other sections.

Following these proposals and recommendations, the entrance via the main staircase and the outdoor areas of the Czechoslovak pavilion featured large exhibits of the Pelton turbine of 10t by Škoda and several heavy objects, like a boiler and pneumatic tools, by Vítkovice Steel Works.[66] These visually impressive objects, however, were placed there against the original plans to show them inside the pavilion in the industries section which was located on the first floor of the pavilion. Because of the elevation and unstable ground, heavy exhibits could be included indoors.[67] Yet moving them in front of the pavilion had a positive effect: the striking objects could serve as an eye-catching prompt inviting visitors to enter the building.

Inside, the individual spaces were designed by different individuals in charge of specific sections, like heavy industries, agriculture, glass, specialized schools, or music. Jaromír Krejcar and Eva Štrimplová were responsible for the design of the Hall of Honour, the entrance space from Quai d'Orsay. [fig. 27] While Krejcar is one of the four main architects/designers of the pavilion, the female architect Štrimplová is rather unknown. She is one of the many names not often mentioned in relation to the design of the pavilion as most literature, primary and secondary, has ascribed the authorship to the four male architects. Her role or extent of work, however, is unclear but her name appears in the main catalogue of the Czechoslovak pavilion as well as in the guide.[68]

The question of authorship could often be a contentious issue, especially in the case of the national pavilion in 1937. In its initial design stages, the pavilion was associated only with Krejcar, which was disputed especially by his colleague Zdeněk Kejř.[69] Apart from Štrimplová and Kejř, many others were involved in the design of the interiors, demonstrating the extent of collaboration and involvement from various individuals. This included, for example, the structural engineer Jaroslav Polívka. His extensive work on the exterior was carried out in Paris and he also had an important role in the construction of the 1939 pavil-

66 "My na světové výstavě v Paříži," *Venkov*, August 22, 1937, 5.

67 Vs, "Viníci ostudy s čs. Pavilonem na pařížské výstavě musí býti potrestáni," *Národní listy*, June 23, 1937, Newspapers Excerpts Archive, MFA, oh8.

68 B. Soumar, ed., *La Tchecoslovaquie* (Prague: Melantrich, 1937), 13; V. Darras, "Praktický průvodce po pařížské výstavě umění a techniky v moderním životě," *L'Indépendance Tchécoslovaque*, no. 11, 1937, 3.

69 Letter from Kejř to the editors of *Přítomnost* of November 12, 1937, Architecture Archive of the National Technical Museum, f.140 Kejř. Kejř reacted to the article by Krejcar, "Poučení z pařížské výstavy," *Přítomnost* 14, no. 40, October 6, 1937, 633–35.

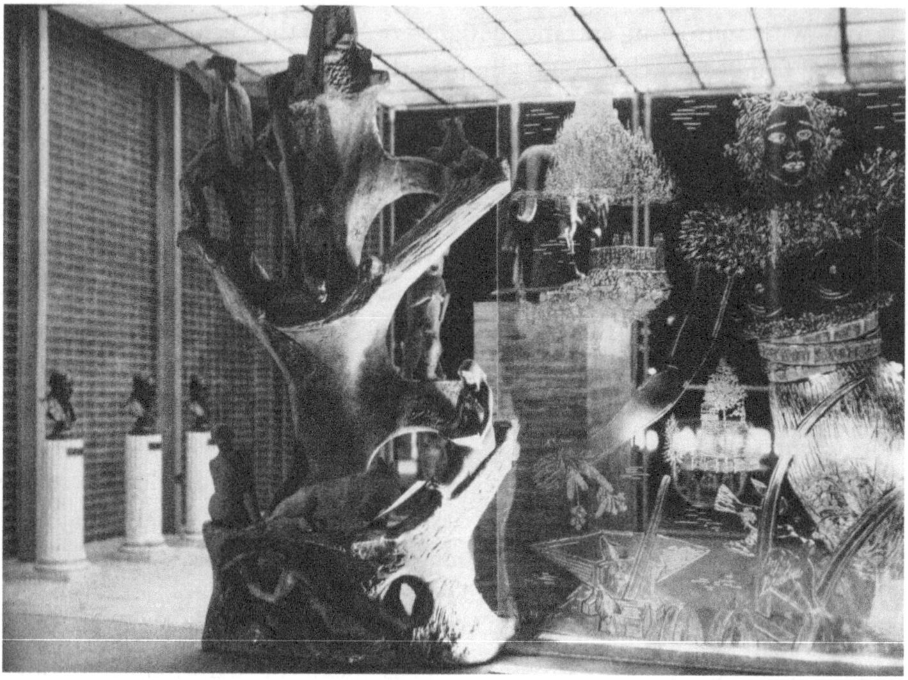

Fig. 27. View of the Hall of Honor, Czechoslovak pavilion, Paris, 1937.

ion in New York.[70] The input of many of the designers and architects—let alone other contributors—has not been properly acknowledged in the contemporary and subsequent accounts of the success of the Czechoslovak pavilion. What such involvement also meant was the fractioning of the exhibition into parts where the interests of individual exhibitors took priority over the overall concept.

Czechoslovakia on display

The internal layout and distribution of exhibits in the interior deserve closer attention because they signify the fractional character of not only this display in Paris but of other Czechoslovak pavilions. The entrance led to Krejcar's and Štrimplová's Hall of Honour which was laid out as deliberately bare and tried

70 Ladislav Jackson, "The Materiality of a Pavilion: The Agency of Materials at the Czecho-Slovak Pavilion in the Turbulent Times of 1938–1940," paper presented at CRAACE conference *Exhibitions, New Nations and the Human Factor, 1873–1939*, Institut national d'historie del'art, Paris, April 4–5, 2022.

to impress with "the simplicity and preciousness of the material."[71] It was no longer the spacious reception area for large crowds and gatherings known from earlier pavilions; such functions were now moved to more luxurious locations and the pavilion restaurants. The space was also designed as large enough to allow the visitor to take a break from more intimate installations.[72] It contained the so-called *Pantheon of the Nation*, a series of sculptures representing Czechoslovakia and its history.

A large bronze sculpture of a lion, the symbol of Czechoslovakia, by Břetislav Kafka (1891–1967) dominated the space. The lion was later moved to Bratislava to complete a monument to Milan Rastislav Štefánik (1880–1919), the Slovak politician and one of the founders of Czechoslovakia who died tragically in 1918. Despite his crucial role in the formation of the state, Štefánik's sculptural representation was absent in Paris, to the critical remarks of Slovak newspapers.[73] Instead, there were four busts of historic figures that were all male and Czech: President Masaryk, the nineteenth-century historian František Palacký (1798–1876), the scientist Jan Evangelista Purkyně (1787–1869) and the composer Bedřich Smetana (1824–1884), placed along the wall in the direction of the visitors' flow.[74]

The stipulated route led visitors from the Hall of Honor to the sections on the political system of Czechoslovakia, its schools, cultures and science first. The School of Decorative Arts section was designed by Otakar Novotný, a professor at the School with various art and craft objects, like glass, textiles, ceramics and graphic design, mainly by František Kysela (1881–1941). His large, engraved glass sheet of the Republic created a transparent division between the Hall of Honor and the rest of the ground floor. Here, Sutnar designed the section of specialized and vocational schools, where more objects of glass, bijou, porcelain and ceramics, as well as embroidery, lace and bookbinding were displayed. They came from the State Graphic School in Prague where Sutnar was the director.[75] Sutnar made good use of textiles and draperies here which served both as shades and to showcase the actual products.

This part was followed by one on industry and trade put together by Ladislav Turnovský from the Ministry of Trade and overseen by Bohuslav Soumar from the designers' team. They emphasized smaller and commercially important com-

71 "Československý státní pavilon na světové výstavě," *Umění* (1936/7): 246.
72 "Vysvětlivky k průvodní zprávě," 2.
73 "Koho representuje čs. pavilion na svetovej výstave v Paríži?" *Slovensky dennik,* August 21, 1937: 5.
74 Purkyně and Smetana were created by sculptors Otakar Španiel (1881–1955) and the rest by Josef Václav Myslbek (1848–1922).
75 Soumar, ed., *Tchecoslovaquie*, 39–41.

modities and again included bijou, especially from the north Bohemian town of Jablonec nad Nisou, ceramics and porcelain. Leather, paper goods, musical instruments, as well as agricultural products were added.[76] The route continued with yet another presentation of glass, now as an individual section of the Glass Research Institute in Hradec Králové, showing a range of objects produced for various purposes and in various techniques, from blown to sheet glass, from technical and engraved glass to miniature figurines.[77] This section was put together by the designer Antonín Heythum (1901–1954) who added a wall of glass rods and illuminated the objects with light emanating from inside the vitrines.[78]

The following sections focused on other branches of export industries. Ceramics and Porcelain showcased the production of the five largest manufacturers from Karlovy Vary and the works of the State Porcelain School from the spa town, arranged by Soumar. The next part on textiles, one of the key Czechoslovak export goods with centers in Prague, Brno and Liberec, was arranged by Alois Wachsman (1898–1942), an architect, scenographer and painter and consisted of products by 21 large companies that manufactured fabrics, textiles, threads, and hats. They included the fezzes from Strakonice, which, at the time, were exported to "the entire Muslim world," signaling the global reach of Czechoslovak products as well as their colonial complicities that I will come back to in the next chapter.[79]

As I mentioned earlier, heavy industries were planned for the Industries Hall on the first floor of the Czechoslovak pavilion. This was another open, well-lit space, originally planned with access by "a mountainous route" with vistas of the Tatras and a car display, yet this idea was later dropped. The space was still meant to impress with its abundance of objects from various companies, supplemented by large-scale photographs and company logos. Suspended under the cupola was the aircraft Praga Air Baby by The Czech-Moravian Company (Českomoravská továrna) a producer of various transportation vehicles including trams, motorcycles and later tanks, that showed here also engines and

76 Soumar, ed., *Tchecoslovaquie*, 45–49.

77 The Bohemian crystal by the sculptor Josef Drahoňovský was shown here alongside the transparent mosaics by Antonín Procházka and Linka Procházková, a mosaic by Jan Zrzavý, an etching on glass based on a design by the painter Emil Filla, and a collection of glass minifigures by Jaroslav Brychta entitled *The Glass Paradise*. Soumar, ed., *Tchecoslovaquie*, 57–59. Václav Čtyroký, "Naše sklářství v Paříži," *Národní politika* (supplement), September 25, 1937, 16.

78 S. Ú., "České sklo na Světové výstavě v Paříži," *Večerní Lidové noviny* 45, July 31, 1937, 3.

79 "My na světové výstavě v Paříži," *Venkov*, August 22, 1937, 5; Marta Filipová, "Colonial Complicities beyond the Empire. Czechoslovakia inbetween Worlds and World's Fairs," in *Routledge Companion to Art and the Formation of Empire*, edited by Emily C. Burns and Alice M. R. Price (London – New York: Routledge, 2024).

models of these vehicles as well as enlarged photographs from its plants and studios. The plane made for an impressive and often photographed object, but it was not the only such exhibit at the exposition. The Air Pavilion also featured a suspended fighter plane amidst rings reminiscent of electron paths in a display designed by Sonia and Robert Delaunay.

The Industries Hall contained exhibits of other heavy industries producers, including Škoda, with their models of a distilling vessel and various engines. An enlarged photograph from the manufacturer and one of the cast iron cruiser Normandie, a "masterpiece of metallurgy," were some of the highlights here.[80] The company Vítkovice, that supplied steel for the construction of the pavilion, was also represented here and displayed models and exhibits of their products alongside other companies and diverse industries. This included Baťa showcasing photographs of the social and work life in Zlín and a display of a single shoe, to which I will return shortly. Agriculture was presented on the ground floor and on the terrace, where it featured agricultural companies growing wheat, hops, seeds of all kinds and processing technologies. The interior part of the terrace housed a painted panorama of a view of Hradčany and Malá Strana in Prague. Authored by the painters Jan Slavíček (1900–1970) and Richard Wiesner (1900–1972), the painting extended for 15m.

The pavilion also contained a folk art section, which demonstrated the difference between the western and eastern folk art of the Bohemian and Slovak regions respectively, as the catalogue noted.[81] Put together by Josef Vydra (1884–1959) of the School of Arts and Crafts in Bratislava with the help of the architect Dušan Jurkovič and others, the exhibit showed items of associations, companies and coops that produced and promoted what was marketed as folk art. This was, for instance, folk textiles, ceramics and lace, which documented the primarily "idyllic life of the people of Slovak mountains ... before their already latent disappearance is accomplished."[82] Vydra and his colleagues visualized the different parts of the state and the split in the geographical and ethnic distribution using folk art and other exhibits.

This was a common narrative and one of the most defining features put forward in all other Czechoslovak participations in the interwar period, as I discuss throughout the book. Slovak regions were put in the role of the traditional countryside and juxtaposed with Czech, industrialized and urbanized regions.

80 Soumar, ed. *Tchecoslovaquie*, 90.
81 "L'art Populaire en Tchecoslovaquie," in Soumar, ed. *Tchecoslovaquie*, 119–120.
82 "L'art Populaire," 124.

In striking contrast to the folk display, abstract kinetic sculptures by Zdeněk Pešánek (1896–1965) were also located on the terrace to represent tourism and the spa industry. Pešánek was a contemporary Czech artist who combined new materials with sound, movement and light in his sculptures and I will return to his work in more detail in Chapter Three in relation to the representation of art and design in Czechoslovak pavilions.[83]

The route through the pavilion was concluded by the tourism section, designed by Krejcar and the painter and graphic designer Zdenek Rykr (1900–1940). Following Krejcar's idea that "every exhibitor makes an effort to first lure the visitor in so that the visitor is satisfied with the content of the pavilion and especially so that the last impression when exiting the pavilion was pleasing and lively," visitors left the pavilion through the so-called Prague street.[84] Adopted from the proposal that Sutnar had entered in the competition with Fuchs, mentioned earlier, the street of shopping windows displayed Bohemian crystal, bijou and jewelry, ceramics, toys and curtains. This was one last attempt to engage the visitors, but also a reminder that the entire pavilion was a showcase of Czechoslovak production meant for sale. Not only heavy industries and agriculture, but also smaller objects of everyday or decorative use were promoted and could be purchased in a nearby shop, located outside the pavilion.

Display techniques

The Czechoslovak pavilion in Paris, like many other national pavilions, was a menagerie of the wishes and desires that various designers, artists, business and government representatives projected into it. They devised an ideal plan, which was always subjected to changes and modifications based on onsite realities and individual decisions. Yet the drive for the desired effect of the displays on visitors remained central. By the late 1930s, the belief that a modern exhibition should "explain, demonstrate and even persuade" the spectator and lead them to a planned reaction in the way advertising did, grew strong.[85] For Czech designers, psychology was an important player in designing exhibitions and they employed a number of innovative techniques, for instance, incorpo-

83 "Československý státní pavilon na světové výstavě," *Umění* (1936/7): 247.
84 "Vysvětlivky k průvodní zprávě projektu," 7.
85 Bayer, "Fundamentals of Exhibition Design," 17.

rated an *interval of silence* and *suggestive presentations* or, as Sutnar did, elicited *shock reactions* from the visitors and controlled their *flow* and the flow of information.

The ideal plan for the Czechoslovak pavilion was to guide the visitor through the pavilion on a route in which the wall became an important feature. It was not just a neutral background or plain divider anymore; the wall could play different roles. It could function not only as a so-called interval of silence, an empty wall space that allowed the visitor a break from the exhibits, but it could also be turned into a self-sufficient display feature with photomontages, epigraphs, and three-dimensional details, which was a technique adopted in many pavilions at Paris, let alone the Czechoslovak one. Especially the Industries Hall made good use of background wall photographs depicting the working life and manufacturing of the respective products in the exhibits of for example Baťa, Vítkovice and Škoda.

The individual displays often made use of another innovative technique, a suggestive presentation, in which an item that was the subject of the display was represented by its graphic abstraction, photography or film.[86] A pertinent example was provided by Baťa, a company that not only produced shoes, but also hosiery, toys, bicycles, airplanes and rubber products for the global market. At the Parisian exposition, Baťa built a separate pavilion of its French branch at

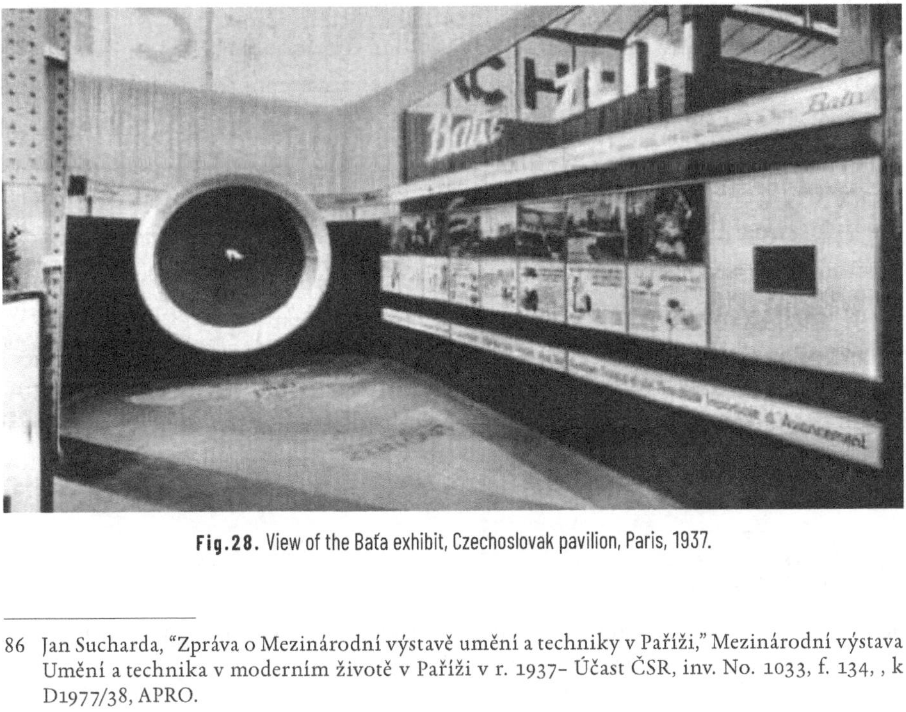

Fig.28. View of the Baťa exhibit, Czechoslovak pavilion, Paris, 1937.

86 Jan Sucharda, "Zpráva o Mezinárodní výstavě umění a techniky v Paříži," Mezinárodní výstava Umění a technika v moderním životě v Paříži v r. 1937- Účast ČSR, inv. No. 1033, f. 134, , k D1977/38, APRO.

Hellocourt where a more detailed insight into the shoe production was offered. It could therefore afford to experiment with its display techniques more in the national pavilion. Reporting on the exhibit, the conservative Czech newspaper *Venkov* noticed that in the Industries Hall, "Baťa exhibits a single shoe. Yet they claim they produce 180k pairs of such shoes a day in their Zlín factory."[87]

This remark referred to an eye-catching installation which indeed featured only one high-heeled shoe in a large lit circle [fig. 28]. The single shoe stood for the extensive production in a display that can be considered as suggestive as well as representative. It was supplemented by the abovementioned large photographs which depicted the different aspects of work and social life in the Czechoslovak town of Zlín where the company was based. They accompanied the overall suggestive presentation by explaining the contribution of the company to modern and healthy life in the city and the country.

The way visual information was distributed, and the concept narrated therefore showed awareness of the effects such presentation had. According to Jan Sucharda, a representative of the Czechoslovak Chamber of Commerce who was commissioned to write up a report on the Czechoslovak pavilion after the exhibition, a key element in the arrangement of the display was movement, both physical, along the planned route, and optical through the visual information.[88] The visitor's eye was made to inspect the different parts of the montage and construct a complete picture, in a way similar to Moholy-Nagy's and Bayer's approach to exhibition space. Enhancing the eye movement in the exhibition was, in Sucharda's view, comparable to theatrical effects and illusions of movement. The comparison with theatre is highly relevant here because there certainly were many overlaps between theatre and exhibition techniques. "The spectator must be a partner in the exhibition theatre," Sucharda suggested and noted that the visitor, even though consuming the exhibition passively, participated in creating its meaning.[89] This view also resonated with El Lissitzky's requests for an active spectator stimulated by means of design he outlined in the Exhibition rooms.[90]

Throughout the interwar period, there were clear parallels between staging exhibitions and theatrical performances.[91] The distribution of space, visual information and people was a shared concern for exhibition and stage designers. It is

87 "My na světové," 5.
88 Sucharda, "Zpráva."
89 Jan Sucharda, "Zpráva," 5.
90 El Lissitzky, "Exhibition Rooms (ca 1926)," in *El Lissitzky: Life, Letters, Texts*, ed. by Sophie Lissitzky-Küppers (London: Thames & Hudson, 1968), 3.
91 Marta Filipová, "The Theatre of Exhibitions."

therefore no wonder that many interwar exhibition designers and architects in Czechoslovakia, including the abovementioned Kroha, Heythum or Sutnar, were not only exhibition, graphic or information designers but also stage designers.

Reception

How successful were the new exhibition techniques and approaches to the exhibition space? While architects and designers were indeed aware of the need to engage the visitor and aimed to create a well-thought-out, well-designed interior, the reality of the pavilion turned out differently. Consistency, or the lack of it, was one of the issues that the pavilion faced. The pavilion, as I mentioned earlier, was split into a series of smaller or larger individual sections each composed by a different designer. The number of architects and designers who were responsible for the various segments of the national pavilion led to the fractionality of the overall exposition. This was seen as a flaw of the interior installation, which then gave a sense that a unified concept was missing, a result of fighting for responsibility between different agents.[92]

In his post-exhibition report, Sucharda identified the inconsistencies between the concepts of the architects and designers on the one hand and the input from the exhibitors and organizers on the other hand. As various ministries were involved in organizing the diverse parts of the pavilion, their goals differed. While some displays aimed at showing the state as modern through the arts, design and architecture, others put primary emphasis on consumerist culture, trade and commerce. Yet, these goals were sometimes not mutually exclusive, and, for example, the showcases of design had both artistic and commercial intentions. This is added to the mixed messages because the same items appeared in different contexts. This was the case, for instance, of glass objects that featured in the trade-oriented section on glass as well as in the section on vocational schools.

The discrepancy between the original concept of the pavilion and its final look was a result of the conflicts between the ideological and materialistic interests of the ministries involved in the display.[93] The ministries of education and public works, on the one hand, and that of industry, on the other, pursued either cultural propaganda or trade and export interests respectively and, as

92 Jan Sucharda, "Zpráva," 7.
93 Sucharda, "Zpráva," 5.

a result, the exhibitors' point of view was often privileged over the visitors' experience. The intention of the designers to lead the visitor through the exhibition on a route that did not overwhelm and exhaust them, ultimately fell short of the actual displays.

The interval of silence, for instance, that would allow rest from the often-overwhelming exhibits, was, in the end, hard to find in the pavilion overcrowded with objects. The intended effects of natural light entering the interior through the Thermolux glass were also diminished in the actual installation of exhibits. Parts of the pavilion were screened off from natural light and lit artificially—these were, again, the glass and porcelain displays and partly the tourism sections. The glass section was, in the end, composed as a magical cave [fig. 29]. The rows of blue columns of light in the "winding hallway with a low ceiling, darkened and narrow" made the space rather oppressive for many.[94] It had an adverse effect on the visitors—it did not let them "breathe out" until the end of this section. This, Sucharda complained, created a strange discord between how the effects of light were envisaged in the exterior and interior and the actual realization.[95]

Many national papers commented on the mishaps in the Czechoslovak pavilion. Further criticism aimed at the overcrowded displays of the heavy industries[96] as well as the "near life-threatening jam" in the narrow, winding curves of the ground floor exposition of glass and ceramics.[97] The Czech theatre director Jiří Frejka, who visited the pavilion, also noticed the conflict between what he described as the "beautiful" exterior and the interior which contained "good and bad objects."[98] He noticed that some of the displays, for instance those of the fez and wood, were overexposed, and despite the theatricality of the overall space, the individual rooms lack a unifying concept.[99]

Regarding the exterior, an anonymous contributor in the newspaper *Venkov* criticized the excessive emphasis on the architecture of the Czechoslovak pavilion at the expense of the interior displays.[100] The author concluded that the architecture of national pavilions in the fairs should be secondary to what was

94 Sucharda, "Zpráva," 5.
95 Sucharda, "Zpráva," 5.
96 "My na Světové výstavě v Paříži," *Venkov*, August 22, 1937, 5.
97 *Rundschau Technischer Arbeit Berlin*, September 22, 1937, quoted in *Národní politika*, October 12, 1937, Excerpts Archive, MFA, oh8.
98 Jiří Frejka, "Děláme československou divadelní exposici," *České slovo*, August 22, 1937, Newspapers Excerpts Archive, MFA, oh8.
99 Frejka, "Děláme."
100 "Náš pavilon na pařížské výstavě," *Venkov*, August 11, 1937, 8.

displayed inside. The temporary pavilions in Paris, including the Soviet, German, and Italian ones, did not represent the best of international architecture and "in terms of the construction, technology, crafts, these [were] improvisations almost like in the theatre. We [the Czechoslovak representation] did not grasp that and that is where the core of all our mistakes lies..."[101] Another daily, the leftist *Právo lidu,* went even further in their criticism and turned attention to ordinary visitors who were left out of the pavilion. The designers and exhibitors failed to address the regular audience because apart from a few excep-

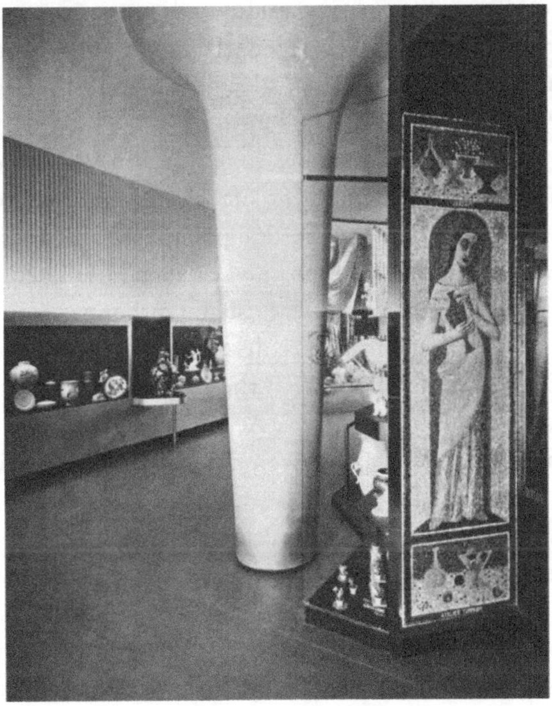

Fig.29. Antonín Heythum, Glass section, Czechoslovak pavilion, Paris, 1937.

tions, there were no displays of ordinary life in Czechoslovakia. Instead, Latin inscriptions in the Hall of Honor gave the impression that the target audience was probably professors of classics, rather than the masses of visitors that normally flow into world's fairs.[102]

101 "Náš pavilon," 8.
102 "Procházky po výstavě," *Právo lidu*, October 19, 1937, Newspaper Excerpts Archive, MFA, oh8.

Conclusion

Krejcar's intent to pitch the display for the average visitor and not the upper ten thousand thus remained wishful thinking. The consideration of the visitors' experience, that Sucharda and others suggested, appeared in a few isolated cases and was almost forgotten following input from different agents involved in the design, composition and content of the display. Why then, would this national pavilion be considered one of the most successful examples of Czechoslovak participation in world's fairs? Despite the frequent criticism in the national newspapers,, its reception changed once prizes and awards were given out over the course of the exhibition. Czechoslovakia received a total of 70 Grand Prixes, 82 Diploms d'honneur, 128 gold medals, 63 silver medals, 11 bronze medals and 354 recognitions.[103]

The top award, the Grand Prix, went for instance to Škoda and Vítkovice in the industries category, to Josef Gočár for his design of a church in Vršovice displayed here, to Vlastislav Hofman for his theatre designs, the Slovak artist Ľudovít Fulla for painting, and to Baťa for the exposition of social living entitled "Woman, child, family." Indeed, world's fairs were usually rather generous in giving out awards. In Paris, over 2,200 prizes were distributed amongst the 52 exhibiting states, but Czechoslovakia still received the third highest number of prizes, after the Soviet Union and Germany.[104] At home, this was eventually considered an enormous success, especially given the fact that both the architecture of the pavilion and the interior layout, as well as the collaborative work of Krejcar, Sutnar and others, received a Grand Prix too.[105]

Many commentators in the international press also appreciated the internal division of the national pavilion and the intimacy of some of the sections. The American architect and historian Talbot Faulkner Hamlin, for instance, noticed that the smaller exhibits in more intimate displays of booths and dioramas on the ground floor contrasted well with the large machines and mechanical parts on the "airy height" on the upper floor.[106] Hamlin also appreciated the visitors' flow, or circulation, as he called it, through the pavilion accessible

103 "Glänzendes Abschneiden der Tschekoslovakei in Paris," *Prager Presse*, November 26, 1937, 5.

104 "Z výstavy v Paříži: na třetím místě," *Večerní České slovo*, November 26, 1937. Full list in "Mezinárodní uznání našim vystavovatelům," *Lidové noviny* 45, November 26, 1937, 4.

105 "Naše propagace cizineckého ruchu získala v Paříži velkou cenu," *Národní střed*, November 27, 1937, Excerpts Archive, MFA, oh8. The news items mention Soumar, Sutnar and only two others—the architects Karel Hiller and Josef Pilař.

106 "Paris 1937. A Critique," 25–34.

by an "ingenious staircase."[107] He found other examples of successful circulation design at the exhibitions in the French Social Service pavilion, Advertising building, and St Gobain glass company building where traffic was directed alongside the displays. The national pavilions of, for instance, Switzerland, Finland and Spain, also designed and controlled the experience where the "element of choice must never be allowed, for that means indecision, delay and crowding" which together with "aimless wandering destroy the effect of [the] exhibits almost entirely."[108]

The detailed discussion of the Czechoslovak pavilion in Paris shows general tendencies in exhibition design of the interwar period. Organizers of national pavilions, including the Czechoslovak ones, ultimately tried to limit the element of choice to a minimum. The visual flow envisaged by the designers directed the movement of the spectator down the pavilion, exposing them to the objects and fighting against exhibition fatigue by introducing innovative techniques like the interval of silence or suggestive presentations. Yet the intended design of the exterior, interior and visual information often clashed with the reality of adjusting the display to the needs and requirements of the individual sectors that were present in the pavilion. The designers' desires for a unified concept of the pavilion and concerns for the visitors' experience were eventually overridden by the ideological aims and commercial motivations of the organizers and exhibitors.

107 "Paris 1937 A Critique," 29.
108 "Paris 1937. A Critique," 29.

Chapter Three

ART AND DESIGN

THE LIMITS OF MODERNISM

Displays of the visual arts, whether they were shown as individual works or in a dedicated section or palace, had been a regular occurrence at world's fairs since their early days. While exhibitions of the art of the so-called old masters became commonplace as of the 1889 Exposition in Paris, the inclusion of contemporary art dates to at least the 1855 Exposition Universelle in Paris.[1] Modernity of these events was formulated by the inclusion of the latest signs of progress found in new technology, inventions and materials, and invited an embrace of the modern in the visual arts too.

Such embrace took many forms; the arts were most often shown in a dedicated pavilion or pavilions which gave most space to the organizing country. National pavilions also habitually incorporated art displays which often counterbalanced the more trade and science-oriented presentations of technology, machinery and statistics in the commercial, industrial and scientific sections and pavilions.[2] Here, exhibitors aimed to promote their economic and political advancements, while showcases of culture, including the fine arts, music and theatre, had a prestigious as well as entertainment added value, presenting the cultural maturity of the respective nation or state.

The range of arts included in world's fairs was also wide and varied from fair to fair. Painting and sculpture were most common both in interiors and exteriors of pavilions or fairgrounds. Outside, murals, statues at buildings and free-standing sculptures and fountains helped to create a unique and festive at-

1 Paul Greenhalgh, *Ephemeral Vistas: The Expositions Universelles, Great Exhibitions, and World's Fairs, 1851–1939* (Manchester, UK – New York: Manchester University Press, 1988), 198 and 200.

2 Greenhalgh, *Ephemeral Vistas*, 198.

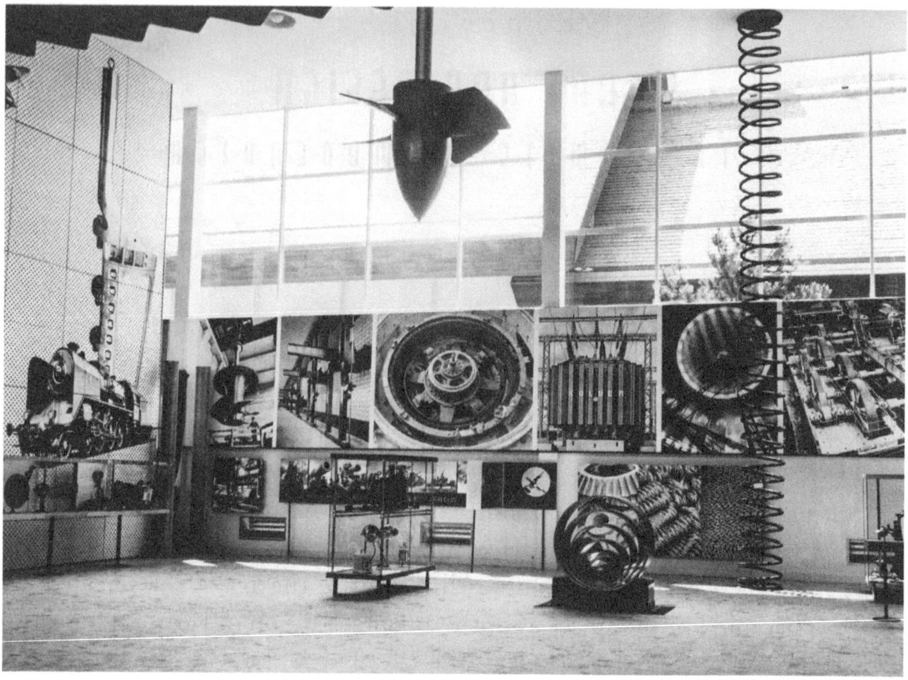

Fig. 30. Czechoslovak pavilion, New York World's Fair, 1939.

mosphere of the visit to the fair ground. Inside, hanging paintings, large scale photographs and free-standing statues complemented the representative character of the overall displays in the respective pavilion. In many cases, art was a part of a carefully conceived narrative, yet there were times when it became a decorative, last-minute addition. These showcases of fine arts were complemented by displays of items that were more functional but also had an important aesthetic element. The objects of craft, such as applied arts and design, and aestheticized parts of machinery like coils, turbines and engines, brought the industrial and commercial aspect of exhibitions back to their original aim [fig. 30].

Throughout the interwar period, Czechoslovakia was looking for the best way of representing its independence, democracy and identity in various exhibits, and fine art and design provided more than a suitable means. Like many other new nation states in a similar position of making their mark through world's fairs, Czechoslovakia continued to build its national institutions that promoted and "allowed national mythologies and the very myth of the nation-state."[3] They in-

3 Donald Preziosi, "The Art of Art History," in *The Art of Art History: A Critical Anthology*, ed. Donald Preziosi (Oxford – New York: Oxford University Press 1998), 508.

cluded not only national museums and galleries, but also art and design schools, and art associations that were showcased at world's fairs.

Yet putting together an official artistic representation presented several problems. While the inclusion of modern art and design could be understood as the state's clear attempt to embrace modernity, there was little consensus amongst the Czechoslovak organizers on which visual form of representation should take precedence and which of the many coexisting, parallel and overlapping modernisms should be prioritized.[4] The interwar period therefore saw an almost ongoing debate about what art was most fitting for international exhibitions organized on such a large scale and with such a multiplicity of aims and about who should be selecting it. Ultimately, too, the question of what constituted the art's modernism was also opened. But indeed, Czechoslovakia was no exception in these considerations, as many other countries, large and small, conducted a similar quest for the most appropriate mode of artistic representation.[5]

France's decision to host the 1925 Exhibition purely on modern decorative arts to construct its modern visual language could be one example. At this exhibition in particular, many participating states attempted the same—construction of their distinctive yet universally accepted modernism. For instance, the Kingdom of Serbs, Croats and Slovenes constructed a pavilion "styled in an expressionist manner, with classical segments and plain facades with Art Deco details."[6] Being in a similar situation to Czechoslovakia in the attempt to forge a single identity for external (and internal) audiences, the modern Yugoslav identity at world's fairs swiped the different ethnic and historic traditions under the umbrella of international modernism.

The entire 1925 Exposition was an endeavor to create a new visual language that later became known as Art Deco.[7] Yet as one Czech commentator noted,

4 Marta Filipová, *Modernity, History and Politics in Czech Art* (New York: Routledge, 2020).

5 For example, Mauricio Tenorio-Trillo, *Mexico at the World's Fairs. Crafting a Modern Nation* (Berkeley – Los Angeles – Oxford: University of California Press, 1996); Greenhalgh, *Ephemeral Vistas*, M. Elizabeth Boone, *"The Spanish Element in Our Nationality": Spain and America at the World's Fairs and Centennial Celebrations, 1876–1915* (University Park, PA: Penn State University Press, 2021); Paul Sigel, *Exponiert. Deutsche Pavillons auf Weltausstellungen* (Berlin: Bauwesen Verlag, 2000); Sven Schuster, *Die Inszenierung der Nation. Das Kaiserreich Brasilien im Zeitalter der Weltausstellungen* (Frankfurt am Main: Peter Lang, 2015).

6 Aleksandar Ignjatović, "Jugoslovenski identitet u arhitekturi izmedu 1904. i 1941. godine," Doctoral dissertation at the University of Belgrade, Faculty of Architecture, 2005, 139. Cited in A. Stamenković, *Yugoslav Pavilions at International Exhibitions*, 304.

7 Makaryk, *April in Paris*; Philip Whalen, "Paris 1925," in *Encyclopaedia*, 236–243; Kostova, *Spectacles of Modernity*; Alain-René Hardy, *Art Deco Textiles: The French Designers* (London: Thames & Hudson 2006); Alastair Duncan, *Art Deco Furniture: The French Designers* (London: Thames and Hudson, 1984); Jared Goss, *French Art Deco* (London: Thames & Hudson 2014).

> the entire composition of the exhibition and its program ... shows a fundamental error: the decorative arts are understood in their old function when it was organically bound to crafts. That is why it is an exhibition of luxury and ornament, not of objects of everyday use. It completely negates the new lifestyle which has been affected by the new element of the machine.[8]

He therefore pointed to the confrontation between the modernity of the new world enabled by technology and the opulence of decorative and often impractical ornaments at the Exhibition. A very simple, yet crucial question was posed there—what is *modern* when one talks about the visual culture? Is it the use of new techniques and technologies, application of geometric ornament, or the use of new materials or art and design that is more accessible and affordable? The Exhibition embraced several ideas of modernity and modernism, found in the introduction of new technologies, materials or ornamentation. Underlying these was a deeper modernist concern about the role of the machine in art and crafts, and its influence on fine art. Artistic displays at world's fairs were not a true representation of the art scene in any given country and they did not construct particular canons of modern art like for instance the Machine Art exhibition at MoMA in 1934.[9] They nevertheless reflected and enforced some of the main tensions between individuals, institutions and apparatuses involved in the formation of the artistic narratives.

Many theorists and practitioners in Czechoslovakia were involved in the lively debate about the artistic content of the national pavilion and reflected these wider concerns. In this chapter, I focus on the attempted selection of an appropriate visual language of representation of the new state. Using a few examples of art and design, I examine the place and importance of the visual arts in the official displays in order to show that the representation of modernity in the visual arts for display in the interwar period was very fluid. The competing motivations and agendas that interfered with the displays, as I argue, explain the seemingly eclectic medley of artefacts that often appeared in the Czechoslovak pavilions.

---

8 Adolf Benš, "Mezinárodní výstava v Paříži dekorativního a průmyslového moderního umění v Paříži 1925," *Stavitel* VI, no. 8–9 (1925): 109–18.

9 Philip Johnson, *Machine Art. March 6 to April 30, 1934* (New York: Museum of Modern Art, 1934); Jennifer Jane Marshall, *Machine Art 1934* (Chicago, Ill. – London: The University of Chicago Press, 2012).

First international exhibitions

The search for an appropriate modern expression in Czechoslovakia was closely bound with the search for the arts that could be representative of the state and nation on the international forums, including world's fairs. Before 1918, art of the future Czechoslovakia made occasional appearance at world's fairs as part of the Austrian and Hungarian pavilions either in the form of individual works or as part of larger exhibits of for example glass. One of the most represented artists was Alfons Mucha (1860–1939), who contributed his works not only to the Exposition Universelle in Paris in 1900, but also to for example the Louisiana Purchase Exposition in St. Louis in 1904 where he created a poster for the French Ministry of Commerce.[10]

Immediately after the creation of independent Czechoslovakia, the visual arts—ranging from folk art and applied arts to fine art—were given the role of representing the new state and its identity. Presentation of fine art could signify cultural maturity and modernity of the new state, displays of folk art its traditions and historicity, while applied and decorative arts showcased the latest designs and production methods combined with a potential commercial opportunity of sale to new markets. On top of that, Czechoslovak presence through art exhibited in specific countries, especially France and the USA, sealed the political and military alliances formed during the First World War.

And it was in France where one of the first occasions for presenting art under the label of "Czechoslovak art" occurred. In 1920, the Louvre hosted an exhibition of Czechoslovak folk art, initiated by General Maurice Pellé (1863-1924) who in 1919 became the leader of the French Military Mission in Czechoslovakia and the chief of staff of the Czechoslovak army during the conflict with Hungary over Slovakia in the same year.[11] Pellé had close ties with the military representatives of Czechoslovakia but also with Tomáš Garrigue Masaryk and many Czech artists whom he met in Paris and during his stationing in Bohemia and Moravia during the war. These included the painters and graphic artists František Kupka, Alfons Mucha, and Max Švabinský, the sculptor Bohumil Kafka,

10 Milan Hlavačka, Jana Orlíková, and Petr Šembera, *Alfons Mucha—Paříž 1900. Pavilon Bosny a Hercegoviny na světové výstavě / Alfons Mucha—Paris 1900. The Pavilion of Bosnia and Herzegovina at the World Exhibition* (Prague: Obecní dům, 2002).

11 The so-called Hungarian-Czechoslovak war was a conflict between April and July 1919 which took place during the disputes over the territories of present-day Slovakia led between 1918 and 1919. The crisis was resolved on a diplomatic level by the Treaty of Trianon in 1920. See for example, David Perman, *The Shaping of the Czechoslovak State: Diplomatic History of the boundaries of Czechoslovakia 1914–1920* (Leiden: E. J. Brill, 1962).

and the female artist Zdenka Braunerová (1858–1934).[12] She not only introduced Pellé to his future wife, her niece Jára, but also to local folk art.[13] As a great admirer of vernacular art that she aimed to revive and preserve, Braunerová helped Pellé organize the Paris exhibition.

The relationship between Braunerová and Pellé is illustrative of the strong personal, cultural and political ties between France and the Czechs and Slovaks. Another such example can be found in the first Minister of War of Czechoslovakia, Milan Rastislav Štefánik, who was of Slovak origin and a French citizen. Before the war, he became an acclaimed astronomer and general of the French army, while during the war, he became a Czechoslovak legionnaire and one of the key members of the resistance government in Paris. The Czech(oslovak)-French relations were also strengthened on an institutional level by the establishment of the French Institute in Prague in 1920 and the Institute of Slavic studies in Paris in 1923, both linked to the French historian of the Slavs Ernest Denis (1849–1921).[14]

The Louvre exhibition therefore fitted into the ongoing cultural exchange and its status can be seen in the fact that the exhibition in Paris was opened by the French president Paul Deschanel. The exhibition, nevertheless, presented a very specific image of the new state, one embedded in folk traditions and art and one that looked at the narratives linking nation's art with folk heritage promoted before the war.[15] The five rooms in the Pavilion Marsau in Rue Rivoli were filled with items loaned from various museums in Czechoslovakia: costumes from Bohemia, Moravia, Slovakia, embroidery, ceramics, household items, furniture, clothes, a painted archway found above the front door (the so-called *žudro*) from the so-called Moravian Slovakia (a region in southeast Moravia), but also contemporary fine art paintings depicting scenic village life by Joža Uprka and the sculpture by his brother František.[16] The mixture of authentic folk art and folk-inspired fine art aimed to create an expression of national sentiments

12 Isabelle Sandiford-Pellé, *Generál Pellé, obrázkový deník* (Prague: Ministerstvo obrany ČR – Prezentační a informační centrum MO, 2010), 155.
13 Sandiford-Pellé, *Generál Pellé*, 155.
14 For example, Tomáš Chrobák and Doubravka Olšáková, *Ernest Denis* (Prague: Eva – Milan Nevole, 2003).
15 The exhibition was organized by Raymond Kœchlin, a French art historian, Lubor Niederle of the Ethnographic Museum in Prague with the help of the ethnographer Karel Chotek, the art historian Renata Tyršová, Madlena Wanklová of the Brno regional museum, and Vlasta Havelková of the Ethnographic Museum. Jaroslav Nauman, "Dojmy z Paříže poválečné," *Volná myšlenka*, May 30, 1920, 5 and Pierre de Quirielle, "Une exposition d'art populaire tchécoslovaque," *Journal des débats politiques et littéraires*, April 28, 1920, 1, https://gallica.bnf.fr/ark:/12148/bpt6k4893140.
16 Quirielle, "Une exposition," 1.

where "the unity of the presented folk art [served as] an affirmation of the unity of the race and the nation."[17]

To this extent, the French journalist and writer Pierre de Quirielle also identified Slovak folk art as a permanent cultural phenomenon that was the "traditional and original, vivid expression of the soul and aspirations" of a country which used it as a weapon against its oppressors, the Magyars.[18] And Slovaks were the people of John Ruskin; their expressions were the manifestation of their "sense of natural and spontaneous" art.[19] De Quirielle's reference to Ruskin and the Arts and Crafts movement stemmed out of the belief that Slovak folk art had a permanent, pre-industrial quality that preserved traditional forms and techniques despite the adverse influences of the outside world.

Czech commentators on the exhibition also stressed a degree of "charming" primitivism contained in the displayed works. The writer Jaroslav Nauman, for instance, reported in the liberal journal *Volná myšlenka* (A Free Thought) that "many visitors were captured by the simple, almost naïve, and still enchanting, beauty of the products of our mothers, standing in such sharp contrast with the refined [art] all around" in the Louvre.[20] Yet the emphasis on the timeless quality of the folk art on display and its pre-industrial roots were key for creating a sense of artistic tradition for foreign viewers. Linking Slovak and Czechoslovak folk art to the ideas of Ruskin and the Arts and Crafts movement made the arts relevant and comprehensible to international viewers. Such art could not be seen as an obsolete, outlived phenomenon, it had to be appreciated for its individual craftsmanship and skill in the face of machine produced art and design.

Searching for a representative style

The dichotomy between traditional visual expressions found in folk art, on the one hand, and machine-made design on the other was an ongoing topic of debate between both organizers of Czechoslovak interwar exhibitions and their critics. The polemic, as well as the search for appropriate representative modern artistic expression, were initiated already at the first substantial participation of Czechoslovakia at a truly international exhibition. It was The Centennial International Exhibition in Rio de Janeiro in 1922 where the

17 Quirielle, "Une exposition," 1.
18 Quirielle, "Une exposition," 1.
19 Quirielle, "Une exposition," 1.
20 Nauman, "Dojmy," 5.

Czechoslovak presence was originally planned as purely commercial, aiming at exploring new trade opportunities and markets of South America. This became more important after it became known that Germany—one of the main competitors—would not take part in this event, which opened more possibilities for the new state.

The artistic side of the presentation nevertheless played an important role, and many objects of art and design were displayed here in the hope of generating sales. This was the first international appearance of the work of František Kysela, a versatile artist and designer from the School of Decorative Arts in Prague, whose works were also featured at the Parisian exhibitions of 1925 and 1937. His stained-glass window in the Czechoslovak pavilion at Rio de Jane depicted the state emblem above the central entrance of a pavilion designed by the architect Pavel Janák as a building with colorful ornament. For Janák the ornament had a purely decorative function, but the entire structure made strong references to folk traditions in the architectural features too. Most prominently, the central entrance was flanked by an archway reminiscent of the painted archway, or *žudro*, of Moravian villages that also featured in the 1920 Paris exhibition.

The search for a suitable national style that architects engaged in in Czechoslovakia and transferred to the national pavilion in Rio de Janeiro was not dissimilar from the local attempts to establish a Brazilian architectural language that would correspond to the modern Brazilian identity.[21] Here, the so-called Brazilianness (*Brasilidade*) in architecture started developing to distinguish national architecture from the Portuguese colonial heritage.[22] It drew on a variety of traditions that included indigenous, African as well as Portuguese influences in which it was similar to the somewhat eclectic nature of modern architecture in Czechoslovakia.[23]

The dialogue between the modern and the traditional in Czechoslovak art and architecture was an important component of the more general search for the new visual identity of the state.[24] Following the end of the war, many architects and designers were aware of the need to find an appropriate visual language which would express the new political situation and serve the new state and people. Janák was one of the architects who experimented with references to local folk tradi-

21 Angela Starita, "Brasilidade in Built Form: Tracing National Identity in Modernist Architecture in Brazil, 1922–1968," Dissertation submitted to the Faculty of New Jersey Institute of Technology at Rutgers, The State University of New Jersey – Newark, May 2020.

22 Starita, "Brasilidade."

23 Hugo Segawa, *Architecture in Brazil. 1900–1990* (New York: Springer, 2013).

24 Filipová, "Traditions," in *Modernity,* 145–177, Lada Hubatová-Vacková, "Užití, využití i zneužití folkloru a lidového umění," in *Budování státu,* 180–244.

tions, which he believed could form the basis of new national architecture.[25] The Czechoslovak pavilion in Rio de Janeiro was an example of such vernacularism that became the expression of "a modern artistic current, characterized by the attempt to break free from the old classical models by introducing elements of national culture."[26] This culture was seen as that of the peasants whose linear motifs from embroidery were effectively translated into the building's ornamental decoration and presented as part of the modern artistic expression.

The state in arts and crafts

It is one of the paradoxes of interwar world's fairs, aimed at showcasing universal modernity and progress, that they enhanced the need for national difference and exceptionalism. The Parisian *International Exposition of Industrial Design and Decorative Art in Modern Life* in 1925 exemplifies such developments, as the official goal of the organizer was to establish a new decorative movement, which would emerge in France and be successful commercially on an international scale. France tried to re-establish itself as a producer of applied arts, having to face increasingly strong competition from Britain and Germany, as well as from the USA. To remove the association that France, at exhibitions or elsewhere, adhered to historicism and old-fashioned décor, all imitations of old styles were banned.[27]

The French organizers and exhibitors, however, struggled to agree on what *modern* meant and they presented varied views and understandings of modern art and design. The poet Paul Géraldy, for example, pointed to France's reluctance to embrace modernity wholeheartedly in his review of the Exhibition; he claimed that the modern here tried to negate "France's history and heritage for the sake of fashion."[28] Expanding the notion of France's unique position in the visual arts, the jeweler and designer René Lalique suggested that the decorative arts of the future should be again elitist. "Elitist art," in his interpretation,

25 Vendula Hnídková, *Národní styl, kultura a politika* (Prague: VŠUO, 2013), Marie Benešová, *Pavel Janák* (Prague: Nakladatelství československých výtvarných umělců, 1959), Norbert Kiesling, *Pavel Janák* (Weitra: Bibliothek der Provinz, 2012).

26 Ruth Lévy, *A Exposição do Centenario. E o meio arquitetônico carioca no início dos anos 1920* (Rio de Janeiro: EBA Publicações, 2010), 231.

27 Julia Kostova, "Spectacles of Modernity. Anxiety and Contradiction at the Interwar Paris Fairs of 1925, 1931 and 1937," PhD Dissertation, Rutgers, The State University of New Jersey, 2011. Makaryk, *April in Paris*; Marc Gaillard, *Paris. Les Expositions Universelles de 1855 à 1937* (Les Presses Franciliennes, Paris, 2003).

28 Paul Géraldy, "L'Architecture vivante," in *L'illustration*, special section on L'exposition internationale des Arts décoratifs et industriels modernes (April 15, 1925), Kostova, "Spectacles of Modernity," 94.

was authentically French and as such could re-establish France as a leader in art and design.[29] At the same time, others called for the new visual language to be more closely linked to French regionalism and local traditions, which—as, for example, Romy Golan has argued—occupied a prime spot at the Exhibition.[30]

Yet despite all the effort to create a uniform new French artistic language, the Exhibition did not come up with any consistent definition or presentation of modernity in art and design. The site of the International Exhibition accommodated a wide array of modernisms that included the various Art Deco pavilions alongside the more restrained architecture as regards the embrace of decoration, like the Lyon and St Etienne pavilions designed by Tony Garnier, the Pavilion of Tourism by Robert Mallet-Stevens, or Le Corbusier's Esprit Nouveau and Konstantin Melnikov's Soviet pavilion. The Exhibition organizers were, nevertheless, far from open to include *all* forms of modernism and for example the non-figurative work of the Bauhaus and De Stijl were excluded from the exhibition grounds. A dissatisfaction with the ban led to a counter-exhibition in Paris, "L'Art d'Aujourd'hui" (The Art of Today), providing an alternative view of what modern, non-narrative art could look like. The exhibition displayed the work of artists from 24 countries, including Fernand Léger, the Delaunays, Natalia Goncharova, Picasso, or Le Corbusier and also several Czech artists living in Paris at the time, such as Josef Šíma, Jindřich Štýrský, and Toyen.[31] On top of that, two artists and art theorists, Karel Teige of Czech origin and Theo van Doesburg from the Netherlands, compiled a call for a protest against the exclusion of De Stijl in the journal *Pásmo* (Zone) published in Prague.[32] Speaking on behalf of the Constructivists and targeting the focus of the Parisian Exhibition on the decorative arts, they called for an organization of a counter-exhibition, congress, international publication and "modern performances" with the aim of eradicating the "applied arts scam" in all areas.[33]

Other Czech critics expressed mixed feelings about the composition of the entire Exhibition as well as about the Czechoslovak participation which for many highlighted the basic issues in the relationship between art and design on the one hand and the diversity of definitions of the modern. The architect Adolf Benš, for example, expressed his skepticism about the displayed decora-

29 Quoted in Kostova, "Spectacles of Modernity," 96.

30 Kostova, "Spectacles of Modernity," 96. Romy Golan, *Modernity and Nostalgia. Art and Politics in France between the Wars* (New Haven and London: Yale University Press, 1995), 58.

31 Anna Pravdová, "V centru pařížského dění: abstrakce 1924–1934," in František Foltýn et al., *František Foltýn 1891–1976, Košice-Paříž-Brno* (Brno: Moravská galerie and Řevnice: Arbor vitae, 2007), 69.

32 "Výzva," *Pásmo* 7-8 (1924–25): 1.

33 "Výzva," 1.

tive arts in the journal *Stavitel* (The Constructor) after the Exhibition closed. He noted that the event was

> highly educational although mostly in the negative sense. [Here] the program had dissolved into an inorganic breadth, the "decorative arts" had been reckoned with in France and in entire Europe and it became clear how unrelated the general state of culture was with the profusion of the decorative arts.[34]

Around 1925, Benš worked in the studio of Josef Gočár, assisted on the design of the Czechoslovak pavilion and was also in charge of the construction of the national pavilion in Paris.[35] His dismissal of the unnecessary decorativeness arts was accompanied by an embrace of "the logical," pure architecture of Le Corbusier which much more corresponded with the contemporary state of mind.[36] The decorative arts at the Exhibition, Benš continued, were therefore reduced to their obsolete function related to craftmanship, which turned the event into a display of luxury and ornament rather than of practical utilities.

Such reading can be applied to parts of the Czechoslovak pavilion too. Overall, the artistic display in Paris was a mixture of decorative ornamentalism, based on craftsmanship, and moderately avant-garde works. The latter could be exemplified by František Drtikol's photographs of female nudes with abstracted backgrounds and the displays of functionalist architecture designs by Jaromír Krejcar and Bohuslav Fuchs.[37] Most of the artwork on display, however, was comfortably conventional and "tame," if not outright conservative.[38] There was a range of sculptural work that included Jan Štursa's *Victory* of a boy waving a laurel branch. The statue celebrated independent Czechoslovakia and was located on the pavilion roof, while Otto Gutfreund's figurative sculptures referenced the visual language of civilism, which commended everyday life and manual work. Otakar Švec's bronze *Motorcyclist* captured modernity through its emphasis of speed in

---

34 Adolf Benš, "Mezinárodní výstava dekorativního a průmyslového moderního umění v Paříži 1925," *Stavitel* VI, no. 8–9 (1925): 110.

35 Vladimír Šlapeta, *Benš: Adolf Benš (1894–1982): architektonické dílo Adolf Benš (1894–1982): architektonické dílo* (Prague: Nadace Charty 77, 2014).

36 Benš, "Mezinárodní výstava," 109.

37 Hana Rousová, "Art Deco," in *Design v Českých zemích 1900–2000: instituce moderního designu*, eds. Iva Knobloch and Radim Vondráček, 177–200 (Prague: Academia, 2016)., 180. *Catalogue officiel de la section Tchécoslovaque. Exposition internationale des arts décoratifs et industriels modernes Paris 1925* (Prague: Comité d'Exposition, 1925).

38 Martina Pachmanová, "Výstavní praktiky státu: export československého umění," in *Budování státu,* 283.

its sleek, abstracted forms emulating movement. This 900kg heavy sculpture was, nevertheless, located somewhat out of the way at the back of the pavilion building, suggesting an ambiguous attitude to this kind of abstracted work [fig. 31].

The large open space on the ground floor was clad in oak panels and featured glass windows by Kysela which depicted references to various crafts like pottering, weaving, painting, or printing and decorative carpets.[39] The space was divided into sections focusing on decorative objects in various materials such as glass, porcelain, embroidery or book binding, with stage design located at the end of the building. The first floor was taken up by the Hall of Honor and equipped with work by teachers and students at the School of Decorative Arts in Prague [fig. 32]. The architect and designer Pavel Janák equipped the space with various sculptural objects in wood and glass depicting Czechoslovak crafts and industries, as well as abstract topics like science, labor, literature or architecture. Janák also supplied glass chandeliers and a model of a fireplace from Slovakia and smaller objects included glass by the designers Josef Drahoňovský (1877-1938) and Jaroslav Horejc (1886–1983) and lace by Emilie Paličková Mildc.[40] Kysela's large wall tapestries with allegories of crafts, executed by the designer Marie Teinitzerová (1879–1960), dominated the space.

The exhibition commissioner for the Czechoslovak pavilion in Paris was V. V. Štech, who had previously helped to organize the national pavilion at Rio de Janeiro. Like many of the architects, designers and artists involved in the pavilion, Štech was linked with the School of Decorative Arts, where he taught. And like many others, he was a member of the Association of Czechoslovak Werkbund. One of the best represented artists in the pavilion was František Kysela, who in 1924 became the rector of the School of Decorative Arts. In many respects, the pavilion—and the Czechoslovak exhibits in other halls—therefore became a showcase of the School and its orientation. In the 1920s, many of the studios at the School taught decorative ornament derived from folk art, because these so-called traditional designs were understood as the expression of the nation.[41] This focus was adopted by the state, which was the official representative of the nation, and various official, public-facing activities and commissions were granted to the School.

39 "Pavillon national," *Catalogue officiel de la Section Tchéchoslovaque. Exposition internationale des arts décoratifs et industriels modernes Paris 1925*, ed. Vilém Dvořák (no publisher, 1925), 13.

40 "Pavillon national," 21.

41 Pavla Pečinková, "Chapters from the History, 1895–1946," in *Vysoká škola uměleckoprůmyslová v Praze/ Academy of Arts, Architecture and Design in Prague, 1885–2005*, eds. Martina Pachmanová and Markéta Pražáková (Prague: VŠUP, 2005), 47.

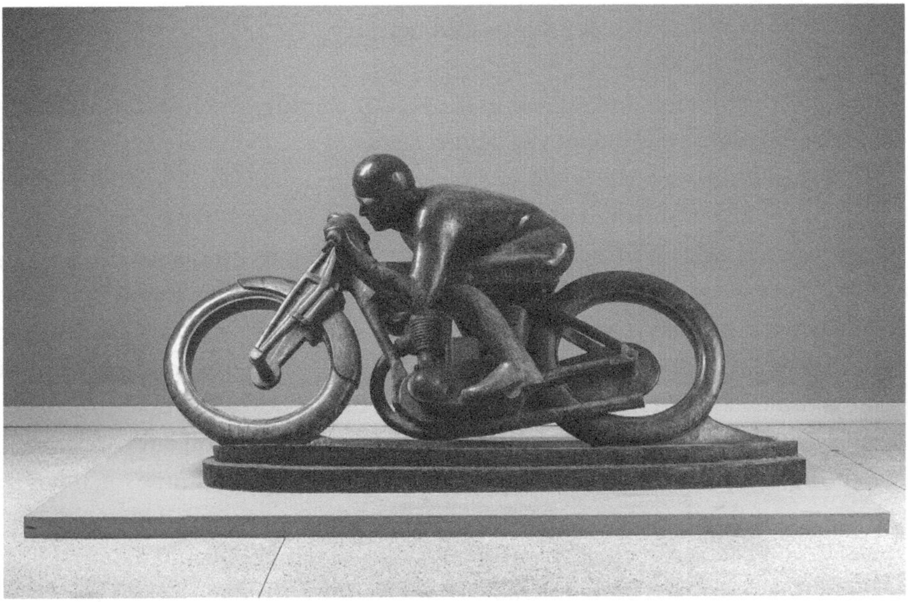

Fig.31. Otakar Švec, Motocyclist, 1924, bronze, 114x226x96cm.

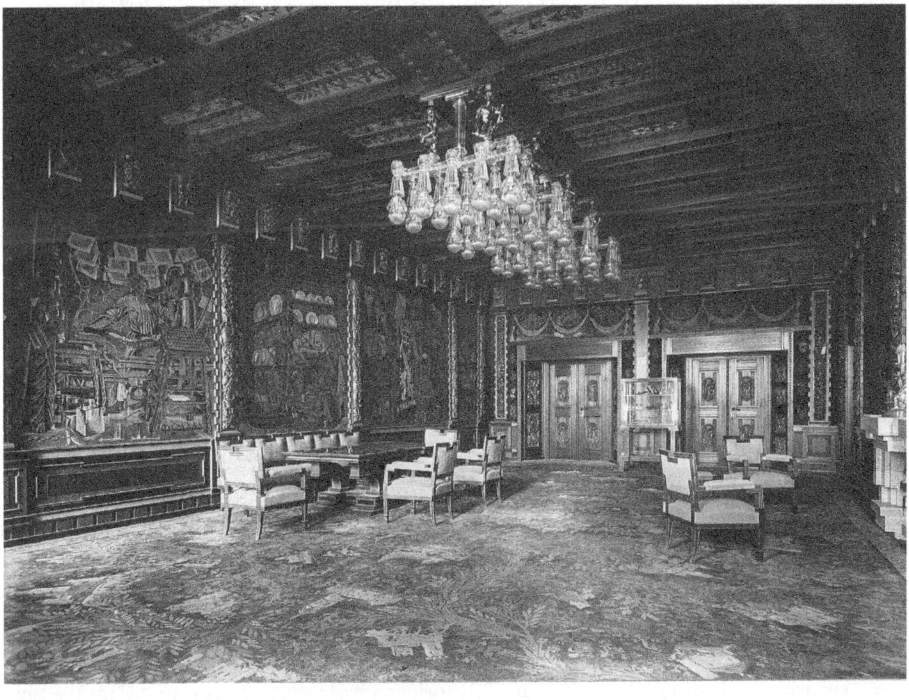

Fig.32. Josef Gočár, Interior of the Czechoslovak pavilion in Paris, 1925.

Crafts and tapestry

Craft, seen as timeless and related to the local (meaning Czech and to an extent Slovak) traditions, was one of the key elements of the national pavilion. It was also featured through the work of Kysela in many places, but especially in his eight large and ten smaller tapestries in the Hall of Honor on the first floor. The work begs for a closer inspection here for its content, technique, and especially for its place within the narrative of modern state on display in Paris. Kysela's endeavor was also noticed by many journalists, designers and art historians in Czechoslovakia who paid a lot of attention to the Czechoslovak national exhibit. One commentator who promoted what can be called local exceptionalism of the display noted that Kysela's "work breathes out [the sentiment of] our country and its spring, it is the artistic transcription of the Czech folk primitivism, it has its rhythm and simple, intimate beauty which does not fade in time."[42] Such rhetoric was embedded in the continued attempts to define specific national qualities of Czech art that would demonstrate its uniqueness on the one hand and importance for national identity on the other.[43] These efforts date back to the late 19th century when revivalists aimed at constructing the idea of the Czech nation in various political and cultural aspects, which included art and architecture. Several decades later, in a completely different political environment of the interwar period, the narratives of national exceptionalism, or uniqueness, were successfully continued and exhibited in the international context.

The tapestries depicting the traditional crafts in the Czechoslovak pavilion played this role very well, as their technique and mode of depiction were rooted in a certain folk naivete [fig. 33]. The traditionally anonymous skill was here adopted by educated designers, with Kysela responsible for the artistic content and the weaver Marie Teinitzerová for executing the tapestries in her studio in the south Bohemian town of Jindřichův Hradec. The smaller tapestries contained decorative motifs like draperies and fruit. The large tapestries covered a total area of 60m^2, and each depicted a different craft. A figure of a craftsman or craftsmen, be it a potter, cabinet maker, printer, glass maker, wall painter, bookbinder, weaver or metalsmith is central to each work. The protagonists are surrounded by their typical tools and machines, showing for instance a pot-

42 "Prof. František Kysela," *Výtvarné snahy* 6, no. 2 (1925): 48.

43 Filipová, *Modernity*; Jindřich Vybíral, "What Is 'Czech' in Art in Bohemia? Alfred Woltmann and Defensive Mechanisms of Czech Artistic Historiography," *Kunstchronik* 59, no. 1 (January 2006): 1–7.

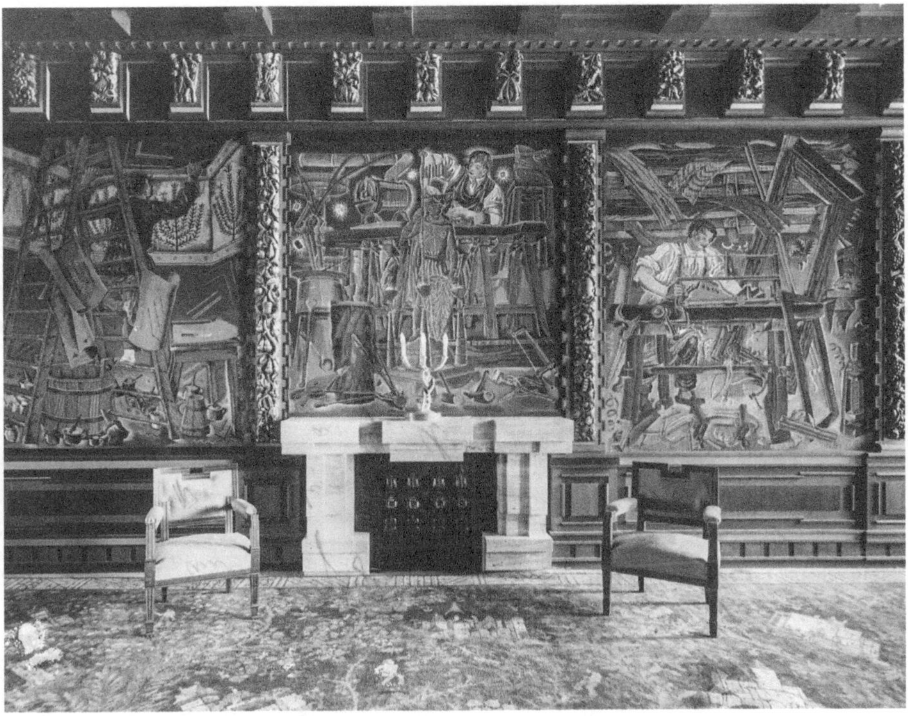

Fig.33. Josef Gočár, interior of the Czechoslovak pavilion in Paris with František Kysela's tapestries, 1925.

tery wheel, chisel, or a weaving machine. In many respects, the tapestries can be considered anachronistic.

The figures, objects and ornaments filled the entire surface of each depiction in a way the area of Renaissance and Baroque tapestries was constructed. In this, the approach to the surface was no different from that of many other works at the exhibition, for instance, the decorative panels by Jean Dupas in the Pavillon du Collectionneur or Domergue-Lagarde's "Pacification-Travail" in the French African Pavilion. They both used figurative narrative which, although abstracted to an extent, was still very didactic about the subject.[44] Kysela, too, combined figurative motives and objects with abstracted ornament that covered the entire area of the panels, framed by the shallow space of the room in which each craft was performed.

44 Llewellyn Smith, *Reports on the present position and tendencies of the industrial arts as indicated at the International Exhibition of Modern Decorative and Industrial Arts, Paris, 1925, with an introductory survey* (London: Harrow-HMSO Press, 1927). Vilém Dvořák, "Gobeliny umělecko-průmyslové školy v Praze na mezinárodní výstavě dekorativního umění v Paříži 195," *Výtvarná práce* 4, nos. 2–3 (1925): 82.

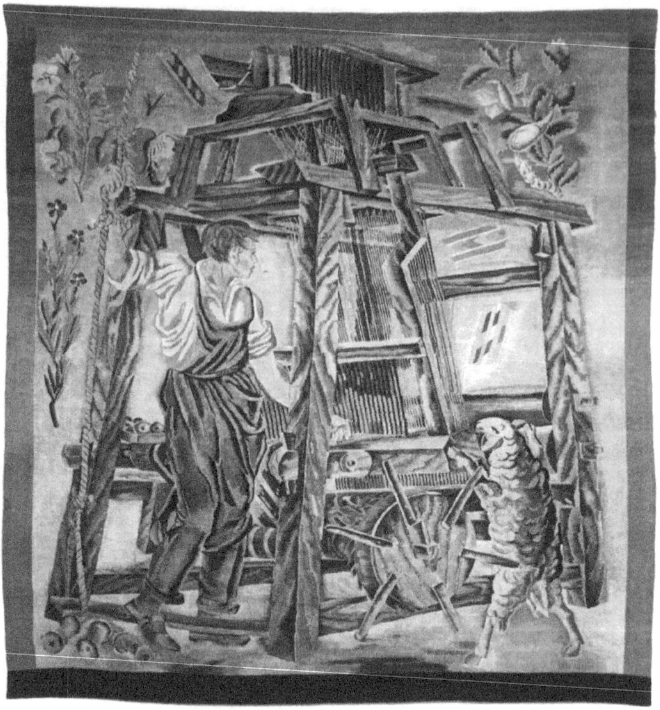

Fig.34. František Kysela, Weaving, tapestry, 1924.

The depiction of the "Weaver" illustrates the designer's attitude to the space, subject matter and the combination of realistic and abstract motifs [fig. 34]. In the "Weaver," a central A shape of the weaving machine, surrounded by abstracted floral ornaments, dominates the composition. The shape re-appears on a few other themes, such as the Painter and the Potter. The heavy wooden weaving machine is making a drape with a geometrical pattern, while abstracted forms also appear as a decoration in the background next to floral motives. Unlike most of the other tapestries which are firmly embedded in hand-made craft-making, the "Weaver's" content can therefore be considered a transition between the traditional and the modern, where the hand-operated device creates a fabric with a modernist pattern.[45] The weaver is located on one side of the machine, which he controls, and looks to the back at the making process. A sheep stands in the bottom right corner on its rear legs curiously looking into the machine. It is also observing what has become of its wool in a detail that adds a uniquely whimsical element to the scene.

45 Ethan Robey, in *Inventing the Modern World: Decorative Arts at the World's Fairs, 1851–1939*, eds. Jason T. Busch and Catherine L. Futter (New York: Skira Rizzoli, 2012), 189.

What all the tapestries share is that the workers and craftsmen (and they all are men) are captured in their everyday activity; they create their products by hand in a seemingly pre-industrial world which are only occasionally disrupted by references to present-day reality. This could be found in the symbol of the Czechoslovak lion with a Slovak cross on its chest in the Glass-maker tapestry, the abstract patterns of the fabric in the "Weaver" or in the furniture in the "national style" that supported the plates in the Potter. Yet the Potter especially, with his bare feet, sand sieve and a wooden bucket firmly pointed to humble, traditional origin of the handicraft. This choice may play into the nostalgia for the pre-industrial techniques in the sense of the Arts and Crafts movement's plea against mechanical industrialization and loss of individual craftsmanship. Or it could also be viewed as an attempt to document the prehistory of contemporary production. In this embrace of the humble and the everyday, Kysela's designs may also be seen in relation to civilism and the daily experience of the worker in history, although devoid of any critical commentary.

This juxtaposition of the traditional handicrafts and the new designs introduced with the modern world can also be read as a trajectory, or one may say evolution, that leads from the craft-making by hand to the modern production enabled by the machine. Such story also well indicated the trajectory of the nation from being dispersed in the Habsburg monarchy to its emancipation in an independent state. Visual depictions of this narrative appeared in many other Czechoslovak pavilions but more explicitly in the national pavilion at the world's fair in New York in 1939. The large painted window dedicated to Hymn to Work, which I will discuss later, focused on the evolution of shoemaking and its ultimate modern pinnacle in the Baťa company.

Kysela's work that featured so prominently in the 1925 pavilion (and in other international exhibitions) aptly complemented the lively debates and considerations on the role of craft, industrial arts, and art in contemporary society that many theorists and practitioners in Czechoslovakia engaged in. The tapestries as well as the entire Czechoslovak representation in 1925 only furthered the ideological split between the vocal associates of the avant-garde like Karel Teige, Josef Čapek, the critic and writer Bohumil Markalous (aka Jaromír John) on the one hand and the representatives of the School of Decorative Arts, Kysela included, on the other. Both groups published a series of texts on their understanding of crafts, applied arts and art industries in 1924 and 1925, during the preparations of the Paris exhibition and the event itself. The radical, avant-garde authors expressed themselves in various articles in journals like *Stavba* (Building), *Bytová kultura* (The Culture of Housing), *Výtvarná práce* (The Artistic Work), and *Drobné umění* (Petit Arts), where they criticized the representatives of the School. They often called them derogatively "umprumáci" (those from the UMPRUM which is short for the Czech translation of the School, Umělecká průmyslovka). Name-calling aside, at the heart of the discussions was the attempt to explain how the visual arts fit in the direction of the modern state and to reflect on their relevance to modern society.

What is modern art and design

One of the most vocal critics of the decorative arts in Czechoslovakia was Karel Teige, previously mentioned, who saw its modern incarnation in the production of the UMPRUM graduates and teachers. He argued that the "artistic industry" represented by the UMPRUM was stuck in the feudal times of primitive manufacture without machinery. Modern times, on the other hand, gave birth to what he called the "industrial arts," which were much more closely linked to modern democracy and liberalism.[46] Teige located the place of origin of the industrial arts in the United States, where the machine was the pioneer of the democratic ideal.

Teige firmly put the democratic industrial arts in contrast with what he understood as the undemocratic, feudal, and pre-industrial artistic industry. Teige challenged what he saw as a misconception that arts and crafts could elevate people's lives and artistic sensibilities.[47] That "artistic handwork, conducted by artistic imagination, can improve our life today, create a unified style, socialize art," was a myth for him because such production was too exclusive and aristocratic.[48] In the discussion about the dynamics between the machine and hand-created works, Teige's views resonated with other leftist artists and theorists in Central Europe, for instance with Moholy-Nagy's praise of the machine as a democratic instrument of political, societal and economic change.[49] The economic aspects of industrial production had also been addressed by Adolf Loos, whose ideas were well known in Czechoslovakia. Loos argued that unnecessary decoration

46 Karel Teige, "Konstruktivismus a likvidace umění," *Disk* II (1925): 4–8, reprinted in *Avantgarda známá a neznámá*, ed. Vladimír Vlašín (Prague: Svoboda, 1972), 123.

47 Teige, "Konstruktivismus," 123.

48 Teige, "Konstruktivismus,"151

49 László Moholy-Nagy, "Constructivism and the Proletariat," 1922, in *Modernism: An Anthology of Sources and Documents* eds. Vassiliki Kolocotroni, Jane Goldman and Olga Taxidou (Chicago: University of Chicago Press, 1998), 299–300.

increased the price of the object and that machine made objects with no or little ornament reaped less expensive products and higher wages for the maker.[50]

The relationship between the creative production and class was an important issue for many left-wing critics. In this regard, despite Kysela's attention to the humble worker, his depictions of the seemingly feudal, exclusive and therefore expensive handcrafts in his tapestries were an easy target for criticism by Teige and others. And so was the entire 1925 exhibition that could be easily criticized for being removed from the everyday realities of the common people and for targeting mainly the upper, wealthier, bourgeois classes by its emphasis on luxury and opulence.

Many of the works in the Czechoslovak pavilion and around the exhibition ground seem to support such conclusions. They were not only removed from the everyday life of the proletariat, but they were also escapist in their embrace of exoticism, othering and occasionally historicism. Before the exhibition opened, Josef Čapek called the event a possible crossroad between ornamentalism of the decorativism of various national styles on the one hand, and more international, social, and democratic, artistic tendencies based on the standard modern needs on the other.[51] Čapek hoped that the 1925 Exhibition would represent the peak of ornamentalism after which ornament would start retreating in favor of less decorative tendencies. He also predicted that after the Exhibition, decorative and ornamental art would find success in the United States, while in countries where it had reached its climax, it would lead to a counterreaction and embrace of non-ornamental and non-decorative art. Čapek was certainly right about the former; the decorative arts of the Parisian Exhibition were soon adopted into the American visual culture. The aesthetics that was born around the turn of the century as socialist, democratic, and interested in wider social classes, became by mid-century the visual language of American consumer capitalism.[52]

Whose modernism

Although ornament and decorativism in the official representations of Czechoslovakia and other countries did somewhat retreat after 1925, just as Čapek pre-

50 Adolf Loos, "Ornament and Crime," in *Modernism*, 80; Adolf Loos, "Umění nebo řemeslo?" in *Navzdory: ornament je zločin, 1900–1930* (Hodkovičky: Pragma, 2015), 141.
51 Josef Čapek, "Na rozcestí," *Bytová kultura* I, no. 6–8 (1924–25): 116.
52 Judith A. Barter, "Designing for Democracy: Modernism and its Utopias," *Art Institute of Chicago Museum Studies* 27, no. 2 (2001): 6–17 and 105.

dicted, it did not disappear completely. At the same time, the influence of the UMPRUM representatives became less prominent in future national pavilions. One can recall Benš's criticism of the 1925 Exhibition as a display of luxury and ornament and not of items of daily use that would represent new, modern lifestyle. In line with it, the coming Czechoslovak exhibits became more conscious of finding modernity in utility. Such turn towards more functional and functionalist displays reflected the continued search for a new visual identity of the state now turning away from the expressions related to the national style.

The Association of the Czechoslovak Werkbund, founded on the model of the German Werkbund that promoted design and the artistic industries, played a key role in this search. Indeed, many members were graduates or teachers of UMPRUM.[53] The Association also had an enormous input in national and international exhibitions of architecture and design that represented Czechoslovakia across Europe.[54] These were for instance the International Book Design Exhibition and the Ausstellung Europäisches Kunstgewerbe (The Exhibition of European Arts and Crafts) in Leipzig in 1927, the Exhibition of Czechoslovak Architecture, the Applied Arts and Building Industries Exhibition in Bucharest, the Stockholm International Exhibition that both took place in 1930 and the 1933 Poznań International Hygiene Exhibition. They were put together by the designer Ladislav Sutnar who was gaining prominence and influence as exhibition designer of the Association, frequently displaying his own work. While many of these exhibitions were promoting the state and its visual arts, the displayed work often prioritized that of Sutnar, his colleagues from the Association, the Association's publisher and manufacturer Družstevní práce (The Cooperative Work) and Krásná jizba (The Beatuful Room), their flagship stores that promoted and sold contemporary design. In his exhibition design, as I have discussed in more detail in Chapter Two, Sutnar put emphasis on clear, functional displays which turn visitors' attention to the object. Books were arranged in show windows, newspapers as mural-like collages, fabrics hung diagonally across a space.

Sutnar's role in shaping the international image of Czechoslovak design is therefore crucial. He also identified a turn after the Paris 1925 Exhibition away from "the superficial decorative design approach of the past" and the displays of

---

53 Iva Knobloch, ed., *Žijeme lidsky? Reforma bydlení 1914–1948. Svaz československého díla. Housing Reform 1914–1948. The Czechoslovak Werkbund* (Prague: UPM – Slovart, 2023).

54 Iva Knobloch, "Krásná jizba a Svaz československého díla jako modernizační hnutí," in *Krásná jizba DP 1927–1948: design pro demokracii*, eds. Lucie Vlčková and Alice Hekrdlová (Prague: Uměleckoprůmyslové museum, 2019), 77.

the national style, which he called flag waving, related to provincialism.[55] Similarly, the entire circle around the Association for the Czechoslovak Work and The Cooperative Work formed a rather closed group of designers and artists who featured in international exhibitions and therefore constructed a specific narrative of the visual arts in the state. Such involvement brings back the question of who should represent the state and how. In other words, it turns focus on who the artists and designers were that were invited to display their work on behalf of the Czechoslovak state. After all, representing the state at such an international forum of a world's fair was both prestigious and commercially attractive and appearing in this setting came with a certain status, important for future careers.

The authority of the exhibition designer over the display nevertheless came second to the official decisions about the focus of the display made at the governmental level. As I outlined in Chapter One, a cross-ministerial committee was put together for each exhibition that decided on the ideological content of the representation. The cultural or economic focus of the display then motivated the choice of exhibitors and participants and the ultimate promotion, or rather propaganda, was not only political and economic but to large extent cultural and diplomatic.[56] Modern design with its commercial potential would appear to fulfil the requirements of both cultural and economically driven representation. Yet, the ideas and products promoted by Sutnar and the Association were sometimes seen as not a great fit.

It was during the preparatory works for the Czechoslovak pavilion at the Chicago Century of Progress in 1933 that individual ministries involved in the organization of the official representations reopened the debate about its focus. In a statement against the direction of the exhibition which could become too avant-garde, the Minister of Trade, Josef Matoušek, emphatically reminded the Board of the Ministerial Committee that

> the Ministry of Trade cannot allow artistic circles to decide on the set up and extent of the industrial (trade) exhibitions, and [it cannot allow] that the right to shape up the installation of their own products is taken away from businesses and instead given to artists of a specific avant-garde di-

---

55 Sutnar, *Early Modern Design Concepts*, c/7-c/1. Also cf. Julia Secklehner, "'Feminine horror' or 'eminent Viennese specialty'?: Vienna's Kunstgewerblerin in Paris, 1925," *Art East Central* 3, no.1 (2023): 13–36, https://doi.org/10.5817/AEC2023-3-2.

56 Joseph S. Nye, "Soft Power," *Foreign Policy* 80 (1990): 153–171; cf. Bartosz Dziewanowski-Stefańczyk, "World's Fairs as Tools of Diplomacy: Interwar Poland," in *World Fairs and the Global Moulding*, eds. Leerssen and Storm, 300–328.

> rection who can remove objects that they do not find likeable from the display... Artistic views often change and stand fundamentally against the practical needs of economic life.[57]

It is not clear who was the exact target of the Minister, but his criticism was partly directed at the 1925 Exhibition which was mostly organized by the Ministry of Education and showcased the work of the Association of the Czechoslovak Werkbund and the UMPRUM circles. Although the display in the national pavilion cannot be seen as avant-garde, the overall focus on the visual and modern (however interpreted) was certainly not aligned with the commercial interests of the Ministry of Trade. Another exhibition Matoušek might have referred to was Sutnar's installation in Barcelona, which as the designer argued captured "Czechoslovak work in the area of new, utility items of high quality and valuable artistic industry through the most typical examples."[58]

Instead of such typical, modern items, the Ministry called for an installation that would promote trade and tourism in the interest of increasing the prestige of the state overseas and which would reap economic results.[59] The Ministry of Education with minister Ivan Dérer, on the other hand, defended a more representative approach that would impress the audiences by its quality and emphasized that especially the Association of Czechoslovak Work was a corporation with "the most experience in the area of exhibition organization," as was visible from its success in Paris in 1925.[60] Despite the harsh criticism of the outdated decorativeness of the national display, the number of awards the participating artists and designers generated was overwhelming. The Ministry therefore recommended the Association to oversee the section of the artistic industries and possibly the entire exhibition in Chicago in 1933.

What modernism?

The discussion about the direction of the official visual representation of the state continued throughout the interwar period. Following Chicago, the exhi-

57 Ministerstvo obchodu Předsednictvu ministerské rady, Věc: Světová výstava v Chicagu, Účast Československa, 31 October 1932, Světová výstava v Chicagu 1933, DM2251, AKPR, 3.

58 Ladislav Sutnar, "Státní odborné školy na Mezinárodní výstavě v Barceloně 1929," *Horizont* no. 25–26 (1930): 47.

59 Ministerstvo obchodu Předsednictvu, 4 and 6.

60 Ministerstvo školství Ministerstvu zahraničních věcí, Věc: Chicago, Světová výstava 1933, Účast Československa, June 21, 1930, Světová výstava v Chicagu 1933, D9475/30, AKPR.

bitions in Spain, the United States and Belgium, indeed contained important exhibits of the arts that attempted to frame modernity of the visual arts and I will return to the American context shortly. It was at the Exhibition of Technology and Art in Modern Life in Paris in 1937 that the place of modern art was hotly debated again on the ministerial as well as artistic level, now in the context of the amplified nationalism and looming new political and military conflicts. This was the year of the *Entartete Kunst* exhibition in Munich, which blacklisted many modernist artists and artworks. The Spanish civil war started in 1936, with Mussolini in charge of the fascists in Italy from 1924. Central Europe, too, was undergoing a radicalization in politics, which included the renewed claims for territorial revisionism in Hungary and the strengthening of the Austrofascist regime in Austria.[61] Czechoslovakia that retained a multi-party system, indeed looked like an island of democracy despite the growing separatist tendencies of Slovaks initiated especially by Andrej Hlinka and his Slovak People's Party.[62]

The place of modern art in this politicized environment was therefore highly charged too. At the Parisian Exposition, the visual arts were again located in a separate building as well as in many national pavilions that oscillated between abstract modernity and conservative traditionalism. The Italian pavilion, for instance, housed several murals by the futurist painter Gino Severini, while Picasso's Guernica, critical of atrocities of the civil war, were featured in the Spanish pavilion alongside his sculptures, Miró's painting and a fountain of mercury by Alexander Calder. Some participating nations took a more conservative path emphasizing the country's historical and religious traditions: the Hungarian pavilion, for instance, showcased paintings on historic, religious topics in the interior while a large wooden statue of Saint Stephen by Pál Pátzay guarded the entrance and received a grand prix.[63] Similarly to the Czechoslovak pavilion here, the Hungarian one contained a mixture of fine art, folk art and crafts, pointing out the historicity, religious character and adherence to traditions of the country.

The Austrian pavilion, by comparison, featured a large photographic montage of the Alps composed by Robert Haas as a view of a landscape modernized by roads. It provided a spectacular effect especially when observed from the outside through the large glass windows of the pavilion designed by Oswald

---

61 Sabrina P. Ramet, ed., *Interwar East Central Europe, 1918–1941: The Failure of Democracy-Building, the Fate of Minorities* (London and New York, NY: Routledge, 2020).

62 Kosatík, *Slovenské století*, 107.

63 Vilmos Gál, "Returning to the Banks of the Seine. Paris, 1937," in *Hungary at the World Fairs* (Budapest: Holnap Kiadó, 2010), 187–90.

Haerdtl.[64] It also fit within the Austrofascist fantasy of idealized rural and traditional Austria combined with modernized progress.[65] Despite this sleek exterior views, the interiors of the Austrian pavilion were fitted with heavy furniture, textiles and decorative objects that used now rather outlived idiom of the Wiener Werkstätte.

The artistic concept of the Czechoslovak pavilion was similarly hybrid. When the plans for the one in 1937 were announced, criticism came from many sides. The members of the Mánes Association, a group of Czech and international artists, architects and art historians, were especially vocal. Josef Gočár and Emil Filla protested the planned vision for the pavilion in their journal *Volné směry* and in a letter to the Minister of Foreign Affairs, Kamil Krofta.[66] They questioned especially the decision that only "a few casts and copies of old masters" were planned for inclusion in an exhibition which was officially focused on modern technology and art.[67] They saw the proposed program of the Czechoslovak display as archaic and outlived, typical of local trade fairs or the Prague ethnographic exhibition in 1895, which was seen as a clear proclamation of 19th century Czech nationalism.

Many commentators called for an appointment of a designated commissioner general who would be responsible for the pavilion's unified content and not act solely on behalf any of the Ministries.[68] For the artistic section of the Exhibition, *Volné směry* also asked for an appointment of a committee of independent experts.[69] This group, they envisaged, would not represent "various local and regional interests and those of some associations," because their attitude to modern art was necessarily negative.[70] *Volné směry* with Emil Filla as the chief editor and the surrealist painter Jindřich Štýrský on the editorial board promoted

---

64 Matthew Rampley, "From Potemkin City to the Estrangement of Vision: Baroque Modernity in Austria, before and after 1918," *Austrian History Yearbook* 47 (2016): 167–87; Elizabeth Cronin, *Heimat Photography in Austria. A Politicized Vision of Peasants and Skiers* (Vienna: Photoinstitut Bonartes: Albertina – Salzburg: Fotohof Edition, 2015); Chris Zintzen, "Shifting Perspectives: Alpine Scenarios in the Work Complex Nach der Natur (Beyond Nature) by Austrian Architectural Photographer Margherita Spiluttini," in *The Draw of the Alps: Alpine Summits and Borderlands in Modern German-Speaking Culture*, ed. Richard McClelland (Berlin and Boston: De Gruyter, 2023), 103–122.

65 Béla Rásky, Artwork of the Month, May 2023: Photomontage at the Paris World's Fair by Robert Haas (1937) https://craace.com/2023/05/30/artwork-of-the-month-may-2023-photomontage-at-the-paris-worlds-fair-by-robert-haas-1937.

66 Ed., "Bude účast Československa na mezinárodní výstavě v Paříži 1937 ostudou?," *Volné směry* 1, no. 32 (1936): 306. Letter to the Ministry of Foreign Affairs, 11 May 1936, Mezinárodní výstava v Paříži 1937, kD5640/36, MFA Archive.

67 Letter to the Ministry of Foreign Affairs.

68 "Účast Československa na mezinárodní výstavě v Paříži r. 1937," *Volné směry* 1, no. 32 (1936): 298.

69 "Účast Československa," 298.

70 "Účast Československa," 299.

their own vision of modern art based on abstraction and the avant-garde. The reference to "some associations" was probably addressed to anyone else who held a different, in Mánes's terms more conservative view of modernism.

Critics of the national pavilions often made suggestive comments but did not explicitly point fingers or named those they criticized. *Volné směry* for example also turned against the author of the exhibition plans whom they identified as "an art critic, known for his outdated mentality and prejudice against anything modern."[71] This "failed art critic" would be Václav Nebeský, who oversaw the fine art presentation. Nebeský was an art critic, collector and art dealer who had lived in Paris for several years and returned to Prague shortly before the exhibition. The modern art display that Nebeský eventually put together did conform to some of the pleas of *Volné směry*. Nebeský included a range of artworks of individuals and artistic associations and his understanding of modern art included more experimental painters like Kremlička, Kubišta, Zrzavý, Josef Čapek but also more conservative authors like Vincenc Beneš, Vlastimil Rada, and the sculptors Kotrba and Lauda. Those that Nebeský excluded for their traditionalism included for instance Max Švabinský, Antonín Slavíček, and Štursa.[72]

Nebeský's selection of art works and artists received some responses in the French press but many in Czechoslovakia. A French critic noticed inspiration of the modern art of Czechoslovakia in "almost primitive" forms and in folk art, which "constantly penetrates [the works] with its energetic rhythms, with its brutal artificiality, with its childlike and cute inventions" of the cubist paintings.[73] Czech press and artists were less accommodating. The journal *Dílo* published a declaration entitled "Criticism of the incompetence in promoting national culture," signed by the representatives of artistic associations like The Union of Fine Artists (that published the journal), the Circle of Female Artists from Prague, the Association Aleš from Brno, and several regional art organizations in Bratislava or Plzeň.[74] They criticized the commissioner general of the exhibition, Dr Krčmář (once a minister of education), whom, together with Nebeský, the authors accused of nepotism and favoritism. Similar to the earlier criticism by Mánes, they thought Nebeský, especially, was out of touch

71 Ed., "Bude účast," 306.

72 Ed., "Československé umění na mezinárodní výstavě v Paříži," *Dílo* 29 (1938–39): 50. Several photographs of the works in the journal *Život* (*Život* 16, no. 1, 1937, 14) give an idea what the selection looked like.

73 Raymond Cogniat, "Moderní československé umění," Život 16, no. 1 (1937): 15.

74 Anonymous, "Kritika neschopnosti v propagaci národní kultury," a copy in "Mezinárodní výstava umění a technika v moderním životě v Paříži v roce 1937 – účast ČSR," inv. no. 1033, file 134, D, APRO.

with the art world in Czechoslovakia and promoted only a specific group of artists—those he knew from Paris and they did not represent Czechoslovak national art.[75] The paintings, that came mostly from Mánes members, "paraphrased Parisian modern schools," were deemed derivative of French art and generally not of a high standard.[76] The modernism that *Dílo* (Work) would have preferred therefore looked more towards essentialist nationalism seen in local traditions, figurative subjects and regionalism.

The artists selected for Paris were, ultimately, labelled as the artistic elite partly based on personal preferences and their personal connections with the exhibition organizers. Yet the editor in *Dílo* identified the elitism in another, often neglected aspect of the selection, namely in the marginal representation of female artists. The editor addressed the absence directly, asking "...if the surrealist Štýrský was invited, why not Toyen too; who is better?" and "why was only the male member invited from the Procházka family of painters, and the female one, equally important, was omitted?"[77] The gender imbalance at the Exhibition adhered to the established division where the fine artists were male, while women were mostly represented by crafts. In this regard, Nebeský did not strike as a particular good choice of a curator. His view from 1919 that a "woman ... was not born to be an artist, she was born to be a work of art" was obviously reapplied at the close of the interwar period. Different areas of the visual arts were reserved to women in his view; women excelled in those fields where art can be performed through their body and soul, like dance and theatre, as well as the decorative and applied arts.[78]

As I argue throughout this book, female voices and women were mostly absent from Czechoslovak national pavilions. The male dominance among pavilion designers and artists also indicates the general state of contemporary society in Czechoslovakia in which women only slowly penetrated more prominent positions.[79] After all, throughout the interwar period, the ministerial organizational committees did not have a single woman involved in the main discussions of the content or direction of the displays. This was certainly true in the

---

75 Ed., "Československé umění na mezinárodní výstavě v Paříži," *Dílo* 29 (1938–39): 50 and "Kritika neschopnosti."

76 "Kritika neschopnosti."

77 Ed., "Československé umění," *Dílo*, 52. In recent literature, Toyen's gender has been reassessed and the artist is often considered as non-binary.

78 Václav Nebeský, "Ženské umění," *Tribuna* 1, no. 23, February 26, 1919, 3. Cf. Martina Pachmanová, *Zrození umělkyně z pěny limonády* (Prague: VŠUP, 2013), 28 or Filipová, *Modernity, History and Politics*, 166–167.

79 Melissa Feinberg, *Elusive Equality: Gender, Citizenship and the Limits of Democracy in Czechoslovakia, 1918–1950* (Pittsburg: University of Pittsburg Press, 2006).

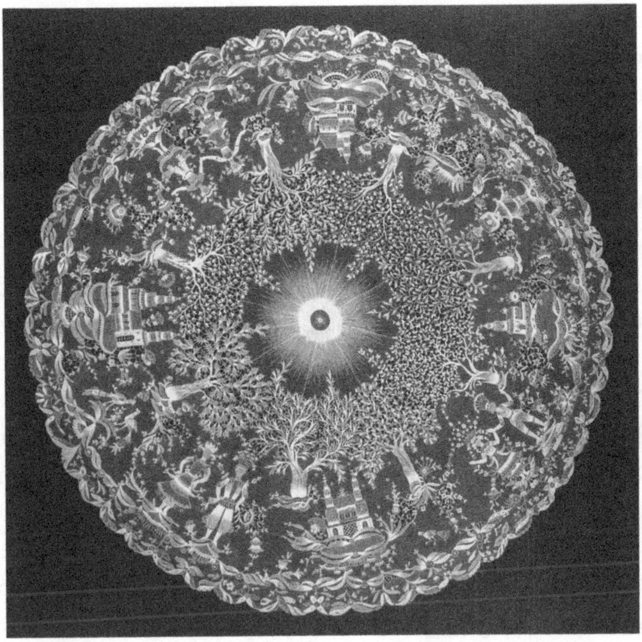

Fig.35. Emilie Paličková Milde, Little Sun, lace doily, 1925.

Czechoslovak pavilions in 1925 and 1937 where some women were represented but their contributions were limited to crafts like lace making and weaving and only occasionally to other techniques. Emilie Paličková Milde, for example, successfully exhibited her lace in many international exhibitions for which she received various awards[80] [fig. 35]. Together with the work of other female artists, such as the ceramicist Helena Johnová, the painter Linka Procházková, or the sculptor Mary Duras, these examples were individual and isolated cases of female participation and came in tiny proportion to the representation of male artists and organizers.

The odd one out

One of these male artists whose work was made prominent in 1937 was Zdeněk Pešánek shown in the national pavilion. His three works were located on the terrace and at least one of them deserves more attention because they demonstrate another aspect of possible artistic modernity that could be associated with

80 Ludmila Kybalová, *Emilie Paličková* (Prague: Nakladatelství československých výtvarných umělců, 1962), 9.

the state. In the Czechoslovak pavilion, Pešánek's work was slightly detached from the rest of the national display in both form and concept, which were not easy to interpret. Yet the choice of his so-called light-kinetic sculptures which combined multiple media, movement, light and sound was rather appropriate in the context of the event and in Pešánek's own words captured "the best the main ideas of the 1937 World Exhibition."[81]

Pešánek was represented here by a sizeable fountain, a series of four sculptures representing 100 years of electricity and a light-kinetic advertisement entitled Radium. The fountain came out as the winner of an open competition of the Ministry of Education, while the four Electricity sculptures were commissioned by the Electric Company and first exhibited in the Museum of Decorative Arts in Prague the previous year. The installation Radium consisted of intersecting geometrical objects displaying two words: Jáchymov and Radium, located next to the fountain. The uranium mined near Jáchymov, in western Bohemia, was the source of radium, separated from uraninite by Marie Skłodowska Curie.[82] The use of radium in new treatments of cancer initiated the development of a local spa that Pešánek's sign promoted. Unfortunately, none of these works have survived as they were either broken in transport or lost; but they are known from models, photographs and descriptions. All three works were in the Tourism section of the national pavilion, yet as they were not part of the original interior design, their size and placement had to be adjusted and crammed into a limited space.

The Fountain, outstanding in its size, experimental execution and application of new materials, promoted the richness and healing effects of Czechoslovak spas and balneology[83] [fig. 36]. The competition instructions called for a glass sculpture to further showcase glassmaking in Czechoslovakia, already quite prominent throughout the pavilion. Yet Pešánek's work consisted less of glass and more of other materials, mostly synthetic resin, in which he frequently worked. Together with wires, neon tubes and bulbs, the use of translucent resin created colorful effects on the water of the round basin. In the center of the fountain were three human torsos, vertical male and female, and one horizontal female made of the plastic with neon tubes and bulbs pulsing light through them.[84] Pešánek wanted to include a mirror at the bottom of the basin, while jets on the rim would spray or jet stream water, with light reflected

81 Untitled document with inscriptions for the displays, AACE NTM, f. 136 Pešánek.
82 "Radium," *Tchecoslovaquie*, 129.
83 Jiří Zemánek, *Zdeněk Pešánek 1896–1965* (Prague: National Gallery - Gema Art, 1996), 176.
84 Zemánek, *Zdeněk Pešánek*, 178.

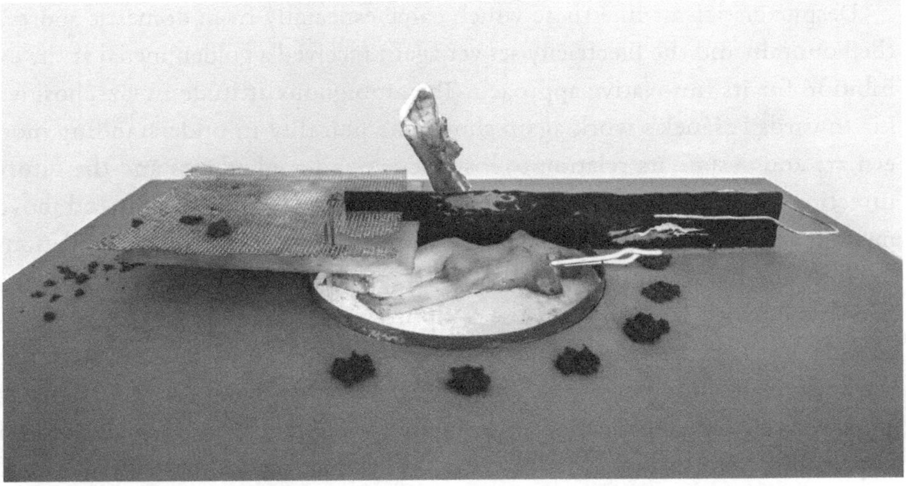

Fig. 36. Zdeněk Pešánek, Spa Fountain, 1936, author's model replica from 1959.

through this. For space limitations this was not executed.[85] The fountain, nevertheless, was supposed to be lit, featuring optical, kinetic and sound effects, similarly to the four Electricity sculptures.

Pešánek's fountain and the other sculptures, however, were received with mixed sentiments. The discrepancy between the artist's original idea and the actual execution had severe consequences for understanding or misunderstanding the work. Both the Electricity series and the Fountain suffered from technical faults. Power supply failed several times and defective engines had to replaced which caused the fountain lights not working for several weeks.[86] The works were also very abstract and combined unusual materials and techniques and as such required extensive additional explanation of the content and artist's intention. And this was not ideal for the context of a quick consumption of information at a world's fair. According to one critic, the fountain came across as a "beginner's attempt with such limited artistic value that it had no right to appear in front of an international audience."[87] And as the painter Josef Čapek noted in relation to an earlier display of Pešánek's work in the Museum of Decorative Arts in Prague, the sculptures made of harsh, unusual materials, were not very flattering in the daylight when the light pulsing in the neon tubes was not visible.

85 Zemánek, *Zdeněk Pešánek*, 178.
86 Letter to the General commissioner's office of the Czechoslovak exposition from Pešánek, August 11, 1937, AACE NTM.
87 Vojtěch Krch, "Československý pavilon v Paříži," *Architekt SIA. Měsíčník pro architekturu, stavbu měst, bytovou péči a umění* 36, no. 9 (1937): 143.

Despite criticisms like these which came especially from domestic sources, the Fountain and the Electricity set yet again received a golden medal at the exhibition for its innovative approach. The ambiguous attitude in Czechoslovakia towards Pešánek's work again shows the plurality in understanding modern art and design, its relation to historic art and applied arts and the future direction. Pešánek thought of the combination of electricity, light and movement as a new kind of artistic industry and new art, yet one with foundations in crafts and hands-on skills. He himself was a classically trained sculptor who had studied with Jan Štursa (whose sculptures often featured in Czechoslovak pavilions) at the Academy of Fine Arts and believed that being good at craft is the first step in an artist's career. That was why teachers taught sculptors how to work with bronze, stone, and ceramics, and students had to learn all the techniques in the workshops. That was why, in Pešánek's view, there was the need for not only schools of drawing and modelling but also for stone-sculpting, ceramic and metal working schools.

Modernity across the ocean—Art in America

A crude trajectory, from Czechoslovakia represented in folk art, decorative applied arts to ultimately being associated with abstract modernism, could be established based on the Parisian exhibitions between 1920 and 1937. The search for representative modernism for display in the American world's fairs took place in parallel but in a less coordinated way. Here, local preferences and political circumstances played a role in constructing the exhibits as seemingly more conservative. In Philadelphia, for example, Švec's bronze sculpture of the "Motorcyclist," which had appeared in Paris in 1925, was left unboxed because the pavilion commissioner did not find it fitting the overall tone of the presentation. The assumption was that the taste in the United States is more conservative, and audiences would appreciate work of more academic nature. Not many large works were transported across the ocean for the high expense transportation incurred and the heavy "Motorcyclist" seems to be an exception. Many works of art were therefore outsourced locally or came from local collections. This applied to displays of fine arts too, where the work was often based on loans from American collectors, which created another picture of modern art.

This was the case at the Century of Progress exhibition, where the art exhibition was organized by the Art Institute of Chicago in both 1933 and 1934. In the first year, the focus was on American collecting and divided into three sections: old masters, outstanding paintings especially French and American and contemporary art, which put emphasis on American artists.[88] With one exception, which was *Whistler's Mother* by James McNeill Whistler, all works came from American galleries and collectors and encompassed quite a wide spread of art history.[89] The International section was also quite diverse and limited to the works available from local lenders. This reflected in the fact that Czechoslovakia was represented by the Jewish painter Georges Kars (1880–1945) who, although born near Prague, spent most of his life outside of Bohemia.[90] At the time of the world's fair (and in fact until today), Kars was not a household name in the art history of Bohemia or Czechoslovakia, having lived in France for most of his life.

Only a few years later in San Francisco, the entire world's fair did not embrace modernism and future-orientation as unambivalently as the one in New York, which I discuss elsewhere, and instead retained the fascination with Art Deco and monumental architecture. Czechoslovakia was not represented here in a separate national pavilion but in a section in the International Hall designed by Antonín Heythum. It consisted of displays in vitrines of craft and design objects, such as glass, crystal, textiles, print, "home industries," folk art, silver ware and shoes. These were combined with large scale photographs of the countryside, vistas of Prague and other towns, the Sokol gatherings and folk dances.

In the climate of changing political landscape which saw the disappearance of Czechoslovakia, Poland and other countries during 1939 and 1940, the works of art chosen for the Czechoslovak section in San Francisco were conservative. Many turned to the safety of landscape and rather nostalgic, rural subjects. Vladimír Hurban, the ambassador to Czechoslovakia in the USA, was responsible for the selection of paintings. Assisted by the artist Andrej Kováčik (1889–1953), both were of Slovak origin which reflected in their choice. Born in Budapest, Kováčik was a rather conservative figure and nationalist. He had already organized an earlier exhibition of about 250 works of Slovak art in New York in 1938, shown at the galleries of the Fine Art Society. He worked on this exhibition together with the Slovak League of America, an organization promoting the welfare of Slovaks in the US.

The choice of paintings in both New York and, a year later, in San Francisco consisted mainly of Slovak landscapes and works inspired by folk art. They

88 *Official Guide: Book of the Fair 1933* (Chicago: Century of Progress Administration Building, 1933), 107.

89 *Official Guide:* Book, 106.

90 *Catalogue of a Century of Progress Exhibition of Painting and Sculpture* (Chicago: The Art Institute of Chicago, 1933), 81.

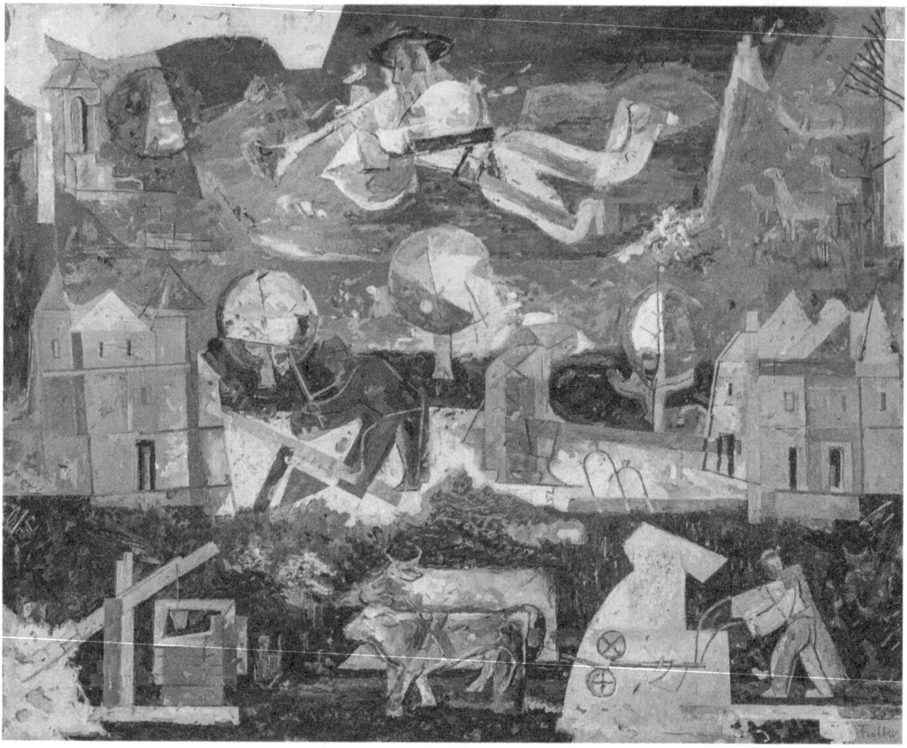

Fig. 37. Ľudovít Fulla, *Song and Work*, 1934-35, painting.

were mainly by Slovak artists, such as Ľudovít Fulla and Martin Benka, as well as by some Czechs, for instance František Malý, who often depicted the Slovak countryside. With works bearing titles such as "Countryside from Terchova," "Wooden Church in Kezmarok" or "Orava Castle," Slovakia was represented as through the rural image even in these cases which were in the hands of Slovak organizers [fig. 37].[91]

Yet even with the focus on Slovakia, the larger picture of art, however modern, in Czechoslovakia was very limited. Missing here were any references to modern architecture, art and design of the Slovak cities of Bratislava and Košice or the achievements of the progressive School of Arts and Crafts in Bratislava, founded in 1928. More experimental and non-figurative works by the Slovak artists represented here, Fulla included, did not feature. Such work, nevertheless, was shown two years earlier in Paris. *The Song and Work* from 1934/35 combined

91 "Czechoslovakia," *Contemporary Art. Official Catalog, Department of Fine Arts* (San Francisco: Golden Gate International Exposition, 1939), 14.

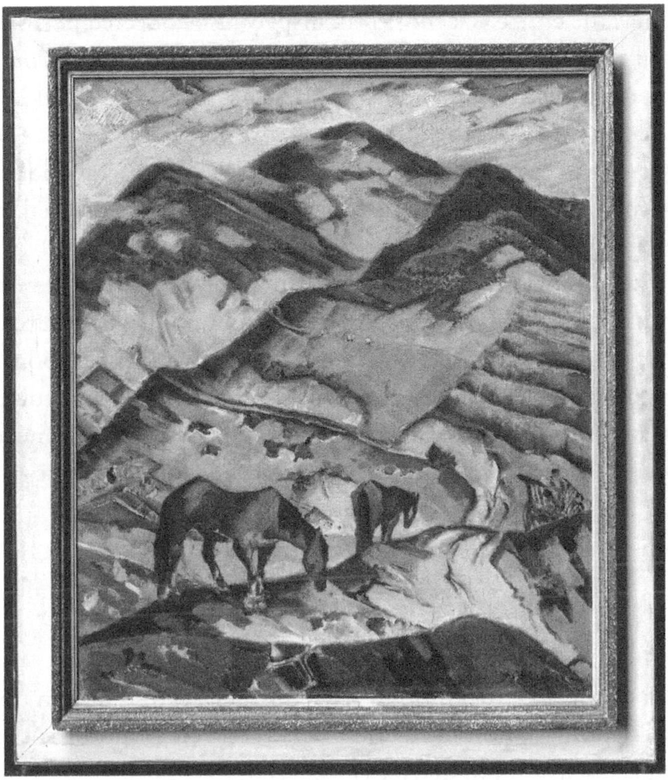

Fig.38. Martin Benka, *Countryside from Terchova*, 1936, painting.

references to traditional culture found in the Slovak peasants and attributes of their life, like musical instruments, the plough or cattle, with a modern approach to space and the distribution of objects in it [fig. 38]. Yet until today, his painting is regarded as aspiring to become "a mythical parable of the Slovak world, [and] its essential foundations in the peasantry and shepherds," replicating the association of Slovaks with the rural, pre-modern environment.[92] Yet the nostalgia for this kind of lost world could be now connected to the direct political threats to the national sovereignty of not only Slovaks but the entire state.

Fulla's more figurative painting of the "Madonna," displayed in San Francisco, also combined folk motives and colors with religious symbols and modernist abstraction, yet in a more moderate way. They were received well in the USA and the artist was named one of the "prominent contemporaries who [were] ambassadors to the Pageant of the Pacific," alongside for example Kokoschka, Klee, Du-

92 KB [Katarína Bajcurová], "Ľudovít Fulla," cat. no. 9/5, in *Budování státu*, 293.

champ, and Paul Nash.[93] The entire Czechoslovak display also collected positive reviews in the American press and exhibition publications. The *New York Times*, for instance, noted that "the picturesqueness [...in] these portrayals of Slovakian countryside and its people" seen in "the glimpses of costume, pastime and the occupations of daily existence" provided a "further insight into racial characteristics of the artists and of their land."[94] Slovakia and its art were reduced to a simplified and selective, yet externally successful image.

The fine art from Czechoslovakia that was displayed in the USA therefore emphasized a more restrained modernist language derived from traditional sources of folk art and realistic painting. This was caused by the fact that those responsible for the selection and supply of the works tried to estimate what works would send out the most favorable picture to the visitors. The art on display had a more illustrative function and communicated basic information about the state—it depicted primarily its landscape and folk traditions.

Conclusion

Much of the criticism of the 1937 pavilion as well as other art and design displays in Czechoslovak representations at world's fairs revolved around the elitism and exclusions in the selection of artists as well as designers. The question of who should represent the state and with what art was therefore not settled during the interwar period. The visual objects chosen for display in the national presentations ranged from fine art and luxury decorative arts to utility design. They could be largely linked to the market economy, whether it was collecting art or commercial sales of design products and operated outside the purely representational status.

With such international exposure that world's fairs offered, the artists selected for the national pavilions and their work would be necessarily associated with the nation and national art. Moreover, these choices to an extent helped to establish the canon of modern art in Czechoslovakia, further validated by numerous awards received at the international exhibitions. This included the members of Mánes, representatives of the UMPRUM, the Association of the Czechoslovak Werkbund as well as the Slovak landscapists.

93 "Cream of Contemporary Europe's Art Exhibited at Golden Gate Fair," *The Art Digest. The News and Opinions of the Art World*, March 15, 1939, 37.

94 "Contemporary Slovakian Art," *New York Times* (November 13, 1938): 185.

Not all artists, however, were accepted in the canon of modern art in the interwar period. Kysela's decorative objects evoked a nostalgia for "primitive" pre-industrial age and premodern crafts displayed in 1925 were replaced by less conservative works in later pavilions. At the same time, Pešánek's avant-garde sculptures represented the other extreme and their appreciation most probably fell victim to the imminent war and insufficient contextualization. The displays of modern art in the Czechoslovak pavilions were therefore very pluralistic, while the presentation of modernism quite restrained.

Chapter Four

COMMODIFYING THE STATE

One of the original aims of world's fairs and international exhibitions had been the presentation of national trade and industries via various companies and businesses. Early exhibitions of industries and arts in the first half of the nineteenth century, often local, were primarily economically driven events, focused on exchanging commodities produced by and for industrial capitalism.[1] From regional and national events they grew into international showcases of comparison and competition that went far beyond the national paradigm.

In their exhibition encounters, large countries like France, Great Britain, and Austria-Hungary had a clearly defined goal of promoting their trade for export and seeking new trading partners. Finding new markets had also been often related to the efforts of colonial expansion and to what was promoted as "spreading civilization" around the world using new technologies, ideas, and products. Czechoslovak goals were not that different even though they did not result in any territorial conquests. However, a degree of colonial aspiration can be detected in Czechoslovakia. Using specific goods such as beads or shoes exported across the globe as a way of entering foreign markets was linked to finding opportunities to acquire affordable raw materials and natural resources for importing back to Czechoslovakia. Such exchanges have recently been explored by Tara Zahra and Zachary Doleshal in relationship to the company Baťa, but many other businesses, including for instance the fez producers in southern Bo-

1 Robey and Raizman, eds., *Expanding Nationalisms*, 4; John R. Davis, "Marginal Exhibition? The All-German Exhibition in Berlin, 1844," in *Cultures of International Exhibitions*, ed. Filipová, 69–90.

hemia and the glassworks of northern Bohemia, were similarly involved in the global trade with all its colonial implications.[2]

The economic motivations for taking part in international exhibitions were therefore the embodiment of increasing globalization, as well as of growing consumerism and capitalism. And as such they were criticized quite early on. Already in 1850 Karl Marx and Friedrich Engels, in advance of the Great Exhibition in London in 1851, described the event as "a striking proof of the concentrated power with which modern large-scale industry is everywhere demolishing national barriers and increasingly blurring local peculiarities of production, society and national character among all peoples."[3] They viewed the planned exhibition and its content as a product and symptom of the consumerist and unjust social system, and this became an established critique of world's fairs.

The emphasis on the consumerist and globalizing aspect of exhibitions is undeniable and has been explored by a range of scholars.[4] Already Walter Benjamin saw nineteenth century large exhibition as pilgrimage sites of commodity fetishism while more recently Burton Benedict addressed the "ritualistic displays of goods and power" of these global events.[5] Yet at the same time, world's fairs retained a high degree of weight given to distinctive national cultures and products. Some of the most ambitious and commercially successful companies with global aspirations, Baťa included, started exhibiting outside of the national pavilions. However, their links to the state and nation often remained intimate. Such companies were still part of the overall diplomatic mission and ideological aims of the respective state.

Ford, General Motors or the Heinz company are illustrative of such a relationship at the fairs in the USA. At the future-oriented World of Tomorrow in New York, 1939, the large Heinz Dome, for instance, showed all its 57 products in a presentation of how "the feasts of yesterday carry over into the world of

2 Tara Zahra, *Against the World: Anti-Globalism and Mass Politics Between the World Wars* (New York: W. W. Norton Company, 2023); Zachary Austin Doleshal, *In the Kingdom of Shoes: Bata, Zlín, Globalization, 1894–1945* (Toronto: University of Toronto Press, 2021); Filipová, "Colonial Complicities."

3 Karl Marx and Frederick Engels, "Review: May-October 1850," *Neue Rheinische Zeitung Revue*, https://www.marxists.org/archive/marx/works/1850/11/01.htm, accessed February 2, 2023.

4 For example, Timothy Mitchell, "The World as Exhibition," *Comparative Studies in Society and History* 31, no. 2 (1989):217–36; Robert W. Rydell, *All the World's a Fair: Visions of Empire at America's International Exhibitions, 1876–1916* (Chicago: University of Chicago Press, 1984); Raizman and Robey, eds., *Expanding Nationalism*.

5 Walter Benjamin, "Paris, Capital of the Nineteenth Century," *Reflections* (New York: Harcourt, Brace, Jovanovich, 1979), 151; Burton Benedict, *The Anthropology of World's Fairs* (Berkeley: University of California Press, 1983). Also for instance, Rosalind H. Williams, *Dream Worlds: Mass Consumptions in Late 19th Century France* (Berkeley: University of California Press, 1982).

tomorrow."[6] At the Golden Gate exhibition in San Francisco, Heinz "dramatically" presented eight period kitchens as "a study of culinary progress from Colonial days to the present time."[7] These showcases were not only aimed at selling ketchup or beans, but they were also ideologically motivated. As the promotional brochure laid out, "One of the miracles of modern civilization ... is the fact that a bottle of Heinz tomato Ketchup looks just the same, tastes the same, is just the same this year, next year and the year after—whether you buy it in Pittsburgh, Pennsylvania, or Melbourne, Australia."[8] Presenting the same product in the global setting of a world's fair helped to create a sense of global appeal, which at the same time referred back to the American culture.

The Czechoslovak participation in international exhibitions of the interwar period displayed a similar interplay of national, global, commercial, and ideological content and emphasis. Moving forward, I will explore the role of a selection of businesses in shaping the image of the state and the motivations that drove them. Private enterprises were involved in the official representation of the state with more than a presentation of company products and production techniques. Using the case of Pilsner beer, Detva textiles, Moser glass and Baťa shoes, whom I hold representative of the main industries included in the Czechoslovak pavilions, I raise the question of what, who, why and how the state and the nation in the commercial sphere were represented , to what extent the national presentations were shaped by their commercial interests and what role the changing political and economic landscape of the interwar period had on such displays and their motivations.

One of the ways of combating the economic downfall of the early 1920s was for large businesses in Czechoslovakia to form close ties with the state administration. Zdeněk Kárník has called the individuals who forged such mutually benefiting relationships "the economic elite," that informed my use of the term "the exhibition elite."[9] The economic elite was composed of individuals with roles in the management of companies, trading and banking associations, as well as in the state administration. Such involvement gave them also a direct influence on the ideological and actual content of the potential national presentations abroad. As I have mentioned in the previous chapter, ministries fre-

6 *Heinz 57 Exhibit Manual Brochure* (New York, 1939), III. Golden Gate International Exposition, Fairs and Expositions Collection, 1893–1967, Environmental Design Archives, University of California Berkeley, 216-X-863.
7 *Heinz 57 Exhibit Manual.*
8 *Heinz 57 Exhibit Manual.*
9 Kárník, *České země I*, 266.

quently quarreled about the focus of the fairs and it was the Ministry of Trade that promoted orientation of the official displays on trade and economic benefits. The frequent involvement of the four companies I discuss here therefore illustrates how such efforts were achieved. The commodities these companies produced, that is textiles, beer, glass and shoes, contributed greatly to the presentation of Czechoslovakia as a wealthy, capitalist country. Putting the Marxist reading aside for a moment, these goods had not only consumer value but also cultural and political value; they became exhibits for display, framed by the state's official position in the world's fairs as well as the global order.[10]

Economy and the new state

After 1918, the new state also had quite a good starting position that could benefit from the concentration of many businesses in Bohemia before the war. In economic terms, the entire Bohemian part of the Habsburg empire before the war was rather prosperous and economically sufficient, representing 65–75% of all industries in the whole of Cisleithania.[11] In 1914, these industries were responsible for 41.2% gross national product of Cisleithania and about 32.1% of the whole monarchy. The situation was rather different in Slovakia and Sub-Carpathian Ruthenia which produced about 16.3% of the gross national product of Transleithania before the war. This increased to 18.5% after the war.[12]

Soon after the creation of the state, the Czechoslovak government introduced new economic measures which included removal of the customs territory and the introduction of new tariffs. The so-called validation of large companies that had a seat in Vienna or Budapest but traded in Czechoslovakia was another important step, which saw the transfer of many such businesses to the new country. Agreement on this was reached with Austria as early as 1920 and with Hungary as late as 1927.[13] One example to illustrate this shift could be the textile manufacturers of fezzes in south Bohemia. Until 1918, the main production for The Joint Stock Company of Austrian Fez Manufacturers was based in and around Strakonice, while the headquarters were in Vienna. With the new

10 Walter Benjamin, "Paris," *Reflections*, 152; Umberto Eco, "A Theory of Expositions," *Travels in Hyperreality. Essays* (San Diego - New York - London: Harcourt, 1990), 458.
11 Kárník, *České země*, 199–200.
12 After the war, it amounted to 18.5% in Czechoslovakia. Karnik, *České země*, 202.
13 Eduard Kubů and Jaroslav Pátek, "Základní charakteristika výchozí situace a vývoje československé ekonomiky v meziválečném období," in *Mýtus a realita hospodářské vyspělosti Československa mezi světovými válkami*, eds. Eduard Kubů and Jaroslav Pátek (Prague: Karolinum, 2000), 13.

Czechoslovak state, the company's seat moved to the south Bohemian town, Vienna became a subsidiary branch and the company's name dropped the reference to Austria.[14]

Generally, though, the period immediately after the First World War created a rather vulnerable environment for new states, in which old economic alliances, customs unions and trading quotas no longer existed. Postwar inflation only exacerbated this situation and affected many countries in and outside Central Europe. One of the solutions of the Czechoslovak government and the minister of finance, Alois Rašín, was a monetary reform performed in 1919. It consisted of separating the Czechoslovak currency from the Austrian one. The quick introduction of a new currency, the Czechoslovak crown, meant that in Czechoslovakia the postwar inflation was not as drastic as in the neighboring countries.[15] The better economic position affected the state's ability and confidence in participating at world's fairs from early on. There was a downside of the relatively strong crown, especially at earlier fairs, like the Centennial Exposition at Rio de Janeiro in 1922. Czechoslovak products were offered at very high prices.

Nevertheless, reaching financial independence was key for future economic self-sufficiency and a stronger position of the country in the restructured global political environment. Institutions and companies that were previously Austrian and Hungarian now became Czechoslovak. This move was furthered by the repatriation of capital to Czechoslovakia in the form of purchasing stocks and capital gains at stock markets by large Czech banks and bringing them to local clients.[16] The general outflow of finances in the form of taxes, tariffs and shares were, before 1918, directed to Vienna and Budapest now stopped. Economic and legal reforms were also motivated by creating a more integrated national economy and legal system, respectively, from two previously unrelated ones. This "Czechoslovakization" concerned not only the financial, legal and industrial aspects, but also the unification of transportation infrastructure and electrification.

On the background of the need to trade independently with external partners, Czechoslovak companies recognized the importance of participating in exhibitions quite soon. They could build on experience from local industrial

14 *150 let textilní výroby výroby FEZKO Strakonice* (Strakonice: FEZKO Strakonice, 1962), 25; Filipová, "Colonial Complicities."

15 Kubů and Pátek, "Základní charakteristika," 12–13; "Rašínova měnová reforma," *Historie ČNB*, https://www.historie.cnb.cz/cs/menova_politika/1_vznik_a_stabilizace_samostatne_ceskoslovenske_meny/rasinova_menova_reforma.html, December 15, 2022.

16 Kubů and Pátek, "Základní charakteristika," 13.

and agricultural fairs that had already been a commonplace in Bohemia, Moravia and Slovakia prior to 1918.[17] In the Bohemian regions, prewar exhibitions were marked by strong nationalist element. Particularly active in hosting trade fairs were the German speaking towns and regions in northern Bohemia, where such exhibitions often replicated the larger events in the monarchy, especially the 1873 Weltausstellung in Vienna. Such exhibitions had a strong German focus even though they did not exclude Czech participation. For the most part, though, they were organized by local Bohemian Germans for local Bohemian Germans seeking recognition for the German-speaking regions.[18]

Exhibitions in for example the north Bohemian Cheb (called Eger in German), Liberec (Reichenberg) and Ústí nad Labem (Aussig) focused on boosting trade and production, fostering business contacts, showcasing goods and innovations, and educating the wider public.[19] As the historian Tomáš Okurka has argued, the earlier nineteenth century events were not yet tinted by nationalistic disputes and splits between Germans and Czechs of Bohemia.[20] These became more prominent after the 1890s Czech and Slavic exhibitions in Prague and especially in the early 20th century. For example, the exhibition that took place in 1906 in Liberec, a city that started fashioning itself as the capital of Germans in Bohemia, thus became an exclusively German promoting German products and achievements.

In terms of the economic divides, businesses across Bohemia and Moravia before 1918 were mostly owned by Bohemian Germans or Viennese and backed by the capital of German and Austrian banks.[21] The most important postwar change for Czechoslovak exhibitions histories was therefore the gradual rise of Czech ownership of local companies and a desire to form new commercial partnerships with the Allies. While Germany continued to be one of the most important trading partners for Czechoslovakia, the political and diplomatic efforts of the Czechoslovak government to create alliances with France, Great

17 Tomáš Okurka, "'Witness to the Momentous Significance of Germ and Labor in Bohemia': Exhibitions in the German-Speaking Regions of Bohemia before the First World War," in *Cultures of International Exhibitions 1840–1940. Great Exhibitions in the Margins*, ed. Marta Filipová (London: Ashgate, 2017), 93. Also Jaroslav Halada and Milan Hlavačka, *Světové výstavy: od Londýna 1851 po Hannover 2000* (Prague: Libri, 2000); Pavol Komora, *Hospodárske a všeobecné výstavy 1842–1940* (Bratislava: Slovenské národné múzeum – Historické múzeum, 2016).

18 Okurka, "'Witness,'" 104.

19 These exhibitions were held in for example Cheb (Eger) in 1871 and 1881, in Litoměřice (Leitmeritz) in 1877, in Česká Lípa (Böhmisch Leipa) in 1880, in Teplice (Teplitz) in 1875, 1879 and 1884, in Ústí nad Labem (Aussig) in 1874 and 1880. Okurka, "'Witness,'" 93.

20 Okurka, "'Witness,'" 94nn.

21 Kárník, *České země* I, 204–206.

Britain, and the United States translated into attempts at establishing trading relationships with these countries. This was, nevertheless, a slow process for an unknown political entity that Czechoslovakia was. World's fairs, nevertheless, were believed to play a key role in enabling the new state to access new partners and markets.

Modern folk industries: Detva

To establish these external relationships, Czechoslovakia displayed the goods that it deemed most suitable for trade. They generally indicated the country's advanced modernity in production and technology, while they also tried to reinforce a consistent brand and identity of the new state. And that had to be consolidated internally too. To do that, objects and commodities representing the historicity and traditions were included. Apart from references to historic architecture in paintings and photographs, or to works of art, folk products, costumes or customs were commonly incorporated to explain how locally embedded contemporary Czechoslovak culture was. Folk art had also other useful functions. As it was often associated with rural regions of the eastern parts of Czechoslovakia, its inclusion in the national pavilions meant that these regions were represented there. While Bohemia and Moravia often displayed what was seen as advanced industries and arts, Slovakia and Sub-Carpathian Ruthenia would be represented by natural resources and folk products.

The political program of Czechoslovakism emphasized and largely invented the joint identity of the Czechoslovaks. At world's fairs, Czechoslovakia showcased the idea of a single nation that consisted of the two main Slavic groups of Czechs and Slovaks while in most cases Hungarians, Poles and Ruthenians were omitted. The only exceptions were German and Jewish businesses and culture that were present through artists and various companies until 1938. At the same time, the cultures of Czechs and Slovaks were represented far from equally, which—after all—reflected the general political, cultural and artistic predominance of the Czechs within the state and its administration.

Each exhibition committee, responsible for the organization of individual pavilions, their design and selection of objects consisted mainly of Czech members who prioritized Czech representation. Yet the committees were conscious of the political need to include Slovakia in some way. During the ministerial discussions of the 1925 exhibition, for example, Václav Vilém Štech of the Ministry of Education concluded that "there is only one way Slovakia and Slovak politics can be represented at a world's fair and it is through folk art."[22] Slovakia was thus frequently reduced to displays related to the traditional cultures of the countryside. This imbalance, however, had also economic motivations. In his comment, Štech also noted that "Slovaks [would] not contribute [financially] towards the exhibition but [would] expect to be represented by their works. [Yet,] Slovak art shall not be omitted if for nothing else but the domestic politics."[23]

A suitable candidate for showcasing Slovakia was found in the company Detva. It fulfilled several requirements of the interwar state's representation at once: it was based in Bratislava and its products combined folk traditions manufactured for use in modern homes.[24] At world's fairs, Detva mostly presented itself by embroideries and lace with floral or geometrical patterns, but it also produced all types of other textiles, carpets, tapestries, glass, ceramics, bijou and provided interior design services. The company was established in 1919 out of the Hungarian association, Izabella, which had been founded in 1895 and had similar goals. When Izabella was dissolved by the Czechoslovak government, Detva was formed to elevate folk art and the home industries in Slovakia and Sub-Carpathian Ruthenia, its production adjusted to the new needs of contemporary people and requirements of the culture of dwelling.[25]

During the interwar period, the full name of the company which had some 2,000 employees, changed a few times to reflect the position of the company and Slovakia within the joint state. In 1927, Detva became a shareholding company Detva,Československý ľudový priemysel / Volks Kunstgewerbe (Detva, the Czechoslovak Folk Art Industries) reflecting the increasing industrial orientation and expansion across the entire state in its title. The change, nevertheless, came with a necessary financial injection, in which the Czechoslovak state gained an almost exclusive majority in shares and helped the company survive an economic collapse. At the end of the interwar period, in 1939, the Czechoslovak company was renamed Detva, slovenský ľudový umelecký priemysel, účastinná spoločnosť v Bratislave (Detva, the Slovak Folk Art Industries, Shareholding Company in Bratislava) to stress its Slovak identity and affiliation with the Slovak State.

Before that, though, Detva expanded across the country and ran several shops across Czechoslovakia, for instance in Prague, Brno, and many spa towns in Bohemia. As was the case with Baťa, Detva's management was entangled with the

22 Štech, quoted in Komora, *Hospodářské*, 204.
23 Štech, quoted in Komora, *Hospodářské*, 204.
24 "Zpráva o zvelebovací a propagační činnosti ústavu 'DETVY' v Bratislavě v období 1924/25 a 1925/26," I.B.1. SNA.
25 *Detva, československý ľudový umelecký priemysel, úč.spol.* (Bratislava: Detva, 1929).

cultural, economic and political elites in the state. The first director was Dušan Jurkovič, an architect known for applying folk-inspired decorativeness and color schemes in his buildings.[26] In later years, the company was chaired by, for example, Frances Crane Masaryková, the American wife of Jan Masaryk, himself a diplomat and the son of the President. Another member of the management, Václav Maule, a long-term vice chair of Detva, was a ministerial advisor for education who participated in Slovak school reforms. At the end of the interwar period, when nationalism in Slovakia was strengthening, Maule was replaced by the Slovak MP Ján Liška, a nationalist politician and businessman. He was closely affiliated with the People's Party of Andrej Hlinka, the Slovak Catholic priest and politician, who called for Slovak autonomy.

The board of the shareholding company also consisted of representatives of various ministries: trade, education, and finance, as well as of "significant personalities of Slovak economic, cultural and political realm."[27] The composition of the board also epitomized the gender imbalance in the state. The lists of the Detva board members consist exclusively of men, with Crane Masaryková being the only woman in an honorary position.

Detva's mission was to "redefine" Slovak folk art, inevitably deemed to extinction, with its characteristic production techniques, motives, and use of domestic materials. The company report from 1926 provides an insight into Detva's approach to production and workforce. It noted that the artistic director at the time, the Czech artist and designer Jaroslav Jareš, was the first to open a workshop in embroidery and weaving with the "humble but talented female workers from the [common] people," and this turned out to be a success.[28] The report also remarked that it was necessary to "educate the rural primitives [in Slovak original *sedliackych primitivov*] in the production meant for the cultivated classes of urban inhabitants in this country and abroad whereby this process of education these simple people" in methods that are new and unusual for them "is very difficult, time-consuming and requires not only a lot of patience but mainly substantial financial sacrifice."[29] The reference to the "primitives" of the countryside evoked somewhat crude character of the people and their work, rather than a belief in their evolutionary inferiority. At the same time, however, the

26 Dana Bořutová, *Architekt Dušan Samuel Jurkovič* (Bratislava: Slovart, 2009).
27 "Výročná zpráva DETVY československý ľudový priemysel, úč.spol. v Bratislave za správný rok od 1. septembera 1930 do 31. augusta 1931," SNA.
28 "Výročná zpráva DETVY československý ľudový priemysel, úč.spol. v Bratislave za správný rok od 1. septembera 1927 do 31. augusta 1928," 1929, inv. no.4, SNA, 4.
29 "Výročná zpráva DETVY 1927–1928," 4.

fact that the production of the "simple people" was meant for the "cultivated" inhabitants of the cities sharpened the contrast between the rural and the urban based on social and class prejudice.

During the interwar period, the eastern regions of the state remained predominantly rural; they did not benefit from the land reforms as much as the western parts and were also subjected to higher taxation.[30] This, combined with the efforts of industrialization and modernization, concentrated mostly in the Czech parts of the state, resulted in a slower economic development in the east. Compared to industries in Bohemia, Slovakia in fact de-industrialized during the interwar period and the divide between Czech and Slovak living standards only deepened.[31] The case of Detva could therefore be seen as symbolic; the company enjoyed the benefits of cheap labor in the countryside while at the same time profiting from intimate connections with the Czech establishment.

This was visible also in the chain of production in the company which followed a top-down trajectory: while the commodities, Detva offered, were made locally, the design was mostly done by artists educated at art and design schools in Czech cities.[32] Detva also cooperated with artists and designers of various institutions, mainly with the Association for Czechoslovak Werkbund, that was intimately involved in the representation of Czechoslovakia at international exhibitions. Detva's commissioning of trained artists was based on the belief that the traditional production of folk art in the countryside, even though technically skilled, was insufficient in terms of taste, and therefore unsuitable for urban consumption. To rectify the tastes (and increase the sales), Detva started running workshops across mostly Slovak regions, where there had been, "until now, primitive and uneconomical forms of production to contribute to the economic development of Slovakia."[33] And while folk art was framed as a tra-

30 Eduard Lukáč, Katarína Mayer, and Martina Lenhardtová, "Development of Adult Education in Interwar Slovakia (1918–1938)," *Ad Alta: Journal of Interdisciplinary Research* (2019): 173–80.

31 Petr Pavlínek, "Regional Development and the Disintegration of Czechoslovakia," *Geoforum* 26, no. 4 (1995): 351–354; and Milan Olejník, "The Development of the Slovak Industries during the First Ten Years of the Czechoslovak Republic," *Človek a spoločnosť: internetový časopis pre pôvodné teoretické a výskumné štúdie z oblasti spoločenských vied* 6, no. 1 (2003): 13–21, https://individualandsociety.org/storage/uploads/casopis/2003/1/the-development-of-the-slovak-industries-during-th/the-development-of-the-slovak-industries-during-the-first-ten-years-of-czechoslovak-republic.pdf; Joseph Rothschild, *East Central Europe between the Two World Wars* (Seattle: University of Washington Press, 1974), 120; Ramet and Skalnik Leff, "Interwar Czechoslovakia – A National State for a Multiethnic Population," in *Interwar East-Central Europe*, ed. Sabrina P. Ramet, 75–108.

32 Vojtěch Merganc (textiles), Jaroslav Jareš, Ľudmila Rambouská-Jarešová and Elena Hollá (embroideries and lace). "Zpráva o zvelebovací a propagační činnosti ústavu Detvy v Bratislavě v období 1924/1925 a 1925/1926," Detva Archive, inv. no. 2, I.B.1, SNA, 1.

33 "Zpráva o zvelebovací a propagační činnosti," 1.

ditional and archaic practice, Detva set out to use it to its own economic benefit and to some extent the benefit of the Slovak region.

The company therefore drew on the available resources and workers in the impoverished parts of the countryside, to whom it provided training and sales opportunities. Such practice was nothing new. In the late nineteenth century, training schools for the so-called house industries, or Fachschule für Hausindustrie in German, were established around the Habsburg monarchy. As Rebecca Houze has argued, they aimed to reintroduce the disappearing folk arts to the peasantry, give them a form refined enough for urban tastes and thus provide peasants with an independent source of income.[34] Detva should therefore be seen as a company that utilized labor in the countryside, portrayed—using the common Czechoslovak stereotypes—as primitive, under-education and unmodern, for commercial gains at national and international markets. At the same time, it found a suitable place in the national pavilions to present Czechoslovakia as a country both traditional and modern.

Detva on display

Detva was a frequent exhibitor at international exhibitions, although it often needed state support and subsidies. Its predecessor Izabella had appeared at, for instance, the Exposition Universelle in Paris in 1900, the Louisiana Purchase Exhibition in St. Louis in 1904, or the International Exhibition in Milan in 1906. After the First World War, Detva continued to be a frequent and successful participant in the Czechoslovak pavilions at interwar world's fairs, winning grand prizes in Paris in 1925, Philadelphia in 1926, and Barcelona in 1929, where it received hors concours prize for embroideries and grand prix for its carpets.[35]

The links with the exhibition elites meant that the presence of Detva in the national pavilions was encouraged by the government. For example, it received subsidies from the Ministries of Trade and Education to promote its activities at future exhibitions and trade fairs.[36] This meant that it could take part not only at the previously mentioned exhibitions, but also at the national exhibition in Brno in 1928 and at the world's fairs in Chicago (1933), Brussels (1935),

34 Rebecca Houze, *Textiles, Fashion, and Design Reform in Austria-Hungary Before the First World War* (London: Routledge, 2017), 80.

35 "Výročná zpráva DETVY československý ľudový priemysel, úč.spol. v Bratislave za správný rok od 1.septembera 1928 do 31.augusta 1929," 1930, Detva, inv. no. 2, I.B.1, SNA, 4.

36 "Výročná zpráva DETVY," 1930, 7.

and at the Empire Exhibition in Johannesburg (1936).[37] Detva also participated in local exhibitions and regular fairs in Prague, Bratislava as well as in all exhibitions of the Association for Czechoslovak Werkbund and Umelecká beseda slovenská (The Slovak Artistic Forum).

It was at the exposition of modern decorative art in Paris in 1925, the one where Štech associated Slovakia primarily with folk art, that Slovakia was for the first time represented "as a producer of artistic objects."[38] The success of the company came in the form of a Grand Prix. Yet here, Detva was also confronted with expositions of other countries that had renewed their folk art for new markets. Sweden, Norway and Finland, for example, showed modern applied arts embedded firmly in their respective local traditions. This confrontation led Detva to open new workshops across Slovakia on embroidery, weaving, and basketry. Market demands, often detectable at world's fairs, also resulted in the change of production focus that favored quality over quantity. This helped to introduce new, more expensive materials like silk and tulle but necessarily increased the price of such products.[39]

By the middle of the interwar period, Detva marketed itself as the producer of luxury items and in the late 1920s and early 1930s became the victim of austerity and the economic crises. Detva felt the impact strongly from the beginning of the 1930s; it suffered losses in revenue especially in the spa towns of Bohemia where it relied on its German customers and tourists.[40] Germany and the USA, the countries especially affected by the economic drop, were for a long time the biggest markets for Detva. Consequently, Detva had to return to the production of less expensive and more utility-oriented items, and even contemplated closing the company down in 1934.

Participation at world's fairs was therefore crucial for Detva to promote its products and target new markets. There was also the opportunity to face the competition, in both positive and negative senses. In the USA, for instance, Detva was first represented in Philadelphia in 1926, where it received a golden medal. This success nevertheless did not translate into the hopes of expanding to the American market, and one of the reasons was seen in the competition. In the USA, it was represented mainly by Hungary, which supplied foreign markets

37 "Výročná zpráva DETVY československý ľudový priemysel, úč.spol. v Bratislave za správný rok 1934," SNA, 6.
38 "Zpráva o zvelebovací," 2.
39 "Výročná zpráva DETVY," 1930.
40 "Výročná zpráva DETVY československý ľudový priemysel, úč.spol. v Bratislave za správný rok od 1. septembera 1930 do 31.augusta 1931," June 11, 1932, inv. no. 4, SNA, 7.

with similar products from their new embroidery companies established after the losses of workshops in the former Greater Hungary, which included Slovakia.[41]

The international success that Detva achieved at the various international exhibitions was nevertheless interpreted as a moral victory over the Hungarian legacy of the company. Before 1918, the products from Slovakia had been presented under the auspices of Hungary and marketed as "the folk art from Upper Hungary" to the displeasure of later Slovak commentators.[42] The rivalry continued into the interwar period and still after the creation of the new states, Karel Knop, the director of Detva in 1926, complained that Hungary won the markets in America with her cheap folkware decorated "with Slavic (Slovak, Croatian, Serbian) motives."[43]

The nationalistic undertones in such claims that tried to affiliate folk production with specific ethnic or national groups worked both in favor of Slovakia and against it. Even though Detva designs came from contemporary artists, the mostly traditional look of the objects defied their modernity in the eyes of some critics. Emanuel Purghart, the director of the press service of the Ministry of Foreign Affairs, is an example of skepticism towards folk art as exhibits and as commodities in the context of national pavilions. Purghart was in charge of several exhibits in the 1925 Czechoslovak pavilion in Paris and this included

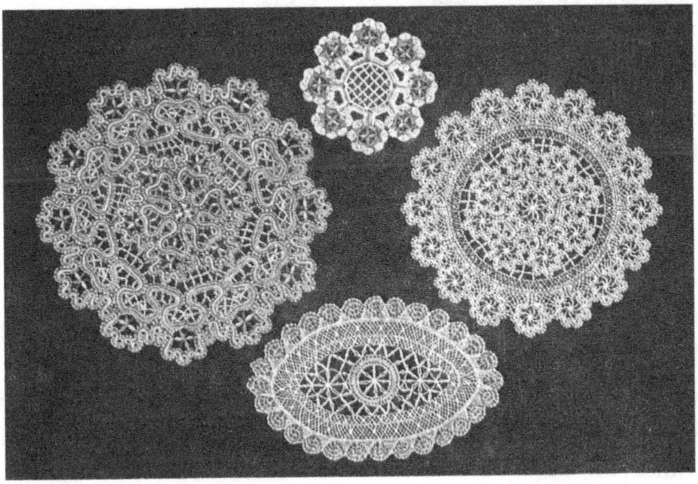

Fig.39. Lace doilies, *Detva. Československý ľudový umelecký priemysel, úč. spol.*

41 Karel Knop, "Organisace lidové výroby na Slovensku i Podkarpatské Rusi a dnešní snahy o její umělecké zvelebení," *Výtvarné snahy* 7 (1926): 138.
42 "Výročná zpráva Detvy," 1931.
43 Knop, "Organisace," 138.

Fig.40. Example of Detva products on display, *Detva. Československý ľudový umelecký priemysel, úč. spol.*

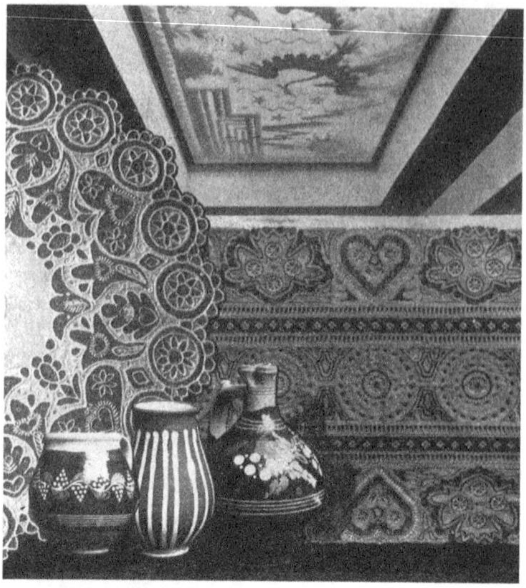

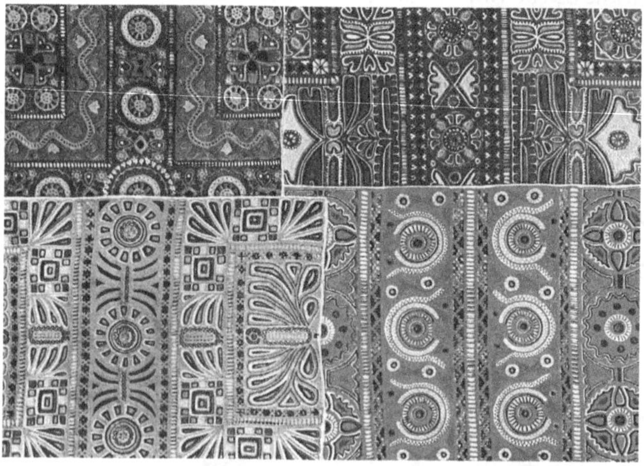

Fig.41. Colorful embroidered pillow cases, *Detva. Československý ľudový umelecký priemysel, úč. spol.*

the folk art section. For him, if Czechoslovakia wanted to be ranked amongst "the most civilized nations," it should avoid displays that may be wrongly seen as originating in "the Balkans or the Orient."[44] As much as the avant-garde, Purghart believed the place of folk art was in museums rather than in modern exhibitions of arts and industries.[45]

44 Quoted in Komora, *Hospodárske*, 213; "Protokol zo schodze správnej rady Detvy 29. aprila 1931," Detva, box 3, inv. no. 30, SNA.

45 Filipová, *Modernity*, 104.

Inevitably, folk art production, however, modernized, continued to be associated with the eastern regions of Czechoslovakia. In Slovakia, folk art also carried surviving anti-Hungarian sentiments which, in the interwar period, translated political animosities into economic and artistic competition. The geopolitical space that Slovakia and Hungary shared before the First World War and the lack of clearly defined borders between the countries made it sometimes impossible to distinguish the ethnic origins of the respective folk production. Such divides, rather fluid before the war, became significant and much more pronounced after 1918 across the whole Czechoslovakia [fig. 39–41].

Nationalized consumption: Pilsner beer

The political re-composition of Central Europe after the First World War changed the status of many ethnic minorities in the new states. In Czechoslovakia, which became a complex multi-ethnic state, the influential position Germans and Hungarians enjoyed before 1918 was redefined in favor of the Czechs. These new conditions had also a stark impact on the confines in which national presentations at world's fairs took place. Before the war, Austria-Hungary either included some of the administrative regions as separate sections (Bohemia or Bosnia, for instance), swiped them under a larger entity (Upper Hungary, aka Slovakia, under Hungary) or simply ignored them. From the moment of the creation of the Dual Monarchy, Hungary enjoyed separate displays under the umbrella of the empire, but separate Austrian and Hungarian pavilions did not occur until the exposition in Paris of 1900.[46]

The Parisian Exposition Universelle in 1878 introduced a clearly defined division into national pavilions in the Rue de nations. This was, nevertheless, preceded by the inclusion of restaurants and bars in fairs and exhibitions that served regional and national food and drink since 1851.[47] These establishments provided visitors with refreshments, helped to introduce them to foreign cuisines and offered to experience foreignness that extended beyond the consumables.[48] To create a better sense of authenticity, customers were served by cos-

46 Vilmos Gál, *Hungary at the World Fairs* (Budapest: Holnap Kiadó, 2010), 108.

47 Thomas Prasch, "Eating the World: London in 1851," *Victorian Literature and Culture* 36, no. 2 (2008): 587–602.

48 Cf. Taka Oshikiri, "Selling Tea as Japanese History: Culture, Consumption and International Expositions, 1873–1910," in *World Fairs and the Global Moulding of National Identities*, eds. Leerssen and Storm, 193–216.

tumed waiters and waitresses, often brought from abroad or recruited from local émigrés. For instance, in the Austrian village at the 1867 exhibition in Paris, there was a sizeable beer hall set up by the Viennese brewer Dreher offering Austrian and Hungarian dishes, beer and wine served by "blue eyed mädchen in national costumes."[49]

The practice of including nationalized food and drink was developed at subsequent fairs. At the Columbian exposition in Chicago in 1893, for instance, the Swedish restaurant offered food items that according to the catalogue: "... no one but a Swede has ever yet succeeded in eating" or indeed drinking in the case of Swedish potato whisky. "Guests may here enjoy, if they can, smoked reindeer, baby sausages, craw-fish tails, raw 'delikatess,' herring, fried stromming, smoked goose breast, reindeer tongues, and 'graflax.'"[50] Venturing into such an outlet therefore provided a safe adventure for the visitor. Food and drink also provided grounds for cultural comparison and for judging other cultures, as well as for creating and sustaining prejudices and stereotypes, whether they were national or colonial. These establishments became generally very popular as most visitors had otherwise no chance of encountering the exotic in person. Food and drink therefore became a useful tool of promoting some of the main aims of these events—progress, modernization as well as globalization, while retaining a sense of national uniqueness. One can only recall the sameness of the Heinz products across international markets I have mentioned at the start.

In this sense, there probably is not a better consumable to represent Czech culture than beer. It had a special place in the national consciousness of the Czechs, carried from Austria-Hungary into interwar Czechoslovakia and in fact until today.[51] The affiliation of beer with national identity dates to the first half of the nineteenth century when the beverage became incorporated into the narrative of nationalism and national self-determination. Beer drinking was turned into a symbolic ceremony of belonging to the Czech and more generally Slavic

---

49 Quote from Eugene Rimmel, *Recollections of the Paris Exhibition of 1867* (London [1868]), 1–2, in Edward N. Kaufman, "The Architectural Museum from World's Fair to Restoration Village," *Assemblage* 9 (1989): 24.

50 *A Week at the Fair, Illustrating the Exhibits and Wonders of the World's Columbian Exposition* (Rand, McNally & Co., 1893), 168. See also Scott, "Tales from the Swedish Café," *World's Fair Chicago, 1893*, 2019, https://worldsfairchicago1893.com/2019/07/20/swedish-cafe; Elizabeth Badger, "The World's Fare: Food and Culture at American World Fairs from 1893–1939," MA Thesis, Western Washington University, 2012.

51 Jana Pařízková and Martina Vlková, "Beer in the Czech Republic," in *Liquid Bread: Beer and Brewing in Cross-Cultural Perspective*, eds. Wulf Schiefenhovel and Helen Macbeth (New York: Berghahn Books, 2011), 101–110.

unity—this elevated beer consumption from a pure material and social experience to an almost spiritual and ideological act.[52]

However, as the geographical region of Bohemia was for centuries shared by ethnic Czechs and Germans, it is impossible to talk about a purely Czech beer tradition in the Czech regions. The history of beer-making in the geopolitical space of Central Europe is interconnected and the German brewing tradition was often indistinguishable from the Czech one. The beer that has been made in the city of Pilsen serves as a great example of this complexity. The pale, heavily hopped beer from bottom-fermented yeast was first brewed in this west Bohemian city in 1842 by the newly appointed head brewer, Josef Groll who was invited to come to Měšťanský pivovar (the Burghers' Brewery) from the German Bavaria. The beer quickly gained popularity throughout Europe and overseas and became second largest beer producer in Austria-Hungary. Subsequently, further breweries were founded in Pilsen, which also brewed beer based on the same technology. The První plzeňský akciový pivovar (the First Pilsen Shareholding Brewery), for example, founded in 1869 became the court supplier for the Kaiser in 1912.

The rise in the popularity of Pilsner beer and more generally in the number of large breweries in Bohemia in the second half of the nineteenth century coincided with two important moments in global economic and political development. First, it was the boom in the number of international exhibitions and world's fairs, especially in Europe and in the USA and second, the increasing migration from Central Europe to North America. In both cases, the beer, presented in the international context as Czech, acted as a symbol of national distinctiveness in the age of continued modernization, industrialization and national self-determination.

With such a close, albeit disputable, link to the Czech nation, Pilsner beer was incorporated into the official presentations of the state at world's fairs in the interwar period. As a company, The First Pilsen Shareholding Brewery was successful already at the Weltausstellung in Vienna in 1873. It built its own pavilion, furnished with Thonet chairs, and an impressive installation in the shape of a pyramid of beer barrels and other beer vessels.[53] The installation was topped by a statue of King Gambrinus, a mythical figure who started making beer in

52 Vladimír Macura, *Znamení zrodu: české obrození jako kulturní typ* (Prague: Československý spisovatel, 1983), 19.
53 Josef Kejha, Jiří Janouškovec and Vladimír Jurina, *Plzeňský Prazdroj, příběh, který nepřestává inspirovat. Plzeňský Prazdroj - od roku 1842* (Plzeň: Nava, 2012), 49.

the sixteenth century.[54] At the Viennese world's fair, the Pilsner pavilion that sold "national" drinks had competition in the American bar or the Hungarian winery, for example, but was awarded a gold medal.

Economies and nationalities

Ethnic restructuring after the war impacted the companies internally, as the major breweries in Plzeň had a large German representation in the management. Until 1918, The First Pilsen Shareholding Brewery was a German enterprise with German management, administration and most employees from Saxony and Bavaria.[55] This started changing towards the end of the war and the first Czech manager, Josef Brych, was appointed in 1919. With him came more Czech employees.[56]

With the break-up of the Habsburg monarchy came the end of the customs union within the Habsburg Monarchy and companies, including breweries, had to re-establish themselves and seek new markets. They also had to tackle the postwar economic crisis by lowering the number of employees and increasing the salaries of the remaining ones. Close trading ties with Germany were affected too, as export to Germany became the victim of the postwar economic drop, later by the crisis of the late 1920s and early 1930s and eventually by the birth of Nazism and the Weimar Republic, which gave preference to German national products.[57]

The fall of the Monarchy temporarily interrupted not only the sales of Pilsner beer but also the successful presentation of the beverage at world's fairs that made only a gradual re-entry into Czechoslovak displays in the interwar period. While prohibition prevented its participation in Philadelphia and partly in Chicago, it was in the late 1930s that beer once again became truly and firmly associated with the national pavilions and Czechoslovak identity. A clear indi-

54 In 1919 Gambrinus gave the name to a new beer made in Plzeň in a separate brewery, to disassociate the original brand, Pilsner Kaiserurquell, from its Austro-Hungarian legacy. Zdeněk Bauer, *Jak vytvořit atraktivní obchodní název firmy, služby, produktu, značky: tvorba názvu - obchodního jména krok za krokem* (Prague: Zdeněk Bauer, 2014).

55 David Totzauer, "Pivovarnictví v Plzni a okolí v meziválečném období," MA Thesis at the University of West Bohemia Plzeň, 2020, 41.

56 The various breweries in Plzeň eventually merged into one shareholding company in 1932, the Plzeňské akciové pivovary (The Pilsen Shareholding Brewery), as another economic crisis was ending.

57 Totzauer, "Pivovarnictví," 38–39.

cation of its status can be found in a letter sent by foreign diplomats to the Ministry of Foreign Affairs in 1935:

> There is a product, the fame of which has been linked by a hundred-year long tradition with the Czech lands, and the quality of which... has gained world recognition, and it is beer. There is no other product that is so characteristically Czechoslovak, so historically famous and close to all the social strata of any nationality as Pilsner beer.... Pilsner beer has become a staple of our republic and the means of [our] promotion![58]

These explicit proclamations were issued in relation to beer export from Czechoslovakia that now became a staple of the entire Czechoslovak state, not just the Czech, or rather Bohemian, part of the country. Eventually, beer also fulfilled the requirements of generating economic benefits, while also promoting and representing the nation culturally. Pilsner beer was therefore included at Brussels (1935) because beer consumption in Belgium was four times higher than in Czechoslovakia.[59] Complaints nevertheless occurred and targeted the non-Czech sounding translation of the brand as Pilsner Urquell that appeared on a large sign above the tap room. Czech press criticized it as an unnecessarily Germanized form of Prazdroj (that translates as "the original source"), the type of Pilsner beer sold here. German press, on the other hand, praised an easier pronunciation of Urquell especially for French speakers.[60] The name, nevertheless, caught on and became part of the brewery's marketing at future exhibitions.

A successful presentation followed in Paris in 1937 and New York in 1939, despite—or because of—the worsening political circumstances in which Pilsner beer continued to play the role of another symbol of Czechoslovak identity and economy. In contrast to the subsidized Detva, breweries contributed to the extensive costs of the national presentation at various fairs. In Paris, for instance, the French branch of Pilsner built a sampling pavilion "to draw the public's attention to the exceptional qualities of Pilsen beer."[61] Originally planned on the

---

58 Letter from Konzulát Československé republiky v Singapore to Ministerstvo zahraničních věcí ČSR o vývozu československého piva do Singapuru, May 29, 1935, Division III, 1918–1939, no. 599, MFA Archive.

59 "Pilsner Urquell auf der Brüsseler Ausstellung," *Prager Presse*, 1 June 1935, Newspaper Excerpt Archive 1916–1944, Oh8, 506/0/43, MFA Archive.

60 "Pilsner Urquell auf der Brüsseler Ausstellung;" "Dojmy českého reportéra ze světové výstavy v Bruselu," *Národní listy*, May 12, 1935, Newspaper Excerpts Archive, MFA Archive.

61 Letter from Biere Urquell Pilsner, Paris, to Legation Tchecoslovaque from May 30, 1936, no. 6328, 1937, MFA Archives, Division III, Section 2.

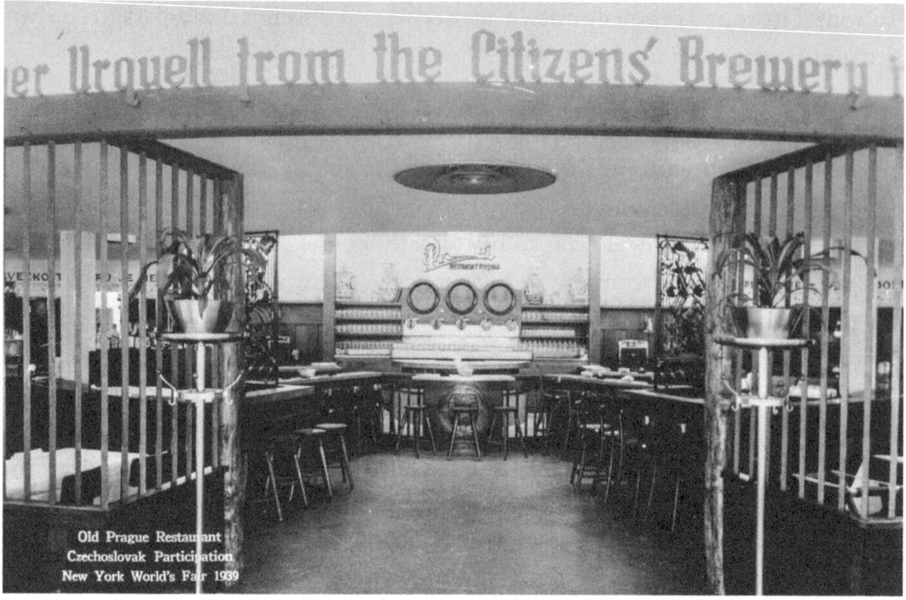

Fig.42. View of the Pilsner restaurant, Czechoslovak pavilion, New York, 1939-40.

pavilion's terrace, the beer hall was built as a wooden structure for 350 guests nearby and designed by the architects Jaromír Krejcar and Eva Štrimplová and paid for by the brewery.[62] The pavilion reportedly sold 1000h of beer at affordable prices and the beverage was awarded another Grand Prix at the Exhibition.[63]

The importance of beer for retaining a sense of national continuity became prominent at the time of the New York world's fair, after Czechoslovakia was first stripped of its borderlands with Germany and then disappeared from the political map completely. Before the Czechoslovak pavilion at this exhibition opened in a reduced form, the American press paid a lot of attention to the political circumstances that affected the display. It also focused on what was exhibited and what represented the surviving Czechoslovak nation without its Czechoslovak state. The *New York Evening Journal* noticed, for instance, that although Germany managed to "gobble up" the little republic, there were still "6,000 half barrels of Pilsner beer that Germany didn't get" and that could represent the nation.[64]

62 V. Darras, "Praktický průvodce po pařížské výstavě umění a techniky v moderním životě," *L'Indépendancee Tchécoslovaque*, no. 11 (1937): 3.
63 Kejha, Janouškovec and Jurina, *Plzeňský Prazdroj, příběh*, 122.
64 J.D. Kerkhoff, "Czech Pavilion, Flags at Half-Staff, Opens," *New York Evening Journal*, May 31, 1939, Series II. Ladislav Sutnar Papers, GRI.

The Old Prague Restaurant here therefore more or less retained the original outline of the national restaurant, which included an interior following the "modern Czecho-Slovak architectural lines," with an outdoor garden in a traditional style. The journalist further remarked that "waiters and waitress [were] dressed in native costumes" served, among others, Pilsner Urquell, "that world-famous beer...—a frothy nectar which tastes especially good with simple, well-cooked meals"[65] [fig. 42]. The combination of the beer, marketed as national, and the folk dress was a common feature in the national pavilions and restaurants. Here, above all, it also served as a reminder of the loss of statehood on the one hand and the perseverance of authenticity and tradition on the other.

Truly Bohemian glass?

By the late 1930s, exhibition organizers, beer producers and political representatives managed to forge an intimate association between the beverage and Czech (more than Czechoslovak) identity that successfully overwrote its history shared with Germans. As I mentioned earlier, Germans, for their substantial numbers, nevertheless continued to be an integral and important part of the population in Czechoslovakia, in many cities across the country and in the border areas between Czechoslovakia, Germany and Austria. Many of these regions had strategic industries and vital natural resources and it was in the borders, that another industry crucial for Austria-Hungary and later Czechoslovakia thrived. It was glassmaking that similarly to brewing had a shared German and Czech ethnic background, yet the inclusion of glass in world's fairs was framed differently from beer. Glass was a material product linked to local natural resources and this, in the understanding of many, gave it a strong affiliation with the land.

Crystal was discovered in Bohemia at the end of the seventeenth century and quickly grew in popularity. Glass making expanded especially in the nineteenth century, by the end of which there were about 100 glassworks in Bohemia, Moravia and Silesia.[66] At the end of the 1920s, there were some 140 glass making companies, producing a range of decorative glass, utility glass, bijou,

65 "Native delicacies to be served in Czecho-Slovakia's outdoor beer garden and restaurant at the New York World's Fair 1939," Release prepared for E. F. Roosevelt, Acting Director, Foreign Government Participation, Series I.A. Administration, I. Central Files, New York World's Fair 1939 and 1940 Incorporated Records 1935–1945, New York Public Library.

66 Petr Nový, Milan Glaveš, and Patrik Illo, *Dva v jednom, 1918–2018. Design českého a slovenského skla* (Jablonec nad Nisou: Muzeum skla a bižuterie v Jablonci nad Nisou, 2018), 15.

sheet glass, mirrors, bottles and construction glass.[67] Until the radical expulsion of Germans from Czechoslovakia after the Second World War, most glass factories in the so-called Sudetenland were owned and run by Germans who controlled the economy in places like Karlovy Vary in eastern Bohemia, and Jablonec nad Nisou, Harrachov, and Kamenický Šenov in the north and northwest of Bohemia.[68]

Just like was the case in breweries, Czechs were often employed as factory workers in glassworks until 1938 when they were pushed inland after the annexation of Sudetenland by Germany. Many companies selling and/or making glass originated in these regions and eventually relocated to Austria or other countries, either during the 1930s or in the mid-1940s. These included famous brands like Rückl and sons from Nižbor in Central Bohemia, Johann Lötz Witwe of Klášterský Mlýn in southwest Bohemia, or the northern Bohemian branch of Lobmeyr in Kamenický Šenov and Harrach glass from Neuwelt or Nový Svět also in the north.

Even though there were many glassmakers in Slovakia, they were often swiped under the umbrella term Czechoslovak glass that was occasionally used. Based mainly under the Tatra mountains, glass production in Slovakia was one of the main industries that suffered from the loss of its traditional markets, especially those in the former Greater Hungary.[69] The entire Czechoslovakia had to solicit new trading opportunities for its extensive glass industry, as it inherited over 90% of all glass production of the former Austria-Hungary.[70] Between the wars, Czechoslovakia was the second largest exporter of glass in Europe behind Germany and before Belgium, with the main markets in the USA, Germany and the United Kingdom including their colonies.[71] The sale and production of glass was, like many other luxury products, harshly affected by the global economic crisis in the late 1920s which caused a huge drop in exports and in employee numbers.

Glass became especially susceptible to nationalized narratives in the interwar state and inevitably in the space of the national pavilions. Previously referred to as Bohemian glass, Czechoslovakia started marketing glass as Czech,

67 Nový, Glaveš, and Illo, *Dva v jednom* 62.
68 Helmut Ricke, ed., *Czech Glass, 1945–1980: Design in an Age of Adversity* (Düsseldorf Stuttgart: Museum Kunst Palast; Arnoldsche, 2005), 27; Marcus Newhall, *Sklo Union: Art before Industry: 20th Century Czech Pressed Glass* (Braintree: Hope Fountain, 2008), 13–28.
69 Nový, Glaveš and Illo, *Dva v jednom*, 49.
70 Antonín Langhamer, *Legenda o českém skle – The Legend of Bohemian Glass – Legende vom böhmischen Glas* (Zlín: Tigris, 1999), 123.
71 Petr Nový, Milan Glaveš, Patrik Illo, *Dva v jednom*, 59, 62.

which was ascribed distinctive national qualities. For example, in 1933, Alois Meteláк, an architect, glass designer and director of a glass-making school in Železný Brod, viewed glass as a material, in which "each nation imprinted some of its soul and its sentiments. Italians [imprinted] their lightness, French their elegance, Swedes their seriousness, Germans their technical perfection."[72] Czech glass was in his view typical of the sense of color, harmony of shapes, liveliness and, like all glassmaking, it grew out of local traditions and the homeland, where the homeland was principally understood as Bohemia.[73]

Exhibiting glass: Moser

These narratives were built with the help of the Glass Institute, which was founded in Hradec Králové in 1922 as a state research institute. One of its roles was to promote glass; it organized several national and international presentations, including a comprehensive National Exhibition of Glass that took place in Prague in 1933 and a very successful display at Brussels in 1935, which featured for instance an aquarium with miniatures of glass flora and fauna by the designer Jaroslav Brychta.[74] Here, as well as in other international exhibitions and world's fairs, Czechoslovakia showcased glass, bijou and pearls either in the exposition of specialized schools, as part of displays of art and design associations, like the Prague-based Association for Czechoslovak Werkbund and Artěl, or as presentations of individual companies.

As I have discussed in Chapter Two in relation to the displays of glass in Paris of 1925 and 1937, glass could be incorporated into both the narrative of decorativeness and modernity of the Czechoslovak state. In the more conservative exhibit at the Exhibition of Modern Decorative Arts and Industries in 1925, Czechoslovakia showcased glass tiles on the exterior of the pavilions, chandeliers in the interior, a collection of glass in the Grand Palais and a section of specialized schools in a separate pavilion. The Czechoslovak glass exhibits faced strong competition here in for example René Lalique's fountain, his interiors, and chandeliers, as well as in the presentations of Sweden, Austria and Belgium.[75]

72 Alois Metalák, "Drobty o našem sklářství," *Národní listy* 73, no. 163, August 23, 1933, 5.

73 Marta Filipová, "Czech Glass or Bohemian Crystal? The Nationality of Design in the Czech Context," in *Designing Worlds: National Design Histories in an Age of Globalization*, eds. by Kjetil Fallan and Grace Lees-Maffei (New York: Berghahn Books, 2016), 141–155,.

74 V[áclav] Čtyroký, "Československé sklářství v Bruselu," *Sklářské rozhledy* XII, no. 6–7 (1935): 82–85.

75 Otakar Webr, "Sklo na mezinárodní výstavě moderních a dekorativních umění v Paříži," *Sklářské rozhledy* III, no. 1 (1926): 8–9.

Germany was not represented at this exhibition, but the Austrian pavilion included Lobmeyr's products like glasses, vases and plates. The displays by other countries provided material for comparison and were often discussed in local newspapers that saw them as a crucial inspiration for further design and content of the Czechoslovak displays.

Until today, one of the most recognizable brands with origins in Bohemia that has been very successful at international exhibitions is the company Moser. Its presence at these events, and its history more generally, was directly affected by political circumstances and ethnic conflicts before the Second World War. The company was founded by Ludwig Moser (1833–1916) in the spa town of Karlovy Vary in western Bohemia in 1857. Moser is one of the many Jewish families that opened their businesses after Jewish inhabitants were granted full civil rights in Austria-Hungary in 1848. Other examples that also frequently appeared in international exhibitions include for instance the Waldes family of the haberdashery Koh-i-noor, the Gutmann brothers and the Rothschild family of the Vítkovice steelworks or the Fürth family of Solo Sušice producing matches[76] [fig. 43–44].

The company was first established as an engraving workshop and a shop and grew into an international business. A glass manufacture was added in 1893 and the company grew into a large enterprise that focused on more affluent cus-

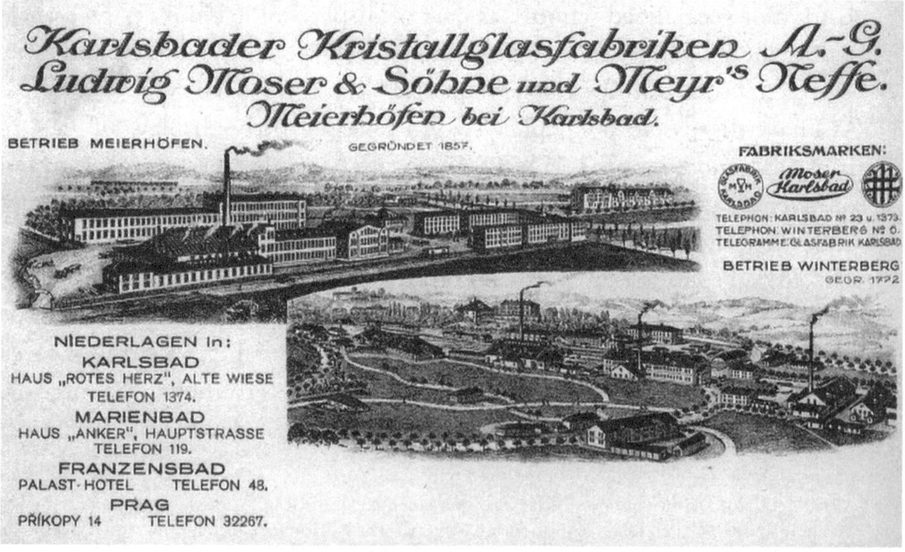

Fig. 43. Advertising of the Glass Company Moser.

76 Tomáš Pěkný, *Historie Židů v Čechách a na Moravě* (Prague: Sefer, 2001), 317.

tomers with their luxurious objects.[77] At the Viennese Weltausstellung in 1873, Moser featured glass engraved with genre scenes based on graphic arts based on templates from Germany. Following its success at Vienna, the company was made the official supplier to the Habsburg imperial court.[78] At the beginning of the twentieth century, it also became the court supplier for the Persian Shah (1901) and to the King of England (1908).[79]

Moser took part in several further exhibitions across the world and at many of these events, the company was awarded high prizes. Moser also featured in much more remote places from the Central European point of view, like Tasmania International Exhibition of 1891–92 and at the International Exhibition on the Isle of Man in 1892.[80] Its geopolitical inclusion at these events was not always consistent; in Tasmania, for instance, Moser was part of the Austrian section, away from a separate Bohemian display of glass and ceramics from the north Bohemian town of Nový Bor.[81]

The company had specialized in luxurious goods and targeted customers that could afford such items. It therefore started working with recognized designers and artists, like Heinrich Hussmann, Ludvika Smrčková and Rudolf Eschler. Smrčková, for instance, provided several designs displayed already at the 1925 exhibition in Paris. She continued to work with the company even after Second World War, when it was nationalized and renamed Karlovarské sklo (The Karlovy Vary Glass). In 1922, Moser started a collaboration with the Wiener Werkstätte, that had retained its visual language of the prewar years into the interwar period and hence was sometimes criticized as outlived.[82] More generally, Moser's high-end products tended to adhere to more conservative and traditional decorations of their objects; with clear crystal being cut, engraved and decorated with their patented oroplastique décor, a deep-surface engraving often with a golden strip.[83]

77 Milan Hlaveš, "Moser," in *Design v českých zemích*, 582.

78 Jan Mergl and Lenka Merglová Pánková, *Moser Crystal Story: 1857–2017: 160 Years* (Karlovy Vary: Moser, 2017), 32.

79 Mergl and Merglová Pánková, *Moser*, 15.

80 Mergl and Merglová Pánková, *Moser*, 52. Such exhibitions included the Centennial Exposition in Philadelphia of 1876, The Third International Exposition in Paris of 1878 and the expositions universelles in the same city of 1889 and 1900, The World's Industrial and Cotton Centennial Exposition in New Orleans of 1884–85, The Melbourne Centennial International Exhibition of 1888, and Louisianna Purchase Exhibition in St. Louis in 1904.

81 *Official Record of the Tasmanian International Exhibition, Held at Launceston, 1891–92* (Tasmania, 1893), 94.

82 Adolf Loos, "Das Wiener Weh," lecture quoted in *Neues Wiener Journal*, April 23, 1927, and *Neues Wiener Tagblatt*, April 21, 1927, in Werner J. Schweiger, *Wiener Werkstätte: Designs in Vienna 1903–1932* (London: Thames and Hudson 1984), 118.

83 Mergl and Merglová, *Moser*, 32.

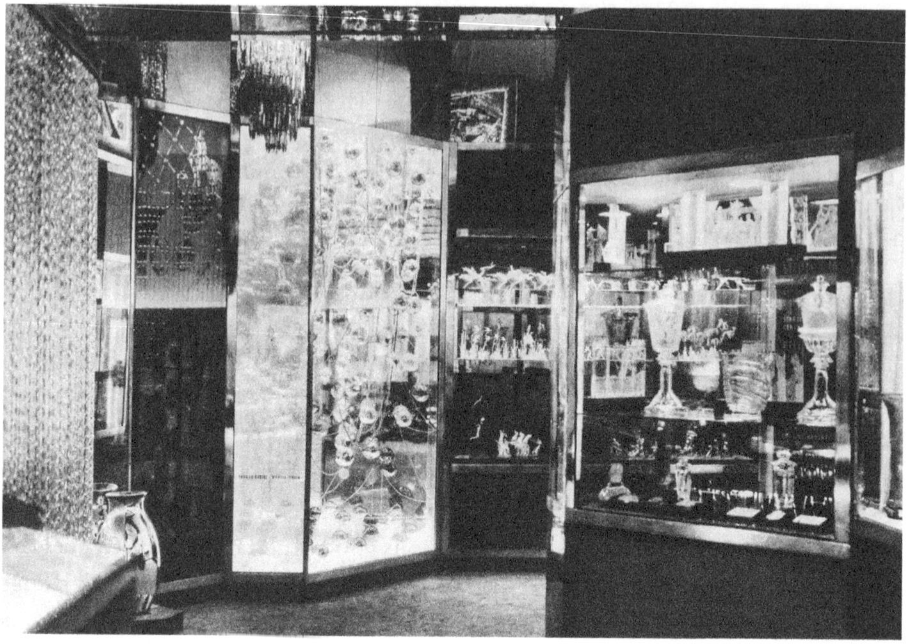

Fig.44. Glass display at the Czechoslovak pavilion in 1937.

Their designs were of the so-called art glass that was primarily meant for decoration and representation, not as everyday utility objects. The exception was Eschler's functionalist designs that came to the fore in the late 1930s and were aimed at middle class consumers. During the interwar period, the company also continued to successfully exhibit their products like engraved or cut, clear or colored glasses and carafes as well as vases decorated by enamel and gold under the label of Czechoslovakia. It was often awarded medals and prizes, and this included a golden medal in Paris in 1925 and Grand Prixes at the exhibitions in Brussels in 1935 and in Paris in 1937 for their "cut, engraved and other glassware and crystals"[84] [fig. 44].

Moser's association with the Czechoslovak state and the national pavilions came to an end when the Sudetenland were ceded to Germany in 1938. Many glass factories found themselves not any more on the territory of Czechoslovakia but in Germany or in the Protectorate of Bohemia and Moravia. Czechoslovakia, or what was left of it, lost the entire production of plate, mirror and bottle glass, alongside about half of the blown glass production, a large part of

84 *Exposition internationale des arts et des techniques dans la vie moderne, Paris 1937, Catalogue Officiel*, vol. II (Paris: M. Déchaux, 1937), 835.

bijou making of Jablonec nad Nisou and several training schools. Out of 93 companies, 60 came under the German Reich together with their employees.[85] First, plans were made to replace and substitute these losses with existing companies and institutions inland, which would also represent the state at the New York world's fair in 1939. The display of specialized schools, art glass, utility glass, bijou, and Christmas decorations, aimed at export to the USA, however, fell victim to further expansion of Germany and the eventual political disappearance of Czechoslovakia.

Moser as a company was gravely affected too. After the Munich Agreement of September 1938, Moser glassworks—based in Karlovy Vary/Karlsbad in the Sudetenland—were nationalized and affiliated with other German glass manufacturers. This was preceded by the departure of the Jewish family, the Moser brothers Leo, Richard and Gustav, from the management of the company already in the early 1930s in response to increasing nationalism in the city and the region.[86] Many other Jewish entrepreneurs had also left central Europe, often for the United States. Moser products nevertheless featured in exhibitions of German industries, and this included the 7th Triennale in Milan that took place in November 1938. Here, the vases of the Prague-born designer Wolfgang von Wersin made by Moser received a Grand Prix.[87] Subsequent administration changed several times.[88] Despite the Jewish origins of the family, their name was kept in the company's title Ludwig Moser & Söhne, Karlsbader Kristallglasfabriken AG until 1941, when it became part of the Staatliche Glasmanufaktur Karlsbad AG with a seat in Berlin.[89]

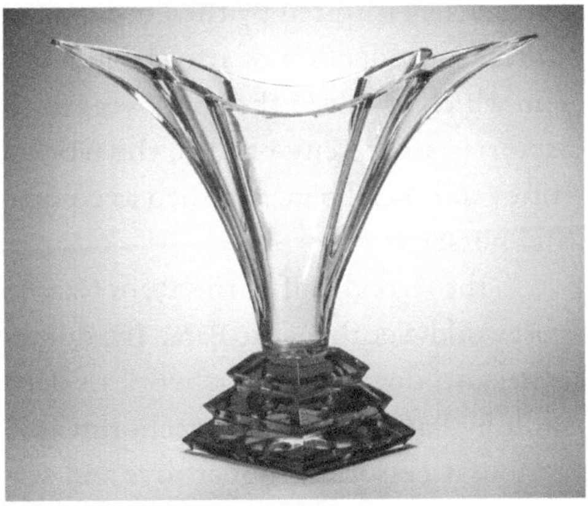

Fig.45. Vase - Fan by Lotte P. Moser, 1925, purple-red royalit glass, height 21.1cm.

The history of Moser glass is therefore a turbulent one, linked to its location in the border regions and Jewish origins, which never-

85 V[áclav] Čtyroký, "Naše sklářství v novém státě," *Sklářské rozhledy* XV, no. 1 (1938): 168.

86 Cf. Martin Wein, *History of the Jews in the Bohemian Lands* (Leiden - Boston: Brill, 2015), Kateřina Čapková, *Czechs, Germans, Jews?: National Identity and the Jews of Bohemia* (New York: Berghahn Books, 2015).

87 Mergl and Merglová, *Moser*, 178.

88 Mergl and Merglová, *Moser*, 176.

89 After the Second World War, the company was nationalized again and became Czechoslovak.

theless did not impact its success at international exhibitions regardless of which national pavilion or section it came under. Glass was able to represent not only specific companies like Moser, but also negotiate the changing political entities and their economic/national interests under the auspices of Austria, Bohemia, Czechoslovakia and Germany.

Building a company, exhibition and the state: Baťa

Detva, Pilsner and Moser represent not only specific branches of important industries in interwar Czechoslovakia but also the way specific products at world's fairs were framed as part of the national culture. One of the most prominent participants in interwar world's fairs was Baťa. The company built a strong presence there alongside a particular view of how such participation should look like. The family business had a seat in the city of Zlín in eastern Moravia where it started its shoe production in 1894. The owners quickly responded to the changing needs of the market, including shortages in the First World War, by using cheaper materials and faster production techniques. Tomáš Baťa, who was one of the founders and the company's director between 1894 and 1932, was particularly inspired by the Fordian and Taylorian ways of running a business. The mass production of affordable shoes was part of an elaborate economic system which involved the employees in the profits (and losses). In the seat of the factories, in the city of Zlín, the workers could also benefit from various amenities and social benefits, such as company housing, training schools, hospitals and nurseries.[90]

In the second half of the 1920s Czechoslovakia became number one in shoe export worldwide thanks to Baťa. The domestic success in shoe production and sales came also down to the network of department stores with an identical image built by Baťa all around Czechoslovakia. They sold not only footwear but also various accessories, rubber toys and services, such as pedicure and shoe repair. However, as the company offered cheaper products and services, they often came at the expense of many local cobblers and small producers.[91]

---

90 Kárník, *České země* I, 251; Jane Pavitt, "The Baťa Project: A Social and Industrial Experiment," *Twentieth Century Architecture no. 1, Industrial Architecture* (1994): 31–44. See also Leda Papastefanaki and Nikos Potamianos, *Labour History in the Semi-Periphery: Southern Europe, 19th-20th Centuries* (Munich and Vienna: De Gruyter Oldenbourg, 2020); Ondřej Ševeček, *Zrození Baťovy průmyslové metropole továrna, městský prostor a společnost ve Zlíně v letech 1900–1938* (Ostrava: Veduta, 2009).

91 Egon Erwin Kisch, *Továrna na boty (Socialisovaný Baťa): K přejmenování nár. podniku Baťa na Svit, nár. podnik* (Gottwaldov: Svit, n.p., 1949), 13.

During the interwar period, Baťa grew into a position of significant influence and confidence which reflected in their boldness to offer their own vision for a national pavilion in the late 1930s that I have already mentioned. It reached this position gradually though, building up its place first as an integral part of the Czechoslovak pavilions. Not much room was given to individual companies to construct their own presentations in the national pavilion space. This job was reserved to the designers and organizers of the different sections who then negotiated the content and placement of the exhibits.

At the same time though, as discussed in the previous chapter, the ideal plan was not always translated into practice, and spatial distribution had to be adjusted based on various demands. In Paris in 1937 where the exhibition had a clear trade focus, for example, company strategies for export occupied three quarters of the pavilion, showcasing glass, ceramics, textiles, fashion, machinery, steel products, musical instruments, bijou and Baťa, which was here synonymous for the national shoe industry.[92] Baťa's position was therefore rather unique; over the interwar period, Baťa participated in the Czechoslovak pavilions most prominently in the 1930s when the company not only promoted its products and services but also its views of social welfare.

One of the first instances of involvement of the company in national displays was the Exhibition of Contemporary Culture that took place in Brno in 1928 that commemorated ten years of the existence of Czechoslovakia emphasizing the greatest achievements of the new entity and its vision for the future. Baťa played an important role in co-financing the reconstitution of the five-meter-tall mammoth, parts of which had been discovered by the local paleontologist and archaeologist Karel Absolon in southern Moravia. The mammoth was displayed in a pavilion entitled Man and Mankind and became a sensation at the Exhibition and I come back to the rationale behind the exhibit more closely in Chapter Six in relation to how the evolution of the Czechoslovak nation was constructed in exhibitions.

Aside from the Brno Exhibition, the company had been active at other fairs across Czechoslovakia too; it frequently took part in the annual Prague trade fairs and regional events, such as the large Slovácko exhibition in Uherské Hradiště in 1937. It was, however, foreign participation that gave the Baťa company the status of a credible, truly international exhibitor. After its first sizeable display at Brussels in 1935, designed by Antonín Heythum and featuring model shoes,

92 "Náčrt programu čsl. oddělení mezinárodní výstavy v Paříži," *Mezinárodní výstava Umění a technika v moderním životě v Paříži 1937*, Division III, Section 2, MFA Archives.

photographs as well as rubber flooring made by the company, Baťa was represented at several venues at the 1937 Paris Exposition.

A memorandum, signed in October 1936 by J. A. Baťa, Ladislav Sutnar and Jaromír Krejcar, amongst others, reveals the approach to exhibiting that the company chief manager and the designers of the Czechoslovak pavilion subscribed to. It is essential, they held,

> that already today we are represented not only in our pavilion but at all national and international departments of the exhibition, especially which concern: sociology, construction, engineering, schools, and demonstration of shoemaking in various European and non-European states.[93]

Exhibitions were no longer about displaying commodities; they branched out to emphasize for example social or educational aspects of national life that the state (and its companies) stood for.

Responding to this call in the Industrial Hall of the Czechoslovak national pavilion in 1937, Baťa showcased a single shoe, which I have already discussed in Chapter Two, alongside large-scale photographs from life in Zlín and work in the factories. The Urbanism pavilion at the fair contained Baťa's display of modern housing and construction in the city of Zlín as well as twenty photographs placed on an illuminated column. Photographs and exhibits of different types of dwellings, social and medical care, youth care and public social facilities featured in the pavilion of Medicine.[94] On top of that, the company built its own pavilion affiliated with the French branch of Baťa in Hellocourt, in which the display cabinets were filled with model shoes complemented by photographs of Zlín that often served as examples of modern and hygienic urban living.

Zlín was a useful example of what the future should look like, including lower taxes, a fully motorized society, better traffic links to the outside world, technical and administrative reforms, and business and vocational training school reforms. J. A. Baťa further specified these visions in a publication *Budujme stát pro 40,000,000 lidí* (*Lets Build a State for 40,000,000 People*) published in 1937.[95] He offered his idea of a state connected by infrastructure of roads, railways and waterways with the aim of building a "strong and economically united Czechoslovak state," which would be "the healthiest, strongest and wealthiest state in

93 "Pp. Karfík, Gahura, Vojta Baťa, Ševčík, Grác, Kadlec, Novosad, Rojt, Meisl." October 24, 1936, Baťa, X. prodejní oddělení, inv. no. 208, MRA in Brno, branch Zlín.
94 "Zlín na světové výstavě v Paříži," *Zlín*, March 8, 1937 1.
95 J. A. Baťa, *Budujme stát pro 40,000,000 lidí* (Zlín: Tisk, 1937).

Europe."[96] Alongside the new infrastructure, Baťa also suggested changes to news reporting, financial and commerce administration, and the legal system. The darker side of his modernist vision, which invited criticism from for example President Beneš, was the suggestion of establishing labor camps for the unemployed, modelled on the New Deal civilian camps in the USA or radical changes in the tax system.[97]

Business with pleasure

The ambition of the company translated also into the design of its displays, whether at exhibitions or shops, which were often commissioned from recognized architects. One of them was Le Corbusier who had devised a template for Baťa stores (never realized though) and who was also a member of the committee for selecting a model house to be used around Zlín.[98] He proposed one of the alternative designs of the Hellocourt pavilion in Paris which was a cube combining a steel frame with glass. Its interior was an ambitiously conceived space which featured an airplane suspended from the ceiling, an attribute that also appeared in the Czechoslovak pavilion as well as in the Air Pavilion at the same exhibition.

In Le Corbusier's design, the Hellocourt pavilion offered more than an exhibition space and included a cinema and a pedicure salon, for which he also designed chairs. It seems that the architect wanted the visitors to have their pedicure done while watching commercials for Baťa from the comfort of lying back in their seats. While the Corbusier's chairs together with the plans for the entire pavilion did not materialize, the pedicure salon and cinema did. To appeal to the visitors, the company showed short films about Zlín and advertisements of Baťa products by recognized Czechoslovak directors. The company could build on its successful involvement in the film industry—it had already cooperated with prominent Czech avant-garde filmmakers Elmar Klos and Alexandr Hack-

96 Baťa, *Budujme stát*, 10.

97 Baťa, *Budujme stát*, 20–21; Zachary Austin Doleshal, "Life and Death in the Kingdom of Shoes: Zlín, Bat'a, and Czechoslovakia, 1923–1941," A Dissertation at the University of Texas at Austin, 2012, 60.

98 J. A. Baťa was professed authority in the company and was not to be challenged. See Martina Hrabová, "Více tváří mistra. Le Corbusier a Československo," *Art and Antiques* 2 (2012): 45–48; Ladislava Horňáková, "Le Corbusierův pobyt v meziválečném Zlíně a jeho vize pro firmu Baťa," in Milan Pitlach and Ladislava Horňáková, *Le Corbusier, Chandigahr* (Ostrava: Kabinet architektury: Galerie výtvarného umění v Ostravě, 2014), 45.

enschmied who made short commercial films for Baťa. Three of them were being shown at the Advertising pavilion in Paris and won one of the three Grand Prixes awarded to the company at the fair.[99]

The experience of the visit in the Baťa pavilion was therefore driven by an attempt to provide not only advertisement and knowledge of the products, but also a bit of relaxation and entertainment for the visitors. The political and ideological aims on the one hand and the entertainment value on the other had been an important aspect of exhibitions from the start and often penetrated exhibits across the exhibition grounds. As Patrick Geddes noted at the close of the 1900 Paris Exposition,

> here, after such unparalleled progress in increasing production, in extending communications, in freer and free trade, was the occasion of inventory and stock-taking; best of all, here was the greatest opportunity that the working world had ever seen of combining business with pleasure, of having the very best of market-days and holidays in one.[100]

How to successfully combine business with pleasure, and education with entertainment were therefore issues that organizers and exhibitors had to address.

Jan Antonín Baťa criticized the Czechoslovak pavilion in 1937 for lacking such engagement in the company newspaper *Zlín*.[101] He pointed out the uneven content and layout of the building, poor visitors' numbers as well as the insufficient representation of local industries. Looking into the future and the planned participation in New York, he proposed a pavilion that would represent the whole state as a vision of Baťa, based on the belief that a company could be synonymous with a state and, to an extent, with a nation.

"What should the pavilion in New York look like?" was a question that the Czechoslovak government and its representatives in charge of presentations of the state abroad asked. It was also posed by a Czech businessman who pondered the image of the Czechoslovak pavilion planned for the New York World's Fair in 1939.[102] He replied to his own question with a proposition that such a pavilion "needs to be technically perfect so that people who visit it could rest [here]

99 *Photo et Cinéma à l'Exposition de Paris 1937: Classe VI, Manifestations Cinématographiques*. Classe XIV, Photographie et Cinématographie (Paris, s.n. 1937), 130.

100 Patrick Geddes, "The Closing Exhibition – Paris 1900," *The Contemporary Review*, November 1900. Quoted in Greenhalgh, *Ephemeral Vistas*, 1.

101 J. A. Baťa, "Dva pavilony Světové výstavy v Paříži," *Zlín* 4, October 1937, 1.

102 J. A. Baťa, "Jak by měl vypadat čsl. pavilon v New Yorku," 26 May 1938, X. prodejní oddělení, inv. No. 208, f. 26, doc. 5493, MRA in Brno, branch Zlín.

and not get more tired by walking around. This could be done with mere wit and technology…"[103] The ensuing description of a possible Czechoslovak pavilion, he offered, went into a lot of detail, describing the internal layout and content. A ramp would lead through all the proposed twelve floors of the pavilions, carts would transport visitors along the representations of the main sites of Czechoslovakia, including Prague, the various mountain ranges, the cave regions, the river Danube or the city of Zlín. The carts would consist of 10 rows of 4 comfortable chairs each and would stop at three designated locations on each floor to allow disembarkation and embarkation of passengers. This resting ride would generate a fee of 50c per person and bring a daily income of USD 60,000. And if anyone did not want to pay, they could simply walk through the pavilion on foot, the proposal suggested [fig. 46].

The plan for the 12-floor building, which would be 40 meters long and 80 meters wide, was drawn up by Jan Antonín Baťa, the head of the company Baťa that manufactured shoes, leather and rubber products, and was also involved in finance, services and transportation industries.[104] By 1939, the Baťa company grew into an

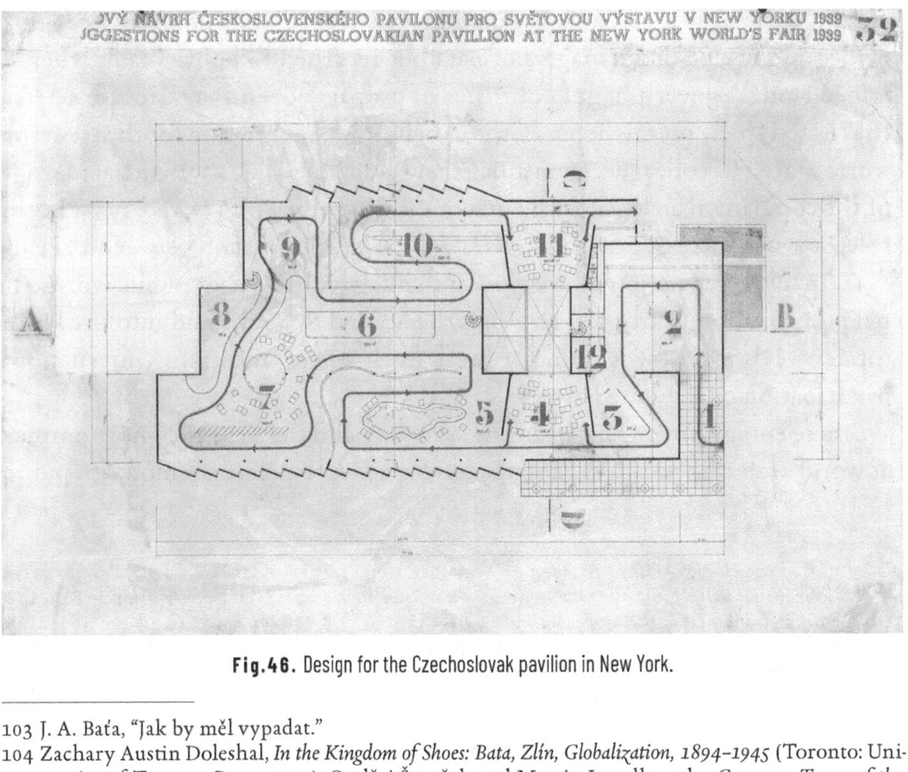

Fig.46. Design for the Czechoslovak pavilion in New York.

103 J. A. Baťa, "Jak by měl vypadat."
104 Zachary Austin Doleshal, *In the Kingdom of Shoes: Bata, Zlín, Globalization, 1894–1945* (Toronto: University of Toronto Press, 2021); Ondřej Ševeček and Martin Jemelka, eds., *Company Towns of the*

enormous enterprise, not only with branches of shoe shops in almost every town in Czechoslovakia but also expanding abroad. Baťa built factories in for example South Africa, Brazil, Canada and India and some of them it also constructed whole towns for its workers.[105] As an establishment with such an international outreach, it cultivated intimate relations with the Czechoslovak government. Equally, the Czechoslovak government was aware of the strategic importance of Baťa which it supported for the benefits to the state. President Masaryk visited Zlín several times and especially his trip in 1928, the year of the tenth anniversary of the state and the Exhibition of Contemporary Culture in Brno, was seen as an endorsement of Baťa activities in business, social life and culture.

The ambitious trajectory the company followed in its business expansion during the interwar period could be easily read alongside the aspirations of Czechoslovakia to become a recognized partner on the international stage of global politics and economy. Besides, many of the Baťa executives were also politically active and had therefore insight of and access to state matters, which included participation in world's fairs. Hugo Vavrečka (1880–1952), for instance, acted as an economy expert, representing Czechoslovakia at the Peace conference in Paris in 1919–20, and as a diplomat in Vienna and Hamburg.[106] In 1932 he became one of the managers of Baťa but returned to politics later when he joined Hodža's government as minister of propaganda in 1938. He also acted as the elected chief commissioner of the Czechoslovak exhibition for the New York world's fair. His colleague, Dominik Čipera (1893–1963) was a financial manager of the company who became the mayor of Zlín between 1932 and 1944. In this role, he was a successor of Tomáš Baťa who held this position between 1923 and 1932. In 1938, Čipera joined the government as a minister of public works and held the position until 1942, through the Second Republic and into the Protectorate.[107] His ministry was one of those in charge of organizing international participation of the state at world's fairs.

These entanglements between the company and the state became prominent at world's fairs especially in the 1930s and the fact that Baťa attempted to put for-

Baťa Concern. History – Cases – Architecture (Stuttgart: Steiner, 2013); Zdeněk Pokluda, *Ze Zlína do světa - příběh Tomáše Bati = From Zlin into the world - the story of Tomas Bata* (Zlín: Nadace Tomáše Bati, 2019).

105 Ševeček and Jemelka, eds. *Company Towns*, 39–44.

106 Vavrečka was also the grandfather of the post-1989 president Václav Havel. Cf. Zdeněk Pokluda, *Baťovi muži* (Zlín: Kovárna VIVA, 2012); Hugo Vavrečka, *Život je spíš román* (Třebíč: Arca JiMfa; Ostrava: Jasmín, 1997).

107 Monika Horáková, "Dominik Čipera - spolupracovník Tomáše Bati," in *Tomáš Baťa: Doba a společnost*, ed. Marek Tomaštík (Brno: Viribus Unitis, 2007); Dominik Čipera, *Ve službách práce a lidu* (Zlín: Tisk, 1944).

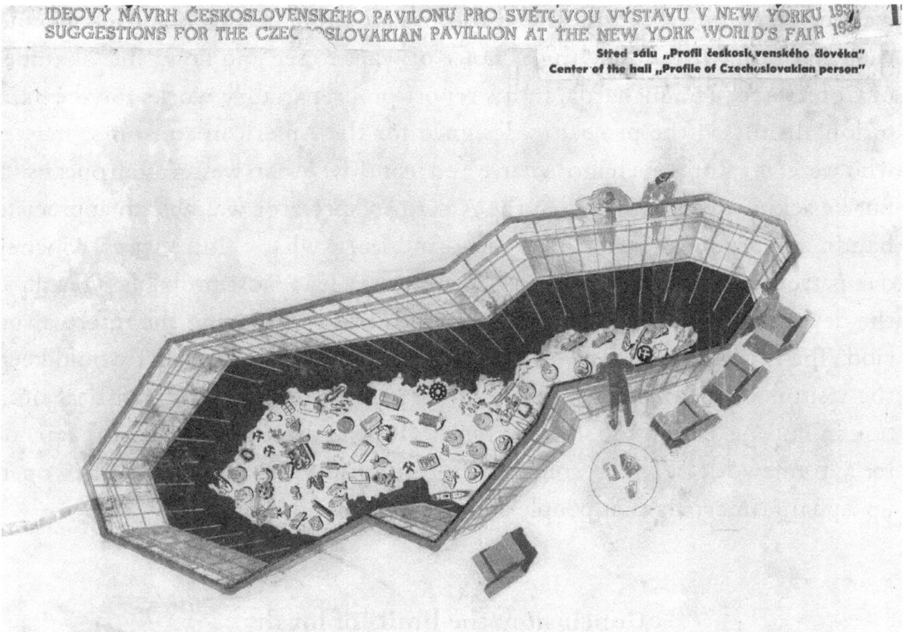

Fig.47. Exhibit Profile of Czechoslovak Person for the Czechoslovak pavilion in New York.

ward their own vision of the national pavilion in 1939 is symptomatic of them. This ambition makes Baťa unique in interwar Czechoslovakia but the access of a company to the state representation using personal contacts was quite common.

This connection was made visible in an exhibit called "The Profile of Czechoslovak Person" [sic], which was part of the Baťa proposal for New York and which represented the map of Czechoslovakia with its typical resources and features. Unlike the collage which was displayed on the wall of the Czechoslovak pavilion and which I discussed in the Introduction, Baťa's map would be inserted into the floor and could be observed from different levels [fig. 47]. It showed the location of key tourist sites like Prague or the spa towns, industries such as mining or extraction of timber or the birthplace of President Masaryk. Zlín was represented by the tallest structure in the country, the Building no. 21 designed by the architect Vladimír Karfík, and placed in the very center of the state.

Baťa's proposal for a pavilion which would represent the entire Czechoslovakia in 1939 was dismissed at the very start by the exhibition committee and especially the Ministry of Public Works.[108] Among the criticisms of the idea

---

108 Dr. Jan Sucharda, "Zpráva o stavu přípravných prací na výstavním pavilonu v New Yorku," April 19, 1938, Světová výstava v New Yorku 1939, inv. no. 1264, box 199, APRO.

was the fact that spectators seated in carts would be passively moved around the exposition without having a choice of what to see and how. The Chamber of Commerce's Jan Sucharda, in his report on preparatory works for the exhibition, dismissed the project as designed for the American consumer masses who were "creatures even more naïve and primitive" than western Europeans.[109] But he acknowledged that even the American spectator was able to appreciate handmade products of artistic value—and learn what culture was.[110] Sucharda's patronizing attitude towards the audiences was most probably a result of the development that world's fairs' grounds had seen during the interwar period. The NYWF's emphasis on attractions and consumerism that would keep the visitors entertained was partly a distraction from the tense political situation in Europe. Partly, too, it was a result of the realization that world's fairs offer a "framework in which commodities' intrinsic value is eclipsed. They open up a phantasmagoria that people enter to be amused."[111]

Conclusion: the limits of loyalty

The intensifying political and ethnic tensions of the late 1930s had a great impact on companies' affiliations with specific nations or states. While in some instances, businesses were subsumed under the new regime, others tried to negotiate the new environment to their economic benefit. This was the case of Baťa. When the company could not build its own vision of Czechoslovakia in 1939, it requested its exhibits to be located where that they would occupy a large proportion of the second floor of the planned national pavilion. The most prominent work here was an enormous stained glass of 1400 square feet, designed by the Czech painter and graphic designer Cyril Bouda. The curved glass depicted the story of the shoe giant Baťa entitled the "Hymn to Work" and I return to it in the concluding chapter. On the floor in front of the window lay a sizeable carpet depicting footsteps, hand-woven by the disabled patients of the Jedlička's Institute in Prague. Comfortable chairs on the side provided means for the visitor to relax.[112]

109 Sucharda, "Zpráva o stavu," 5.
110 Sucharda, "Zpráva o stavu," 5.
111 Benjamin, "Paris," 152.
112 "Čs. pavilon na svetovej výstave znázorňuje republiku aká bola," *Souvenir Programme of the Eleventh Slovak National Day* (Chicago: Mally Press, 1939).

While these plans were drawn up and fulfilled, Czechoslovakia as a state was experiencing a gradual political disappearance. This presented challenges also to the official, national display as well as to the exhibiting companies. In the setting where Czechoslovakia no longer existed as an independent state, the question of who should represent it and by what became more pressing than ever. While earlier in the 1930s, Baťa was happy to provide its own vision of the Czechoslovak state, and to an extent the nation, after the creation of the Protectorate and the Slovak republic, it quickly backed down from such associations.

A state that had vanished presented many problems for the organizers and the individual participating companies. The political restructuring raised many legal, economic and moral questions. The contract of participation at the world's fair became invalid, as the legal entity that entered it no longer existed. It became unclear who should pay the customs fees, shipping and storage, and who could provide insurance for objects from the now alien country. Similarly, ownership of objects that originated from a country that no longer existed was questioned. Yet such a situation was not new and repeated dilemmas that had presented themselves during the world's fair in 1915 in San Francisco. The First World War in Europe resulted in a radical political restructuring that led to uncertainties about many exhibits at the fair. Those from for example Austria, Germany or Finland could not be returned to the states they came from first because of the war as these states no longer existed in their prewar composition. Many exhibits from these countries were therefore withheld in the United States well into the 1920s.[113]

For Baťa, the situation around the New York participation was equally alarming. Right after the German occupation in March 1939, the new government in Prague, represented by Dominik Čipera, a Baťa manager, mayor of Zlín and a minister of public works, decided over the liquidation of the national hall and reconfirmed stopping all the works on the pavilion soon after.[114] However, the authority of the Protectorate government was not recognized in the USA and the national exhibition continued in honor of the state that vanished.

Following the orders from Prague, the chief manager of the department for exhibiting of Baťa, Josef Černovský, explicitly requested all the company exhibits to be removed from the pavilion. In an attempt to disassociate the exhibits

113 Records in Manuscripts Collections. Panama Pacific International Exposition, Records 1911-1929, BANC MSS C-A 190, carton 167, U.C. Berkley Bancroft Library.

114 Telegram from Malota to Černovský of April 21, 1939, Korespondence ředitelny fy Baťa s USA CZ z Čs. pavilonu v New Yorku – Světová výstava. I/4 Ředitelna inv.no. 412, MRA in Brno, branch Zlín.

from the Czechoslovak legacy, he described the hand-woven carpet and the glass windows as "purely private property" which should therefore be taken out of the Czechoslovak pavilion.[115] When his request for removal was not successful, Černovský suggested to the headquarters in Zlín to continue the exhibit, but not in what he described as an anti-German spirit.[116] The company Baťa, nevertheless, insisted on the disassociation from the pavilion, which the exhibition organizers working with the American authorities and the Czechoslovak embassy refused. The "German sovereignty over the provinces of Bohemia, Moravia and Slovakia" was not recognized by the American government and the Fair Corporation who continued to work with the "accredited representatives of the former government of Czechoslovakia," especially the former Czechoslovak ambassador Hurban.[117]

Baťa was not the only company that declined to participate in the pavilion as many other exhibitors and companies from former Czechoslovakia expressed their wish not to be included. Yet the quick separation of Baťa from being seen as representative of Czechoslovakia indicates that it decided to prioritize its business interests. They took priority over the company's identification with the state, which it had previously cultivated, and which culminated in their own vision of the 1939 Czechoslovak pavilion.

This course of events also indicates the vulnerable relationship between businesses and the political establishment they come to represent. While Baťa tried to steer away from being associated with Czechoslovakia, other companies were more directly exposed to the changing political circumstances. Beer remained a useful symbol of Czechoslovak identity, while glass from the border regions became incorporated into the German state. And Detva, after the struggles with the Hungarian competition in the interwar period, came to eventually represent the Slovak Republic after 1939. All these individual case studies, selected out of a broad range of Czechoslovak industries that had appeared regularly at interwar world's fairs, show that participation of companies in national pavilions was never innocent but always framed by affiliations to national groups and political systems.

115 Letter from Josef Černovský to George Janeček, 26 April 26, 1939, New York World's Fair 1939 and 1940 Incorporated records 1935–1945, Series I, P03.Foreign, b. 306, f. 2, entry no. 492, NYPL.

116 Telegram from Josef Černovský to Vavrečka, Hlavnička of April 28, 1939, The MRA in Brno, branch Zlín.

117 Letter from J. C. Holmes to Oliver M. Sayler, of 12 April 1939, New York World's Fair 1939 and 1940 Incorporated records 1935–1945, Series I, P03.Foreign, b. 306, f. 2, NYPL.

Chapter Five

PEOPLE OF THE FAIR

Most planning of the official presentation at world's fairs was done in Prague by committees of representatives from various ministries. After the effort to establish a permanent body that would oversee preparations of all international participation and secure the continuity of the presentations failed, each individual national pavilion created its own committee with its own vision, agenda and disputes. The committee would devise the narrative for the national pavilion, but the careful plans were often disrupted at different moments of the exhibition process, most often by political circumstances or on-the-spot circumstances.

Adjustments to layouts, location or inclusion of items for display, or the very nature of the exhibits often had to change according to the specific situation and conditions in the place of the world's fair. This could be the economic or political circumstances, like the Great Depression that had a direct impact on the fairs in Barcelona (1929) and Chicago (1933) or the start of the German expansion in Central Europe in 1938. The March Anschluss of Austria was followed by German annexation of the borderlands of Czechoslovakia, known as the Sudetenland, in September 1938. Czechoslovakia ultimately was split in March 1939 and the invasion of Poland took place in September 1939, which marked the start of the Second World War. The most extreme adjustments of national displays therefore took place at the last two world's fairs of the interwar period in New York and San Francisco in 1939/40, which were directly affected by these political circumstances.

From the moment of the first idea to participate in a world's fair, national pavilions had been organically developing and transforming. And while their modifications were often caused by external political and economic causes, it was individuals who enabled them at various stages, starting with the initial idea for an exhibit and ending with experiencing the exhibits. In the name of economy of expression, I have inevitably been referring to the state or the national pavilion as if they were a unanimous and active agent in many places throughout the book. These bodies, indeed, are mere umbrella terms for an array of individuals who represented them. Alexander T. Geppert identified five main types of actors whose agency can be detected in exhibitions. They were the initiators who proposed an exhibition or participation, the official organizers (including commissioners, the respective country's representatives and exhibitors), the local and foreign active participants, like employees and performers, reviewers and critics, and finally the audiences.[1] These categories, or rather roles included within them, would sometimes overlap and in the case of Czechoslovakia, such intersections took original forms, as I will show further.

I have already sketched out the mechanics of the organization process of national pavilions in Czechoslovakia in the first chapter. The long minutes of frequent meetings testify to a protruded initial stage, during which the committee composed of the initiators and organizers would meet regularly before eventually settling on the concept and content of the display. And while they attempted to construct a specific narrative of the display, it was often the various participants, reviewers and audiences that formulated the ultimate look and meaning of the pavilions and entire exhibitions. It is, after all, people who give meaning to exhibition spaces; they use them either according to or against the intended design, both spatial and behavioral. Exhibition space, like any other space, can be conceived of both as a physical and mental category, constructed by human agency.[2] Inhabiting and experiencing exhibition spaces does not have to be passive and visitors may actively "use" and interact with them or challenge the prescribed behavior. Agency can therefore be uncovered at all stages of exhibition projects.

A crucial role in creating the meaning of national displays and of entire fairs was therefore played by visitors and audiences. Their personal accounts focus on aspects of the exhibitions that differ from official reports for governments or companies; they highlight different aspects of exhibitions including the interplay of colors, range of smells or what one could buy.[3] This chapter is partly

1 Alexander C. T. Geppert, "Introduction: How to Read an Exposition," *Fleeting Cities. Imperial Expositions in Fin-de-Siecle Europe* (Palgrave Macmillan, 2010), 5–6.

2 Henri Lefebvre, *The Production of Space*, trans. Donald Nicholson-Smith (Oxford: Blackwell, 1991), 1–11.

3 Letter from Jiřina Václavek to her grandfather in Czechoslovakia of February 1, 1934, courtesy of Peggy Hazard.

built on some of these testimonies, where they survived, which reveal agency in the interaction between the human element and the physical or abstract concept of the exhibition. Examination of the involvement and participation of the different agents in various aspects of the national pavilion, agents that have often been excluded from official exhibition histories, also reveal broader social and political issues that come to the fore here.

My discussion so far has focused mostly on the physical layout and the content of the pavilions and now I would like to shift my attention to the less visible, yet equally crucial aspects of national representation. I argue that apart from the so-called exhibition elites, identified as the initiators, organizers and exhibitors, less prominent individuals played a crucial, yet not always recognized, role in building the national displays. One example from the 1922 exhibition in Rio de Janeiro gives a general flavor of the issue. During the planning stage, the Czechoslovak foreign office decided to send out several workers from Czechoslovakia to Brazil to help with the installation. Its memo set out what skills were solicited and further stipulated that "it would be highly desirable if one of these workers would bring his wife who could be used for minor works in the exhibition, such as dusting the exhibits."[4] A carpenter named Jaroslav Bohdalecký, who was contracted for the work was therefore able to bring his wife Františka to Rio de Janeiro. With a passport, a health certificate and a vaccination card, Jaroslav and his wife left for Brazil at the expense of the Czechoslovak government. The state also redeemed Jaroslav's loss of income at home, wear and tear of his tools and the cost of acquiring a cap with an inscription "Pavilhão Tchecoslovaco" against presenting respective receipts.[5]

The anecdote is significant in several respects. First, it shows that employees of the Czechoslovak national pavilions were paid to travel from Czechoslovakia across the world to do work that local workers could easily perform. Second, the primary emphasis was on male labor; women, who came as wives, were secondary benefits that could help with the maintenance. It may be assumed that because the Rio de Janeiro exhibition took place so soon after the end of the First World War and the political reconstruction of Europe, contacts with distant countries had not been re-established and it was simply easier to source a workforce from home. Rio de Janeiro may not have had such a rich pool of émigrés from Central Europe like the USA that the Czechoslovak government could

4 "Výstava v Rio de Janeiro. Jaroslav Bohdalecký, pracovní síla pro pavilon ČSR," Presidium pro zahraniční obchod. *Rio de Janeiro* no. 232 1921–22, NACR, no. 10128.

5 "Výstava v Rio de Janeiro."

easily call upon. Yet the auxiliary role that Františka was given is indicative of a larger issue of gender imbalance in the Czechoslovak (and many other) pavilions.

As I will show further, women were involved in world's fairs mainly as providers of services rather than initiators or leaders. In this chapter, I turn my attention to agents of exhibitions that could be considered as other, marginal or circumstantial. The notion of "otherness" in world's fairs has been fruitfully explored with attention to various ethnic minorities that were most often seen as subjects of display and less often as initiators, contesters of the master narratives of the fairs or their visitors.[6] Here, I expand the notion of otherness to include the so-called internal and external others to Czechoslovakia, in this case women, ethnic minorities and the diaspora. While they form groups that are impossible to compare, they share the feature that they were instrumental, yet little acknowledged, in the ideological building project of the Czechoslovak state and in the physical building of the national pavilions.

"Czechoslovaks" in the USA

Although this chapter examines several world's fairs to demonstrate how different actors interacted with the displays, my focus is The Century of Progress Exposition which took place in Chicago in 1933 and 1934. It was one of the most successful events in the history of world's fairs and Czechoslovak participations. Chicago was also the location of a very active Czech and Slovak diaspora and the place where the involvement of diverse groups, like women's organizations and ethnic communities, in the fair was very prominent. More than any other large exhibition in the interwar period, The Century of Progress provides a great case study for the exploration of how migration, gender and class affected national pavilions and world's fairs.

The Czech and Slovak diaspora in the United States was sizeable; in 1920 it amounted to 600,000 of Czechs and Slovaks each. Prior to the First World War,

---

6 Dagnosław Demski and Dominika Czarnecka, eds., *Staged Otherness: Ethnic Shows in Central and Eastern Europe, 1850–1939* (Budapest – New York: Central European University Press, 2021); Guido Abbattista, *Moving Bodies, Displaying Nations. National Cultures, Race and Gender in World Expositions 19th to 21st Century* (EUT Edizioni Università di Trieste, Trieste, 2014); Pascal Blanchard et al, eds., *Human Zoos: Science and Spectacle in the Age of Colonial Empires* (Liverpool: Liverpool University Press 2008); Debra Hanson, "Re-presenting the Arab-Islamic World at the Nineteenth-Century World's Fairs," in *Expanding Nationalism*, 15–32; Katelyn K. Know, *Race on Display in the 20th- and 21st Century France* (Liverpool: Liverpool University Press, 2016); Tracy Teslow, *Constructing Race: The Science of Bodies and Cultures in American Anthropology* (Cambridge: Cambridge University Press 2014).

Czechs had been settling primarily in Illinois, Ohio, Nebraska, New York, Texas, and Wisconsin, while Slovaks found homes in Pennsylvania, Ohio, New Jersey, New York, and Illinois.[7] Geographically detached from Central Europe, the diaspora played a significant role in both building the independent state in the war and constructing ideas of the nation during the interwar period.[8] However, their image of the homeland, which these communities had often not seen for decades or ever, differed from the interwar reality, as much as the image of the life abroad differed from reality in the minds of those who did not leave.[9]

The first generation left Austria-Hungary in the late 1800s and only experienced the creation of the nation-state of Czechoslovakia remotely. There were the exceptions of those who were involved in the efforts of Masaryk and his allies to raise American awareness about the calls for Czech—and eventually Czechoslovak—independence towards the end of the war. Yet the single Czechoslovak identity felt also remote to many and was not always part of the everyday narrative. The Czechoslovak National Council of America, for instance, with a seat in Chicago was founded to support the efforts to create an independent state but inconsistently referred in its communications to the inhabitants with different names. They were called Bohemians, Czechs and Slovaks, Czechoslovaks, Czechs and Slavs, or "Czecho-slovaks."[10]

There were also great generational divides among the émigrés that were important to how Czechoslovakia was viewed after 1918. Most of the first-generation Czechs in the United States self-identified as Czech based on their language. They maintained Czech associations such as Sokol, women's clubs, legionaries' clubs, community buildings, pubs and cooperatives.[11] Already at the end of the nineteenth century, the Czech Americans were engaged in lively cultural as well as trade affairs that involved the homeland. Taking part in world's fairs and exhibitions at both ends was an indispensable part of retaining the contacts. At the Columbian Exposition in Chicago in 1893, for instance, the Czechs invited the composer Antonín Dvořák—who was at the time a director of the National Conservatory of Music of America in New York—to conduct several pieces at

7 Thomas Čapek Jr., *Czechoslovak Immigration.* Classroom material. Service Bureau for Intercultural Education (New York) Chicago Public Library, 8.

8 Jan Auerhahn, "Pokus o demografii zahraničních Čechů a Slováků," in *Československá vlastivěda II. Národopis*, ed. Jiří Horák, Karel Chotek and Jindřich Matiegka (Prague: Sfinx 1936), 107.

9 Cf. Tara Zahra, *The Great Departure. Mass Migration from Eastern Europe and the Making of the Free World* (New York: W.W. Norton & Company, 2016), especially chapter 3, Happy and Unhappy Returns, 120-163.

10 Dagmar Hájková, *"Naše česká věc," Češi v Americe za první světové války* (Prague: NLN, 2011), 113.

11 Alena Jaklová, *Proměny etnické a kulturní identity českých přistěhovalců ve Spojených státech amerických. Interpretace analýz českoamerického periodického tisku 19. a 20. století* (Prague: ČVUT, 2014), 58.

the Bohemian Day on August 12.[12] In Europe, exhibits of Czech Americans appeared at the 1891 Jubilee Exhibition and the 1895 Czechoslavic Ethnographic Exhibition in Prague where the life of the émigrés was featured in the reconstruction of townhouses, farms, a church and a saloon.[13]

As many émigrés who originally came from Bohemia and upper Hungary lacked direct or indirect experience of the homeland, it was therefore the Czechoslovak pavilion built at American world's fairs that *externalized* it for them. The national pavilion became a crucial substitute for the country that the local diaspora would probably never see. This translated into an active involvement of many Czech and Slovak émigrés in various aspects of the Czechoslovak pavilions in the interwar period or the Bohemian sections before the First World War. Apart from the Columbian Exposition, they took part in the Louisiana Purchase Exhibition in St. Louis in 1904, the Panama-Pacific International Exposition in San Francisco in 1915 and, after the war at the Sesqui-Centennial Exhibition in Philadelphia in 1926, Chicago's Century of Progress in 1933 and the two fairs of 1939/40 in New York and San Francisco.

A painting under the arm

Philadelphia's Sesqui-Centennial Exhibition was the first world's fair in the USA after the war and after the introduction of new immigration laws in 1924 that restricted the number of immigrants into the country. The Czech and Slovak diaspora participated in the organization of the Czechoslovak pavilion and many individuals prominently intervened with the exhibit [fig. 48]. During the preparations of the Czechoslovak pavilion for Philadelphia, a Mr. Příbramský of the local Sokol branch, turned up at the exhibitions grounds, a large framed picture of President Masaryk under his arm.[14] Masaryk was depicted reading out the Declaration of Independence in front of the Independence Hall in Philadelphia on October 26, 1918. Stanislav Špaček, the building works supervisor of the Czechoslovak pavilion, reported in this way to Prague that the chief of the local Sokol organization turned up at the almost completed pavilion with the

12 "Slováci, zastúpení na Columbickej výstave 1893. Súčastnení na Českom dni 12. augusta 1893. Z pamäti Boženy Kosmanovej, rod. Mallý," *The Fifth Slovak National Day*, 1933, 77.

13 Marta Filipová, "Peasants on Display. The Czechoslavic Ethnographic Exhibition of 1895," *Journal of Design History* 24, no. 1 (2011): 15–36.

14 Stanislav Špaček, "XV. Zpráva. V. Konečná zpráva o instalaci československého pavilonu na výstavě ve Philadelphii," September 16, 1926, AKPR, file Světová výstava ve Filadelfii, D5531/26, 2

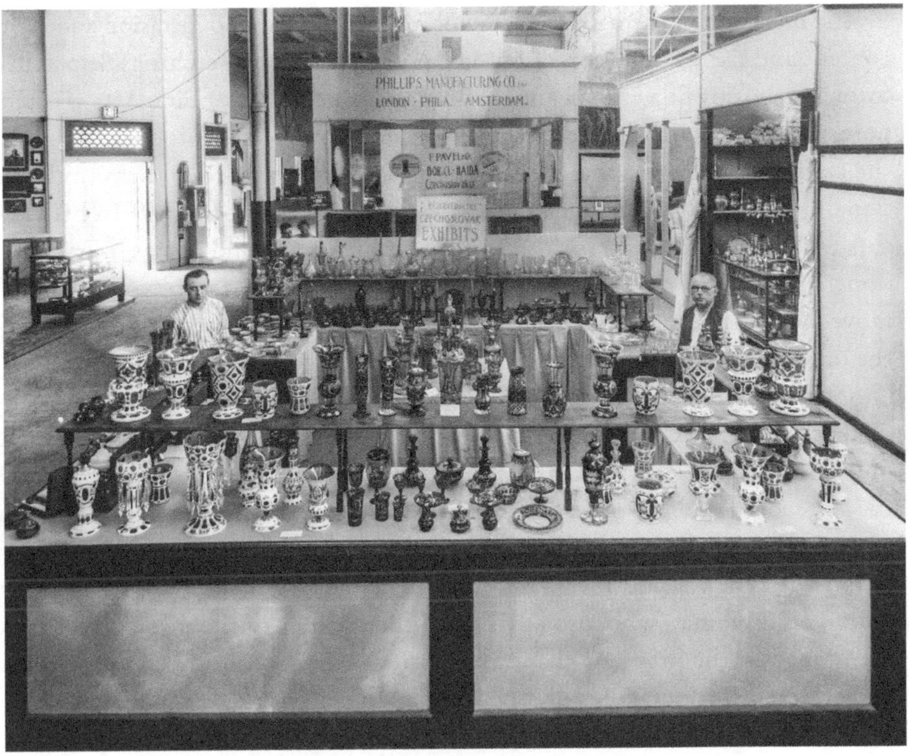

Fig.48. Interior. Czechoslovak exposition, Sesqui-Centennial Exposition, 1926.

artwork to be included in the display. Špaček accepted the painting and found a place for it in the pavilion.

One of the arguments for the Czechoslovak participation at this fair was that in October 1918 Tomáš Garrigue Masaryk had signed "The Declaration of Common Aims of the Independent Mid-European Nations" in Philadelphia. The document set out a collaboration of nations between the Baltic, the Adriatic and the Black Seas, e.g., Czechoslovakia, Yugoslavia, Poland, Ukraine as well as various ethnic groups like the Uhro-Rusyns and Zionists, in supporting their independence and preventing further wars.[15] Masaryk had then read out the Declaration bearing signatures of 11 other representatives on the stairs of the Independence Hall in the city. This was most probably the moment captured in the painting which was delivered by Příbramský and which is unfortunately unknown today.

15 "Declaration of Common Aims of the Independent Mid-European Nations," http://www.carpatho-rusyn.org/fame/proc.htm, accessed on July 3, 2023.

The inclusion of the painting was not his only on-the-spot decision about the content and the distribution of objects. A box of over 100 etchings dispatched from Czechoslovakia and meant for display in the Palace of Fine Arts was temporarily lost in transit. When it was eventually found, it was too late to install the etchings in the original location so Špaček made space for them in the national pavilion in an arrangement that he described as "pleasing." In his view, this was a better location for the works because they could receive more attention. To accommodate them, he had paintings of the Ministry of Railways moved to a different location because he saw them as too large and crowding the original space.

It was also the supervisor's decision to leave unboxed the heavy *Motorcyclist* by Otakar Švec sent here from the 1925 exposition in Paris, where it was awarded one of the prizes. Špaček explained that the sculpture "caused too much trouble, the cast would not give out a good impression to the public, hence ... the 'Motorcyclist' was boxed up and taken to storage."[16] He took these steps upon consultation with the Czechoslovak ambassador Zdeněk Fierlinger, a former legionnaire and diplomat, who also advised on the content of the display. As I have mentioned in the previous chapter, the sculpture was considered too modern and abstract in comparison to other objects in the national exhibit.

In the larger scheme of things, the interventions of Příbramský, Špaček and his fellow expatriates into the Czechoslovak display in Philadelphia may come across as insignificant. The additions, removals and relocations of a few objects did not affect the overall effect the display had. Yet such input highlights an important aspect of the exhibits, especially those organized in the United States. Their content and ideology were often shaped, or intended to be shaped, by the official or unofficial involvement of the local émigré communities which went far beyond their participation as visitors or performers.

Czechs and Slovaks of Chicago

As I mentioned, Chicago was one of the main centers of the Czech and Slovak diaspora that, in the interwar period, was most prominently engaged with the Century of Progress Exposition in 1933. The city had a community of about 200,000 Czechs around the time of the world's fair who arrived to work in the

16 Špaček, "XV. Zpráva," 1.

garment factories, the docks, meatpacking and steelmaking industries.[17] Here, the first substantial Czech neighborhood was symptomatically called Pilsen after the west-Bohemian beer producing city or rather, a pub referring to it, established there in the mid-1870s.[18] The Bohemian National Cemetery was founded here in 1877 and featured the first "Czech" crematorium dedicated in 1919. Émigrés also founded many clubs and associations and organized cultural and social events here. The social club Slovanská Lípa (The Slavic Linden) was founded in 1861 and sponsored Czech language theatre plays, Lyra, the workingmen's singing society in 1890. Catholic, Protestant and Freethinkers' schools were established, the latter dedicated to the Czech ethnographer and traveler Vojta Náprstek and one to T. G. Masaryk.[19] Active Sokol branches opened at the end of the 19th century and organized massive gatherings of the American Sokol Union gymnasts in 1925, 1929, 1933 and 1937.[20]

Chicago had also long been an important hub for politicians and when Masaryk arrived in the city in May 1918, he received a "royal" welcome from a parade of 20,000 American Czechs and Slovaks in national and Sokol costumes.[21] It was also the center of exile nationalism during the war not only for the Czechs but for other Slavs too.[22] Before 1918, the Poles of Chicago, for instance, also campaigned for the creation of independent Poland. And after the Polish state was founded, they kept their Polishness by printing Polish-language publications, being united in various organizations and by observing traditional customs.[23] The Polish national heroes like Tadeusz Kościuszko, the general who fought in the American War of Independence alongside Kazimierz Pulaski, were fondly commemorated by the community together with non-military personalities like the astronomer Copernicus. There were also frequent exchanges between the Polish and Czech communities; the Czechs as an older and more established group in the city often served as a model for the Poles and their institutions.[24]

17 Malynne M. Sternstein, *Czechs of Chicagoland* (Arcadia Publishing, 2008), 23.
18 Sternstein, *Czechs of Chicagoland*, 23.
19 John J. Reichman, *The Czechoslovaks of Chicago* (Chicago: Czechoslovak Historical Society of Illinois, 1937), 11.
20 Joseph Čermák, "Physical Education among Czechs and Slovaks in America," 42–45, in *Czechs and Slovaks: World's Fair Memorial of the Czechoslovak Group* (Chicago, 1933).
21 J. J. Zmrhal, "Masaryk v Americe," in *TGM 1850–1950 sborník*, ed. Otakar Odložník (Chicago: Národní jednota čsl. protestantů v Americe a v Kanadě. 1950), Chicago Public Library, 19.
22 Tara Zahra, *The Great Departure: Mass Migration from Eastern Europe and the Making of the Free World* (New York: W.W.Norton, 2017), 106–107.
23 Dominic A. Pacyga, *American Warsaw: The Rise, Fall and Rebirth of Polish Chicago* (Chicago: Chicago University Press, 2021).
24 Pacyga, *American Warsaw*, 105.

Judging from the vast number of Czech and Slovak publications printed in Chicago and Illinois in the interwar period, the community was thriving and keen on preserving its languages. The range of publications included titles like *Svornost* (Unity), "mouthpiece of liberalism," the socialist *Spravedlnost* (Justice), the Catholic daily *Národ* (Nation), the semi-weekly *Katolík* (The Catholic), and many others. The daily *Denní Hlasatel* (The Daily Tribune) and its semi-weekly agricultural version the *Hlasatel* (The Tribune) were also linked to a radio broadcast in Czech and had a wide reach.[25] This variety also reflected the political, religious and social diversity of the communities, catering for freethinkers, Catholics, Protestants, legionnaires, and women's clubs.[26] The newspapers, weeklies, calendars and magazines were also full of advertisements for local businesses run by Czechs and Slovaks. They were comprised of cafés and pubs, groceries and breweries like the Atlas, fashion and accessories shops, including branches of Baťa, various investment companies or repair services. These lists indicate that the diaspora expanded from manual work to owning businesses, surgeries and law firms with better economic self-sufficiency.[27]

Many members of the local Czech and Slovak diaspora welcomed the prospect of a Czechoslovak pavilion at the 1933 world's fair and engaged with the official Czechoslovak representation in Chicago in several ways. While the design of the pavilion and most of the content came from Czechoslovakia, the diaspora was involved as organizers, employees, performers and, indeed, visitors, to recall Geppert's division of the different actors in world's fairs. Initially, the participation of the émigrés in Chicago was planned independently of the governmental plans in Czechoslovakia. They had been actively calling for a Czechoslovak pavilion in the early stages of the fair, by which they fought the initial reluctance of the Czechoslovak government wary of the difficult economic climate of the early 1930s. This included a personal visit to Czechoslovakia by the Chicago mayor, Antonín Čermák. Having Czech roots, Čermák left for the USA with his family when he was a boy. In America, he became a businessman and a Democratic politician and eventually was elected mayor of Chicago in 1931. Czechoslovakia became the first stop on his tour of Europe to lobby for the participation of countries and companies in the planned world's fair. The primacy of Czechoslovakia on the itinerary was interpreted in his homeland as proof

25 Reichman, *The Czechoslovaks of Chicago*, 60.
26 Reichman, *The Czechoslovaks*, 8
27 Jaroslav E. S. Vojan, "How the Czechs Helped Build the American Civilization," 28–34, in *The Czechoslovaks of Chicago*, ed. by John R. Reichman (Chicago, 1937).

that the country's presence at the fair was important not only for the extensive émigré community but for the global audience too.

After the Czechoslovak officials reached the final decision to build a national pavilion, the diaspora formed the Czechoslovak Group (which in Czech appeared as Československý výbor) to contribute to it. Čermák became the honorary chair until his demise in early 1933. The Group had a committee consisting of twenty Czechs and twenty Slovaks, which was a split quite different from the practice in Czechoslovakia, where organization was reserved primarily to Czechs. Nevertheless, the joint venture in Chicago was a result of a compromise that the organizers had to arrive at.[28] Views on the union of the Czechs and Slovaks in Europe differed amongst the communities which were far from unanimous in their embrace of Czechoslovakism. As early as 1931, the *Slovak* Committee for the World's Fair was established to represent all Slovaks of America in "a unified and united way" at the exposition.[29] Yet soon it became clear that the Slovaks of America could not have their own representation separate from Czechoslovakia, because only recognized governments could have their pavilions. Therefore, they joined forces with the American Czechs.[30] They still sought to be presented as two equal national groups, not as Czechoslovaks but rather as hyphenated Czecho-Slovaks.

The Slovaks in the USA that held nationalist views were represented by the Slovak League of America, the same organization that was behind the emancipatory efforts of Slovaks in America before 1918. American Slovaks perceived themselves as the champions of the Slovak cause, unlike the Slovaks in Upper Hungary, and in position to take political decisions.[31] In 1915, representatives of the Slovak League had signed the Cleveland Agreement with the Czech National Alliance and both groups agreed on working jointly towards an independent state in which the two nationalities would have an equally independent standing. This goal was later slightly revised in the Pittsburgh Czecho-Slovak Agreement of 1918, signed by the Slovak League, the Czech National Alliance, the Alliance of the Czech Catholics and T. G. Masaryk. This document envisaged Slovakia as an autonomous part of the joint state with its own administration, parliament and judicial system.[32] This promise was, however, also never de-

28 J. E. S. Vojan, "Století pokroku, světová výstava v Chicagu 1933," *Kalendář New Yorských listů na rok 1933*, vol. 12, 1933, 96.
29 "Slováci vystúpili z výboru svetovej výstavy," Náš svet, January 6, 1933, 3.
30 Slováci vystúpili, 3.
31 Kosatík, *Slovenské století*, 86; Michael Cude, "The Imagined Exiles: Slovak-Americans and the Slovak Question during the First Czechoslovak Republic," *Studia Historica Gedanensia* V (2014): 297.
32 Kosatík, *Slovenské století*, 85–86.

livered to the dissatisfaction of Slovaks both in the USA and in Czechoslovakia. Later, Masaryk claimed that the Pittsburgh document was not legally binding as it was drawn up in haste and not by those living in the land.[33]

The early attempt of American Czechs and American Slovaks to be presented separately from the new state at the Chicago world's fair suggests that the local organizational committee acted independently from the official representation of the Czechoslovak government. It perceived itself as composed of "local American citizens of Czech and Slovak descent."[34] Moreover, when the American Czechs and Slovaks settled on cooperation, the hyphen in the title of The Czecho-Slovak Group was of special importance; it suggested that the two groups were autonomous, not subsumed under a joint Czechoslovak identity. This, indeed, reflected the political situation in Czechoslovakia outside of Prague. While the state, represented officially at fairs by various ministries, institutions and companies, tried to consciously show itself as unambiguously unified with a single Czechoslovak identity, by 1933 the separatist tendencies grew stronger.

In Slovakia at this time, the political landscape was divided between autonomists and centralists, the latter consisting of the Agrarian and Social Democratic parties. The former included the Nationalists and the People's Party. It was especially the People's Party that was highly critical of the ideology of Czechoslovakism and anti-clericalism associated with Prague.[35] Moreover, throughout the interwar period, the radical representatives realized that Slovak autonomy was not limited to the Czechoslovak state. The Slovak League of America was especially vocal in its demands for a recognition of Slovaks as an independent nation and kept contacts with the nationalists in Slovakia.[36]

After the Slovak plans for an autonomous display failed, The Czechoslovak Group continued its efforts of putting together its own exhibit for a while. The Group aimed to show to the American audiences what the Czech and Slovak diaspora achieved "both in cities and in the countryside" during the previous century of progress.[37] This vision included presenting the first settlements against the current look of neighborhoods and farms, local businesses, schools, clubs and associations, and journalism.[38] Eventually, however, the presence of the di-

33 Kosatík, *Slovenské století.*

34 "Slováci vystúpili," 3.

35 Róbert Arpáš and Matej Hanula, "Postoje hlavných slovenských politických prúdov k čechoslovakizmu v medzivojnovom období," in *Čecho/slovakismus*, eds. Adam Hudek, Michal Kopeček and Jan Mervart (Prague: NLN, 2019), 182.

36 Cude, "The Imagined Exiles," 303.

37 Vojan, "Naše účast na světové výstavě 1933," *Most* no. 5, October–November 1931, 4.

38 Vojan, "Naše účast," 4.

aspora was reduced to a small section within the pavilion built by the Czechoslovak government, a topic I will come back to shortly. Despite that, the diaspora remained actively involved and organized the Czechoslovak national day, Sokol gathering, concerts and theatre performances and published *The World's Fair Memorial of the Czechoslovak Group.*[39]

While the American Czechs and Slovaks were engaged in the pavilion, it is difficult to see the same level of enthusiasm in those who became the easternmost minority of the new state—the Ruthenians. While they were described as one of the oldest emigration groups from central Europe with about 250,000 Ruthenians living across the USA in the interwar period, it seems that the involvement of the Ruthenian diaspora in the 1933 fair was minimal.[40] The representation of Ruthenia in the Czechoslovak pavilions was overall marginal, reduced to references to the forested landscape and natural resources of the region. In the same way, the American Ruthenians were not part of the Czech and Slovak diasporic communities or the program of Czechoslovakism. As much as the Czechoslovak nation consisted of many disparate ethnicities bound by tentative unity, the diaspora cannot be considered as a single community. In the case of the Ruthenians, the failed promise of autonomy within the Czechoslovak state and the growing enthusiasm for closer cultural and political affiliation with Russia affected the dwindling feelings of allegiance with Czechoslovakia.

Where is my homeland

The initial reluctance of the Czechoslovak government to invest in a national pavilion at Chicago in the economically unstable times eventually waned, partly due to the lobbying and personal intervention of Mayor Čermák. In Czechoslovakia, Čermák partly represented the Czechs of Chicago (and less the Slovaks) as he saw the exhibition as an excellent opportunity for showing the best Czechoslovak industries and arts to Chicago, the whole United States and the world.[41] Before the exhibition, the Czechoslovak consul general in the USA was quite explicit about this when he stated that the Czechoslovak participation was seen as important for the Czechs (and Slovaks) in Chicago. He identi-

39 *The World's Fair Memorial of the Czechoslovak Group* (Chicago, 1933).

40 Dušan Kubány to Central committee of Century of Progress, Government Correspondence, no.2, box 54 Century of Progress, University of Illinois at Chicago; Magosci, *With their Backs to the Mountains*, 244.

41 F. L. Vlček, *Náš lid v Americe (Sebrané články a verše)* (Týn nad Vltavou: O. Tunek, 1935).

fied the young generation of Czech Americans, to whom "we entrust the legacy of the national endeavor, and in which we try to preserve the love and pride in their Czechoslovak roots."[42] Most probably aware of the distancing of the second generation from the homeland, he appealed to the shared cultural roots: at the exhibition, "... we can show ... to those who mostly never stepped on the Czechoslovak ground the achievements and cultural prowess of the Czech and Slovak race [nation]."[43]

Moreover, many Czechs and Slovaks of the second and third generations in the USA were more distanced from the homeland of their parents and grandparents, and according to some accounts were losing interest in their heritage, which was often reduced to lace and sheep. The younger Czechs, for instance, had more in common with Americans or other ethnicities than the Slovaks with whom a joint state was created. The feeling of detachment of many young Czech Americans from the land of their parents and grandparents was often exacerbated by the image of Czechoslovakia that was projected in the United States through ethnographic films.

For instance, the popular documentary film *Zem spieva* (The Earth Sings) by Karol Plicka from 1933, depicting the seasonal cycle of rural life in Slovakia, was shown around USA as part of the tour that a few members of Matica slovenská, a patriotic organization supporting Slovak culture, language and education, made in 1935/36. The delegation consisted of, for example, the writer J. C. Hronský who lectured on modern Slovak literature, the photographer and journalist Jozef Cincík who delivered talks on Slovak art and architecture, the politician and promoter of Slovak independence Konštantín Čulen on Slovak history, and Plicka who not only showed his film but also talked about Slovak music to the Slovak communities in America.[44] They all emphasized the authentic and original character of Slovakia and its people.[45] Yet not everyone subscribed to the images promoted by these talks and the film. In Indiana, for instance, the former inhabitants of Moravian Slovakia, which is until today considered a region with a strong folklore identity between Moravia and Slovakia, complained that the depiction of the hardship and poverty of life in the rural regions harmed

42 "General consulate of the Czechoslovak Republic in Chicago to the Ministry of Foreign Affairs," May 25, 1932, APRO.

43 "General consulate."

44 "Delegácia a jej poslanie," *Zájazd Delegácie Matice Slovenskej do Ameriky v r. 1935–1936* (Chicago: Výbor pre Zájazd Matice Slovenskej v Chicagu, 1936). Cf. also Konštantín Čulen, *Slováci v Amerike* (Turčianský Sv. Martin: Matica slovenská, 1938).

45 "Delegácia a jej poslanie," *Zájazd Delegácie Matice Slovenskej do Ameriky v r. 1935–1936* (Chicago: Výbor pre Zájazd Matice Slovenskej v Chicagu, 1936), 9.

their efforts to present Czechoslovakia as a modern and affluent country. Instead, they recommended showing "films with modern buildings and factories, interiors of large plants, banks and department stores, sports grounds, Sokol gatherings, skiing and other sports, in short, present the second generation with evidence that Czechoslovakia was a progressive country matching the American lifestyle."[46] The image of the homeland among the Czech and Slovak diaspora was far from a unified one and it was based on generational, political and religious allegiances.

The diaspora at the fair

While the younger generation might have not been so keen on the folk heritage of their ancestors, the modernity of the Czechoslovak pavilion was more important. The diaspora in Chicago therefore found it crucial to be actively involved in constructing what they perceived as the correct image of Czechs and Slovaks in America. The Czechoslovak pavilion thus featured some works by local artists that were offered in a way like Mr Příbramský's contribution in Philadelphia. The large entrance hall in Chicago, for instance, was fitted with a bust of Antonín Čermák, the late Chicago mayor of Czech origin, created by Anton Vožech, a sculptor from New York. Many other artworks were supplied by the local organizations including the local Czechoslovak Arts Club.[47]

Members of the Czech and Slovak communities in the USA also ran various stalls in the pavilion. This included embroidery sold by Josef Šimek from Chicago, or wicker baskets by Frank Tuma from Philadelphia, who not only resold goods imported from Czechoslovakia but made his own baskets.[48] Other goods for sale included glass decorations, toys, ceramics, and textiles, and for their abundance (there were about 60 stalls indoors and outdoors), the pavilion was criticized as crammed and reminiscent of bazaar with its focus on cheap trinkets[49] (fig. 12).

The diaspora also planned to include a comprehensive exhibit on Czechoslovak America in the Czechoslovak pavilion, with key information about the

46 Jaroslav E. S. Vojan, "Budoucí forma češství v Americe," *Kalendář New Yorských listů* 18 (1939): 29.
47 Letter of the Czechoslovak American Chamber of Commerce to the Publicity Department of August 28, 1934, Century of Progress Government Correspondence, no. 2, box 54, Special Collections, University of Illinois at Chicago.
48 "Košíkářství na výstavě 'Století pokroku'," *Spravedlnost*, June 16, 1933, 10.
49 Bohumil Klír, "Věc tapná a nechutná, *Svornost*, October 6, 1933, 10.

communities, like statistical information on the number of Czechs and Slovaks in the USA and their organizations. It intended to showcase books on Czechoslovakia and newspapers published in the States, historic coins called *tolar* (the precursors of the American dollar), as well as a ship propeller, an invention that "had direct and long-lasting impact on the development of the American civilization," as the pamphlet on the pavilion claimed.[50] Originally envisaged on a much larger scale, this section was ultimately reduced to a small room towards the pavilion exit. For that, it was harshly criticized by Czech-language papers in the USA as too small, rushed, and not comprehensive enough.[51]

In the light of not enough photographic evidence, such descriptive criticism also helps to build a picture of what the exhibit looked like. The disapproving remarks came especially from the newspaper *Věstník československé obce legionářské v Americe* (Bulletin of the Association of Czechoslovak legionnaires in America) published by the legionnaires settled in the USA. They represented a substantial and vocal community of the volunteer armed forces active in the First World War. The legionaries were recruited either from individuals born outside of Austria-Hungary, including the USA, or those who deserted the Austrian army during the war. They fought in France, Russia, Italy and Serbia in the First World War as well as in conflicts shortly after its end in Slovakia and the Těšín region. After 1918, the legionnaires from the USA settled back in America and were joined by many volunteers from France.[52]

One of them was Bohumil Klír who expressed his disappointment with the space ultimately left to the diaspora, because it was easily missed by visitors. He described that the exhibits were partial, in bad condition and their "arrangement was truly sloppy."[53] This was the case, for example, of a map of the United States which was supposed to show locations of Czech and Slovak communities; but these were only represented by pins. There were also "stained and wind torn" photographs of prominent personalities of Czech decent (that included the anthropologist Aleš Hrdlička and the Illinois attorney general and federal judge Otto Kerner) and incomplete lists of clubs and associations. Printed publications like magazines were "weathered and smudged, corners loose," and were attached to the wall by carpet tacks.[54] The entire Czech and Slovak America ex-

---

50 J. H. Hruška, "Výstavka československé Ameriky," *The World's Fair Memorial of the Czechoslovak Group* (Chicago, 1933), 106.

51 "Čsl. legionář píše," *Věstník českoslov. obce legionářské v Americe*, March 1934, 1.

52 Dagmar Hájková, *"Naše česká věc" Češi v Americe za první světové války* (Prague: NLN, 2011), 107.

53 Klír, "Čsl. legionář píše svému bratrstvu—z Chicaga!" *Věstník československé obce legionářské v Americe*, March 1934, 1.

54 Klír, "Čsl. legionář," 1.

hibit, overseen by the Czechoslovak Ministry of Trade and not the diaspora, looked therefore more like an afterthought or a last-minute addition.

Apart from its description of the actual presentation, the *Věstník* also highlighted another issue related to the diaspora, namely the involvement of émigrés as employees, or rather, the lack of their involvement. Left leaning Klír pointed for example to the fact that people of Czech and Slovak decent found it hard to obtain work at the national pavilion. The reason was the assumption of the Czechoslovak Ministry of Trade that the émigrés lived in a rich country and assumed to be well-off. They were, therefore, seen as part of the "ignorant, uncultured and dark crowd that is called in Czechoslovakia the 'colonies' or 'minorities'," Klír complained.[55] Such understanding was in line with the distorted image of the life in the USA that many in the home country held. On the one hand, American Czechs and Slovaks were seen as ignorant, on the other their lives were idealized and romanticized as economically comfortable. These discrepancies were captured in many contemporary novels too, famously in Marie Majerová's *Siréna* (The Syren) on the disenchantment with American capitalism and individualism, or in Karel Čapek's *Hordubal* about a farmer who returns from America to the Sub-Carpathian Ruthenia in many ways alienated by the experience from the rest of the people and his family.[56]

There was a further critical point Klír made that indicates that some members of the diaspora did find work at the pavilion. The exhibition's labor office opened three years prior to the event, recruiting various skilled and non-skilled workers for contractors, builders, exhibitors or concessioners.[57] The office closed down in mid-May 1933 when all vacancies were filled up, although workers kept turning up on the off chance of being hired. In connection with employment at the fair issues, Klír described a case of mistreatment of some sixty female sales assistants whose working conditions at the stalls in the Czechoslovak pavilion were near appalling and who were treated, in his view, "without human or social sensitivity."[58]

This meant that they were underpaid despite the high profits generated by the sales of souvenirs in the pavilion, and their social welfare, health and gen-

---

55 Klír, "Čsl. legionář píše svému bratrstvu," *Věstník československé obce legionářské v Americe*, October 1934, 2

56 Marie Majerová, *Siréna* (Editura de stat, 1918); Karel Čapek, *Hordubal* (Prague: František Borový, 1933); cf. Josef Švéda, *Země zaslíbená. Antologie o obraze Ameriky v české literatuře a kultuře* (Příbram: Pistorius & Olšanská, 2016).

57 "Na výstavě se lidi už nepřijímají do práce," *Svornost*, May 20, 1933, 2.

58 Klír, "Čsl. Legionář píše svému bratrstvu – z Chicaga!" *Věstník československé obce legionářské v Americe*, November 1933, 2.

eral wellbeing was not looked-after.[59] They worked eight hour shifts seven days a week under insufficient health conditions; only one toilet was provided for all the employees who worked "in the cramped space ... [behind] the glass walls of the pavilion" without much ventilation.[60] Such practices were, unfortunately, not exclusive to the Czechoslovak pavilion. The Franco-British Exhibition, supervised by Commissioner General of Hungarian origin Imre Kiraly, for example, employed large numbers of women who worked in infamously poor conditions of twelve-hour shifts, very few breaks and only one toilet.[61]

For Klír, the conditions in the Czechoslovak pavilion had dire consequences for the whole Czech (and Slovak) community in Chicago and the USA, who came out disenchanted with the Czechoslovak pavilion and the image it offered: "...after all the learning of Czech and ... emphasizing the importance of awareness and the sense of national belonging to the old country of their parents, the second generation encountered here the 'democratic' representatives of Czechoslovakia."[62] The "democratic" representatives were meant as a sarcastic reference to the bureaucrats responsible for the pavilion, the commissioner Ladislav Turnovský who left Chicago shortly after the national pavilion opened, and, Bohumil Soumar who was in charge of managing the Czechoslovak exhibit. Soumar's only interest, as Klír saw it, was in doing business, not in human and social sensitivity.

By highlighting the problems that ordinary workers faced not only in Chicago but at other world's fairs, Klír touched upon an underlying issue with such events related to workers' conditions and rights. In Chicago, the situation was particularly dire, as the fair took place at the end of the Great Depression that caused a massive wave of unemployment and all kinds of related difficulties with people's housing, health or access to food. The Century of Progress exhibition was, indeed, organized to overcome these struggles and set out to promote a fresh start for the capitalist society and American life.[63] But as the Czech critics pointed out, social injustice persisted and was only permeated by the unfair work conditions at the fair. L. I. Janík, a communist commentator for *Spravedlnost* called for a workers' movement that would "strive for removal of the pres-

59 Klír, "Čsl. legionář píše," *Věstník*, October 1933, 3.
60 Klír, "Čsl. legionář píše," *Věstník*, October 1933, 2.
61 Paul Greenhalgh, *Ephemeral Vistas: The Expositions Universelles, Great Exhibitions, and World's Fairs, 1851–1939* (Manchester – New York: Manchester University Press, 1988), 186.
62 Klír, "Čsl. legionář píše svému bratrstvu – z Chicaga!" *Věstník*, October 1933, 3.
63 Carla Leshne, "Progress is More than Just a Word: The Workers Film and Photo League at 1933 Chicago World's Fair," *Meet Me at the Fair: A World's Fair Reader*, eds. Celia Pearce, Bobby Schweizer, Laura Hollengreen, and Rebecca Rouse (Pittsburg: ETC Press, 2014), 372.

ent order in which world's fairs are organized to daze and exploit the people."[64]

Elsewhere, unrest and dissatisfaction with various aspects of the organization of fairs, exclusions and expenditure were common but not part of the mainstream narratives that are often reiterated about these events. Many protests came from disadvantaged groups. To give an example, strikes occurred amongst casual workers at the 1893 World's Columbian Exposition calling for decent pay. Frequent was mistreatment of minorities, especially of the African Americans. In 1926, they stood up against the convention of Ku Klux Klan planned and approved for the Sesquicentennial Exposition.[65] The Century of Progress also saw protests of black Americans against discrimination at some establishments, mostly restaurants.[66] It was reported that "some colored people" were refused service at the Czechoslovak restaurant too.[67]

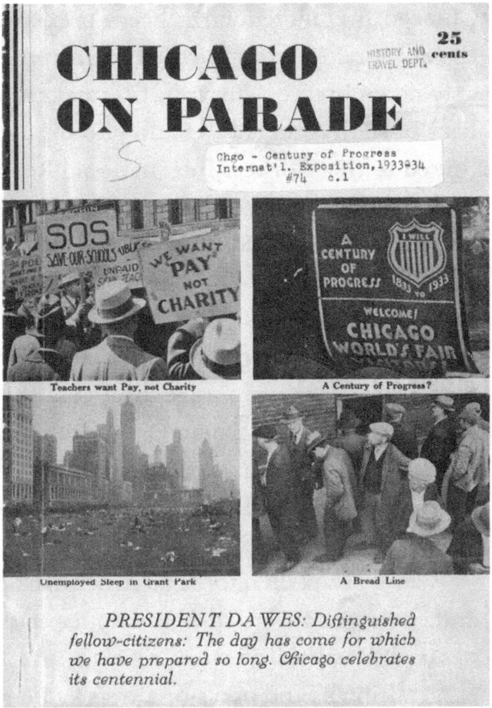

Fig.49. *Chicago on Parade* pamphlet.

A lot of dissatisfaction with fairs came from the fact that financing such a costly event was seen as an unnecessary extravaganza. This was particularly prominent with the Chicago world's fair, the organization of which was tainted by the Great Depression. This grim side of the fair was captured, for instance, in the pamphlet *Chicago on Parade* published in 1933[68] (fig. 49). The small booklet contained images and caricatures from the last three years of hardship in

64 L. I. Janik, "Konec výstavy," *Svornost*, November 17, 1933, 4.

65 Thomas H. Keels, *Sesqui! Greed, Graft, and the Forgotten World's Fair of 1926* (Philadelphia: Temple University Press, 2017), 176–83.

66 August Meier and Elliott M. Rudwick, "Negro Protest at the Chicago World's Fair, 1933-1934," *Journal of the Illinois State Historical Society (1908–1984)* 59, no. 2 (Summer, 1966): 161–71.

67 A letter from Colonel Finch of 28 September 1933, Century of Progress Government Correspondence, no.2, box 54, Czechoslovakia-Denmark, Special Collections, University of Illinois at Chicago.

68 *Chicago on Parade*, 1933, Century of Progress Collection, box 5.13, CPL.

Chicago, juxtaposed with cheery proclamations about the joys and successes of the event taken from the opening ceremony. The photographs were most probably taken by the American filmmaker, photographer and activist Conrad Friberg of the Workers Film and Photo League. Founded in 1930, the League was linked to the Workers International Relief and a part of the Communist International which channeled relief for working class organizations.

The images in *Chicago on Parade* depicted the shacks many men lived in, rough sleepers in Grant Park, "one of the largest hotels in the world," protests and picket lines calling for reasonable wages, calls for improvement of the public school system, hunger marches, and breadlines.[69] The whole idea of the century of progress and future direction of the city was uprooted by this booklet. Put together by a Herman O. Duncan, the booklet was addressed to the representatives of the Fair, especially the president of the fair, Rufus Dawes, and the mayor of Chicago, Edward J. Kelly, whose quotes appear throughout against the poverty photographs.

World's fairs therefore magnified class divides. Habitually, they were designed and envisaged for middle- and upper-class visitors, whether they came for business or pleasure. The working class was considered as workforce, providers of entertainment and occasionally as audience who were often encouraged to visit the fair at specific days or times. As education and instruction played an important role at exhibitions, the working-class visitors were targeted too. These attempts ranged from displays of diagrams and the latest technological inventions to more entertaining demonstrations of manufacturing techniques for new products, which could include hands on trials or tastings of food samples.

In the USA, world's fairs were primarily attended by audiences local to the hosting city while in Europe they were more linked with tourism and visits to attractive locations like Paris.[70] The fairgoers, of whichever class, were supervised and helped by exhibition attendants and the police. In Chicago, the police at the fair could be easily recognized by their new uniforms, described by one astonished observer from Czechoslovakia as "very fetching, with black trousers, a jacket and colonial helmets which made them visible from afar."[71] The visitor portrayed the police as well-built and tall men, noted that there was about 500 of them and that they "invited respect further increased by a large revolver."[72]

69 Carla Leshne, "Progress is More," 373; *Chicago on Parade.*
70 Leerssen and Storm, eds., *World Fairs and the Global Moulding*, 9.
71 R. Šticha, "Něco o Americe a výstavě v Chicagu," *Československá Paříž*, March 4, 1934, no. 28, 2.
72 Šticha, "Něco o Americe," 2.

The police kept the grounds in order by patrolling them. Observers however complained about their reluctance to deal with petty theft at the pavilion stalls.[73] Plain-clothed police were also tasked with collecting dues from the concessioners who were supposed to supply 10% of their proceeds to the organizers. The police, and this included uniformed policemen and women, kept an eye on the popular performance and their law abidance. The most famous case was the Sally Rand fan dance, performed first in the Streets of Paris, later in the Italian and finally the Oriental Village. She was arrested for obscenity and displaying nudity and was fined $25. This was, nevertheless, a rather negligible amount given the lucrative proceeds from her performance.[74] Rand's case, nevertheless, is characteristic of the disruptions that often took place at world's fairs. While order was expected not only in the presentation of nations, people and objects, disruption necessarily occurred in many forms, including petty crime like theft, as well as corruption, illegal sale of alcohol or prostitution.

"Native" experience

Sally Rand was not the only case of a subversive element in a world's fair but most prominent and talked about one. In Chicago, there were other partly or fully nude female dancers in the Dance Ship or the various "exotic" reimaginations of the Oriental Village, Cuban Rhumba or Old Mexico.[75] The women dancers certainly managed to find a profitable source of income at the fair, but they were an exception. Employment of women across the fair took various forms, but very often female attendants were exploited, as the case of the sales assistants from the Czechoslovak pavilion showed. In his critical account, Klír also noticed the waitresses in the Czechoslovak pavilion who, he described scornfully, had to wear "national costumes or something that is reminiscent of them."[76]

The Czechoslovak restaurant in Chicago, as much as all other Czechoslovak food outlets at interwar world's fairs, was equipped with beer and some light refreshments, like Prague ham. The consumption was accompanied by entertainment provided by musicians, singers and to an extent by waitresses in national dresses. In Chicago, adverts in newspapers announced for example the Bohe-

73 Šticha, "Něco o Americe," 2.
74 Cheryl Ganz, *The 1933 Chicago World's Fair: A Century of Progress* (Urbana, Chicago: University of Illinois Press, 2012), 20.
75 Ganz, *The 1933 Chicago*, 24.
76 Klír, "Čsl. Legionář píše svému bratrstvu – z Chicaga!" *Věstník*, October 1933, 2.

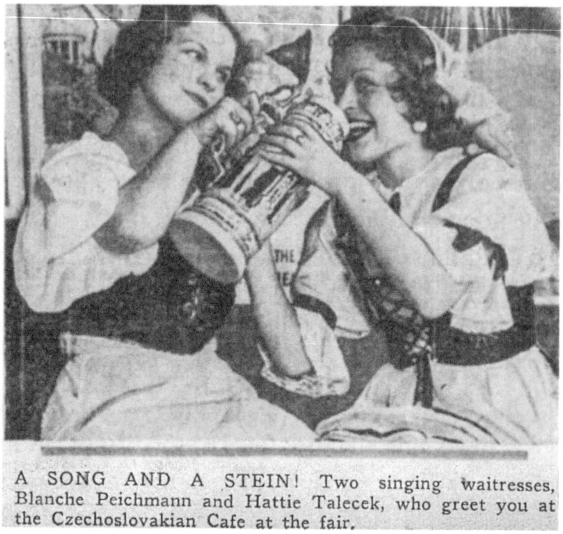

Fig. 50. Two singing waitresses at the Czechoslovak restaurant.

mian String Quintet that featured the "light opera star of international reputation," Mr. Anton J. Wagner who emigrated to the USA in his teens and sang popular melodies at the restaurant together with a Mrs. Frances Trnka.[77] Further, there were also two singing waitresses Blanche Peichmann and Hattie Talecek, who indeed wore something reminiscent of national costumes and greeted and entertained guests at the restaurant (fig. 50).

The entire food and drink establishment also met with harsh criticism, as it was seen as not Czechoslovak enough. Klír understood the restaurant as unpatriotic because, as another commentator remarked disparagingly, it was run by a "Austrian Jew" from Vienna and a German from the German Reich.[78] Nationalistic prejudice and racist stereotypes were therefore nothing alien to at least some members of the diaspora, like Klír. Part of the disillusion with the restaurant came from the fact that originally, the Slovak diaspora suggested to build an "authentic," originally furnished Slovak "búda," or a village hut which where wine from the southwestern town of Pezinok would be served in Modra jugs, food would include biscuits with pork rinds, and sausages with horseradish while "a gypsy band playing Slovak melodies would make everyone merry."[79] The idea was nevertheless turned down by the pavilion organizers from Czechoslovakia and a Czechoslovak restaurant was built next to the national pavilion.

77 "Wagner Sings for Czeko-Slovakians (sic)," *Century of Progress Scrapbooks*, June 26 – July 22, 1933, box 1, folder 3, CPL.
78 "Skutok, ktorý sme veru neočakávali," *Náš svet*, 4, no. 25, 1933, 2.
79 Jarábek, "Slovenské zprávy o svetovej výstave," *Svornost*, May 3, 1933, 2.

The refusal of the Slovak proposal was interpreted as another instance where the American Czechs and Slovaks were ignored.

More generally, the Czechoslovak restaurants and national pavilions provided a "native" experience by serving local food and drink, entertaining visitors by music and, indeed, dressing female attendants in national costumes. This practice also arose criticism which pointed out that such clothes were removed from everyday reality and gave out the wrong idea about modern Czechoslovakia. These comments appeared not only at Chicago, but for instance in relation to the preparatory works for the Czechoslovak pavilion in Paris in 1937 where women in folk costumes were to appear. In their disapproval of this inclusion, the Czechoslovak consulate in Strasburg referred to the earlier Czechoslovak pavilion in Brussels in 1935. It criticized the fact that the waitresses in the restaurant there were dressed in what *appeared* as national "Czechoslovak costumes" which was not an appropriate way of representing the country.

In this connection, the consulate also recalled protests about the planned Alsatian village for the 1937 fair that came from Alsace.[80] The emphatically entitled article "Elsässer als Ersatz für die Neger" (Alsatians as substitute for the Negro) published in *Elsässer Bote* compared the planned costumed participants in the Alsatian village to the recreations of Black villages at French colonial exhibitions.[81] Following the paper's disapproval of any possible connotations that costumed Alsatians could have with black people, the Czechoslovak consulate warned against exoticizing Czech and Slovak folk culture.

The Ministry of Trade that organized the Czechoslovak exhibit refuted these remarks and disputed the comparison between what the article called "negrocising" of the regional culture. It pointed out that national costumes were occasional, festive wear which increased attractiveness of the display when used in the context of a national pavilion. This was, according to the Ministry, the case in Brussels where the costumes on the waitresses contributed to the popularity and success of the Czechoslovak restaurant and to increased sales of beer. The Ministry also explained that the costumes were not meant as national, rather they were "stylized colorful dress for serving" beer, designed by Antonín Heythum for this purpose.[82] For the Czech commentators, whether from the consulate or the Ministry of Trade, there were different levels of primitivism that were

80 "Výstava v Paříži 1937, protest Alsasanů, aby jejich svéráz sloužil jako atrakce," letter from The Consulate of the Czechoslovak Republic in Strasburg to the Ministry of Foreign Affairs and the Ministry of Trade of December 5, 1936, section III, part V, MFA Archive.

81 Quoted in "Výstava v Paříži 1937," 2.

82 "Výstava v Paříži 1937," 2.

stressed in the Czech and Slovak people in the context of world's fairs. Yet none was fully comparable to the exoticism of African and other indigenous villages and their "native" inhabitants in the colonial displays.

While indigenous people from outside of Europe were shown as *exotic* through their native attire and dwellings, the same attributes applied to Europeans had different meanings. Dressing female attendants in folk costumes was a common trope not only for the Czechoslovak pavilion but for other European exhibits too, including Poland, Sweden or Italy. The main reason for the inclusion of such attire was to bring up the country's historicity as well as authenticity, which folk art indicated to many, with the added value of attractiveness. Individuals and groups in native costumes had been included from the onset of these events and provided the markers of ethnic origin and difference.[83] Placed in a juxtaposition with the attributes of progress and modernity, ranging from scientific and technological inventions, man-made building materials, to hygiene displays, they were more than an attractive spectacle. They also showed human evolution from the somewhat primitive origins of local cultures to civilized modernity.

Yet as I have discussed in the chapter on Art and Design, folk culture held this position of ambiguity when it came to international displays and representation of Czechs and Slovaks abroad. While folk culture had played a major role in the late 19th century exhibitions that took place in Prague, during the interwar period, it was often maneuvered into a manifestation of lost traditions and nostalgia in the interwar period. As such, folk culture stayed an important and integral part of émigrés' expression of belonging to the home country articulated in special events at world's fairs. This included official openings and closures of national pavilions, parades, Sokol gatherings and national days, which I will explore in more detail.

Gendered fairs

First, however, I would like to address the issue of gendered culture that I have mentioned a few times and place it in the context of the more general questions of gender presence and absence at world's fairs. Recently, scholars of world's fairs and large exhibitions have been paying more attention to gender, with publications focusing on women's roles and spaces in individual case studies, and those

83 Sadiah Qureshi, *Peoples on Parade: Exhibitions, Empire and Anthropology in Nineteenth Century Britain* (Chicago–London: University of Chicago Press, 2011), 119.

that integrate gender into their broader narratives.[84] Women held a wide range of roles at world's fairs, not always visible or acknowledged. They were waitresses and sales attendants, they were performers and objects of display through their products or bodies, they came as visitors and in a few cases, they were also organizers and designers. In Czechoslovak pavilions, women's presence beyond the services like sales and waitressing was marginal.

As I have mentioned earlier, the original concept of the national pavilions was devised in Prague by cross-ministerial committees that consisted exclusively of men. The architecture of the Czechoslovak pavilions was also designed solely by men, with only one known entry from a female architect, Marie Jansová, into the competition for the 1937 pavilion in Paris. Some female artists and designers, like Minka Podhajská, Helena Johnová, and Emilie Paličková Milde did appear in the Czechoslovak displays throughout the interwar period, most prominently as part of the decorative displays in Paris in 1925.

Paličková Milde was, in fact, one of the most successful designers judging from the number of prizes her lace received at the fairs. Interwar exhibitions also included a few female artists, e.g., the painter Linka Procházková and the sculptors Mary Duras and Marta Jirásková appeared in the modern arts exhibit at the 1937 exposition in Paris. In the larger scheme of the Czechoslovak displays in this period, though, their total number and representation was miniscule. On top of that, there were also women artists and designers whose level of involvement has been difficult to unfold because they cooperated with their husbands or partners, and this would overshadow their contribution. This was the case of the previously mentioned Eva Štrimplová, who collaborated with Krejcar on parts of the Czechoslovak pavilion in 1937, or the case of Charlotte and Antonín Heythums. In their case, the extent of the contribution of Charlotte to the designs bearing the name of her husband Antonín Heythum is still unknown today.

This invisibility of women in Czechoslovak displays at international fairs seems to contradict the political climate in the new republic that was favorable to women's emancipation. Already before the war, there had been various liberal women's societies and clubs that called for gender equality and suffrage, which women gained in 1920. In the interwar period, the Women's National Council

84 The first group is represented, for example, by Myriam Boussahba-Bravard and Rebecca Rogers, eds., *Women in International and Universal Exhibitions, 1876–1937* (London: Routledge, 2018); T. J. Boisseau and Abigail M. Markwyn, eds., *Gendering the Fair: Histories of Women and Gender at World's Fairs* (Urbana: University of Illinois Press, 2010) , the second by for instance Ganz, *The 1933 Chicago* and Greenhalgh, *Fair World.*

was established as the central body in which various women's rights organizations were united and fought for further rights for women.[85] They were supported also by President Masaryk who often lectured and published on the topic of gender and women's rights, therefore being seen as a promoter of feminism.[86]

Masaryk called for equality and the removal of privileges as a feature of humanity and modernity because being modern for him was being equal.[87] His appeals, however, did not translate universally to all aspects of social life. While soon after the end of the First World War and the collapse of the old monarchic order, women's access to education and employment in Czechoslovakia was set to improve, prevailing views still operated within the mindset that women belonged to the domestic sphere. And women as independent earners were exceptions. Even Masaryk did not go as far as to claim that women could do the same jobs as men and recognized a level of biological determination in gender roles.[88]

Melissa Feinberg has identified this contradiction between the government's official stance and the reality, as the right to vote did not come with the right to work.[89] During the interwar period and more markedly in the 1930s, married women were discouraged or often prevented from working in civil service, and this limited their access to decision making. The reason for this exclusion was framed as economic fairness; within a family unit, one earner should suffice and that should be the man. As much as during the Habsburg monarchy prior to 1918, individual rights in Czechoslovakia were suspended in favor of women's "responsibilities to their families and by extension to the nation."[90] And this included being part of teams that organized participation at international exhibitions.

The committees responsible for international exhibitions, women in Czechoslovakia most often developed spaces separate from the main exhibition realm. Women's organizations put together many exhibitions of women's art for national and international audiences in, for example, The Exhibition of Czech Female Painters in Paris (1927), The Exhibition of Czechoslovak Women's Art

85 Jana Burešová, *Proměny společenského postavení českých žen v první polovině 20. století* (Olomouc: Univerzita Palackého v Olomouci, 2001), 42.

86 T.G. Masaryk, "Moderní názor na ženu," in *Masaryk a ženy* (Prague: Ženská národní rada, 1930), 63–64.

87 T.G. Masaryk, "Modern opinions about women," quoted in Melissa Feinberg, *Elusive Equality: Gender, Citizenship, and the Limits of Democracy in Czechoslovakia, 1918–1950* (Pittsburgh: University of Pittsburgh Press, 2006), 11.

88 Feinberg, *Elusive Equality*, 18. Masaryk, "Dodatek k Několika poznámkám o práci československých žen," in *Masaryk a ženy* (Prague: Ženská národní rada, 1930), 22.

89 Feinberg, "Women in the Civil Service," *Elusive Equality*, 99–128.

90 Feinberg, *Elusive Equality*, 126.

in Buenos Aires (1929) or The Exhibition of Modern Art of the Czechoslovak Women in Sofia (1936).[91] These efforts often received official political backing that promoted women's exhibitions as a sign of the country's democracy.[92] On a national level, two exhibitions that focused on women that took place in Brno between 1929 and 1930 can be seen as part of this trend. The Exhibition of Modern Woman and the Civilized Woman centered around women and modernity.[93] They challenged in their own ways the existing image of women and suggested improvements to various areas of women's lives, including dressing, home interiors and work. Yet in the end the exhibitions had no larger impact on translating this message into the Czechoslovak pavilions abroad where women continued to play a peripheral role.

Outside the Czechoslovak pavilions, women at world's fairs occupied separate spaces that were inhabited and/or organized by women. These only slowly developed into their more integrated involvement in the exhibitions. From the 19th century, many fairs included women's buildings and women's pavilions which showcased women's work across professions, from material production to philanthropic deeds.[94] Most prominent of these were the Woman's Building at the Columbian Exhibition in Chicago 1893. In the USA, seven woman's buildings had been built at world's fairs by the beginning of the Second World War, and fourteen appeared in the main European fairs.[95]

In Europe, women's buildings and exhibitions were also many and date back to the end of the 19th century. They included the "Pavillon der Frauenarbeiten" in Vienna, in 1873, that was proposed by Rudolf von Eitelberger, whose activities as the director of the Kunstgewerbemuseum and elsewhere aimed at improving the taste of middle-class women, partly by promoting their work in applied arts.[96] One of the largest events for women's exhibitions was the Werkbund Exhibition in Cologne in 1914 which featured the Woman's House (The Haus der Frau) by Margarethe Knüppelholz-Roeser. Set amongst the buildings by Walter Gropius, Peter Behrens and Henry van der Velde, for instance, the pavilion

91 Martina Pachmanová, *Z Prahy až do Buenos Aires: "ženské umění" a mezinárodní reprezentace meziválečného Československa* (Prague: VŠUP, 2014), 63.
92 Pachmanová, *Z Prahy*, 60.
93 Martina Pachmanová, *Civilizovaná žena: ideál i paradox prvorepublikové vizuální kultury* (Prague: VŠUP, 2021).
94 Myriam Boussahba-Bravard and Rebecca Rogers, "Introduction. Positioning Women in World's Fairs," *Women in International and Universal Exhibitions, 1876–1937* (London: Routledge, 2017), 3.
95 Mary Pepchinski, "Woman's Buildings at European and American World's Fairs, 1893–1939," in *Gendering the Fair. Histories of Women and Gender at World's Fairs*, eds. Tracey Jean Boisseau and Abigail Markwyn (Urbana: University of Illinois Press, 2010), 187–207.
96 Pepchinski, "The Woman's Building," 194.

tried to emphasize women had an important role in design and the design reform.[97] In the same year, Emilie Winkelmann designed her own Woman's House with 25 rooms at the International Book and Graphic Exhibition in Leipzig.[98]

After the First World War, like the Exhibition of Contemporary Culture in Brno in 1928, the Polish General Exhibition (Powszechna wystawa krajowa) commemorated ten years of the existence of the Polish republic. It took place in Poznań in 1929 and with 112 exhibition buildings it became the largest fair in Poland.[99] Here, the Pavilion of Women's Work was initiated by the journalist Maria Ruszczyńska with the first lady, Michalina Mościcka, as the patron. The women involved here were connected with the progressive circles of Poland and their pavilion aimed at showing the various activities of Polish women, regardless their class, profession or religion.[100]

Designed by Anatolia Hryniewicka-Piotrowska (1896–1989), a modernist architect, color played an important role here as the façade was a striking combination of yellow and grey.[101] Yet in the same exhibition, women oversaw another pavilion, the Pavilion of Female Farmers. It was put together as an alternative to the Pavilion of Women's Work which the female farmers saw as initiated by "sisters from cities" who pursued different goals.[102] The female farmers pavilion was therefore based more around the traditional values related to living in the countryside. The reductive construction of identity in exhibitions and fairs therefore applied to women's spaces too. This was the case of the Polish pavilions, as much as of the Czechoslovak national presentations which gave women a very specific role.

The interwar period therefore opened new questions that concerned women's roles in society and hence at world's fairs. These could be the recognition of the advantages of modernity, that included consumerism; adherence to traditions whether found in the countryside or the family, activism in politics and cultural life. All of them were addressed at four consecutive Woman's World's Fairs which were held between 1925 and 1928 in Chicago, were organized by women

97 Despina Stratigakos, "Women and the Werkbund: Gender Politics and German Design Reform, 1907–14," *Journal of the Society of Architectural Historians* 62, no. 4 (2003): 499.

98 Mary Pepchinski, "Women's Buildings," 194.

99 *Przewodnik po Powszechnej Wystawie Krajowej* (Poznań: Biuro Ogłoszeń "PAR," 1929), https://cyryl.poznan.pl/katalog.php?baza=obiekty&id_obiektu=83563.

100 Magdalena Heruday-Kiełczewska, *Wystawa Pracy Kobiet na Powszechnej Wystawie Krajowej*, https://tytus.edu.pl/2018/09/23/wystawa-pracy-kobiet-na-powszechnej-wystawie-krajowej/ and *Przewodnik*

101 Olgierd Czerner and Hieronim Listowski, eds. *The Polish Avant-garde: Architecture and Town Planning, 1918–1933* (Paris: Editions du Moniteur, 1981), 86.

102 Rada Naczelna Ziemianek, eds., *Pawilon Ziemianek i Włościanek na Powszechnej Wystawie Krajowej w Poznaniu 1929 r.* (Warszawa: Komitet Pawilonu Ziemianek i Włościanek, 1930), 5.

and showcased women's ability to free themselves through work.[103] They were held on a small scale in comparison with the grand world's fairs that lasted several months, but their ambition was to provide women's view of women's work.

During the four years, however, the fairs increasingly moved away from putting emphasis on modern technology and its potential to liberate women towards more conservative displays of home improvements, decoration and traditional crafts.[104] The 1927 instalment of the Woman's World's Fair also contained The Czechoslovak Folk-Art Exhibit, which firmly aligned material culture of the homeland with peasant and female production. It indicates the move of the women's event to more traditional displays, as well as the persisting embrace of folk art as a reminder of the "ancestral home" amongst the Czechoslovak émigré women.

The Exhibit consisted of a display of costumes, folk products and techniques that were recognized as disappearing in Czechoslovakia. A brochure with articles and pictures was also published, acknowledging that many folk practices were dying out in a process that was seen as an inevitable result of progress and the change of lifestyle.[105] One of the texts reminded the reader that "cooperative societies have been organized [in remote areas of Czechoslovakia] to foster the making of laces and embroideries and to find markets for them as a means of livelihood to women who... produce delicate ornaments for the dresses of American ladies."[106] One way of preserving folk culture was thus to turn it into a commodity produced for national and international markets, with a potential to inform the American public (and markets) of the existence of Czechoslovakia. In the Czechoslovak Folk-Art Exhibit, the question of female involvement in world's fairs was successfully merged with the practical issues of preserving folk cultures in a modern world.

National days

The examples of the gender imbalance, the lack of recognition of casual workers, and the marginalized involvement of the external Czechs and Slovaks have a common denominator in othering the aspects of Czechoslovak identity that

103 Tracey Jean Boisseau, "Once Again in Chicago: Revisioning Women as Workers at the Chicago Woman's World's Fairs of 1925–1928," *Women's History Review* 18, no. 2 (2009): 265–291.

104 Boisseau, "Once Again," 268.

105 Vojan, "Lidové umění československé," in *The Woman's World's Fair, Coliseum – Chicago – May – 1927. The Czechoslovak Folk-Art Exhibit*, The University of Chicago Library, 8.

106 "Folk art of Czechoslovak Women," *Czechoslovak Folk-Art Exhibit*, 9.

stand outside the mainstream society and culture. The internal and external others of the Czechoslovak state, nation and their presentations included also ethnic minorities, like Germans, Hungarians, Poles, or Roma that I discuss throughout the book. In the previous chapter, I also argued that folk art and cultures were subject to exoticization which placed folk culture into a role of the other too, because it created a specific juxtaposition with high culture emphasized in the Czechoslovak pavilions. The contrast between the traditional, countryside-related image of local culture with the modern, urban and international focus of the exhibitions took many forms. As I already discussed, folk culture played an important, albeit often contradictory, part in the feelings of affiliation with the homeland of the diaspora. This became prominent also in the national days in which Czech and Slovak émigrés actively participated.

National days were special events that ran on a set date and condensed what was conceived as the best of the country's culture into one day. Generally, the national days in the USA drew from the local diasporas and this was the case of Czechoslovakia where Czechs and Slovak émigré communities could execute their ideas and efforts of relating to their home country. A Bohemian day was held already in Chicago in 1893, as I mentioned previously, and each subsequent fair in the USA hosted Bohemians, Slovaks and/or Czechoslovaks.

In the interwar period, the Philadelphia Sesqui-Centennial Exhibition included a Czechoslovak national day alongside days that provided dedicated time and space for other foreign countries, commemorations of various occasions like the country's independence days, or diverse institutions. In Philadelphia, such days were dedicated to an array of causes that ranged from the Masonic Clubs to Colonial Dames as well as to countries and American states. The Czechoslovak Day took place on 28 October to coincide with the anniversary of the foundation of the republic. Another event on November 6 was organized to remember the American-Czechoslovaks and the day was managed by the diaspora with the local Sokol organization in the lead [fig. 51]. Attended by some six thousand members of the community from Pittsburgh, Chicago, New York, and Baltimore, the day consisted of speeches, Sokol and workers gymnastic drills, dances in national costumes and songs of the "Carpatho-Russian choir."[107] This was one of the very few occasions where Sub-Carpathian Ruthenia, whose diaspora in Pennsylvania was substantial, had some presence in the Czechoslovak representation. Subsequent pavilions in the USA did not seem to involve or engage the Ruthenian community.

107 Austin and Hauser, *The Sesqui-Centennial*, 411 and 412.

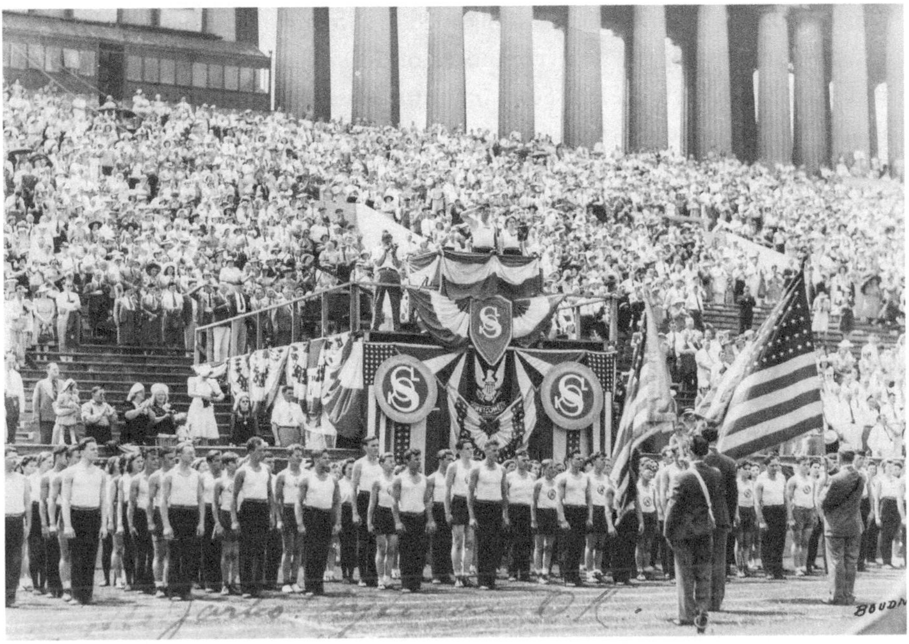

Fig.51. Sokol gathering at Century of Progress, Chicago, 1933.

The Czechoslovak national day in Chicago was a much larger event, organized by the local diaspora as a visually strong mixture of processions, dramatic, music and dance performances, and speeches. It did not differ too much from other national days of other countries or at other world's fairs. The Polish Day in Chicago, for instance, fell on August 26, 1934, and was organized by the Polish Americans. Their focus was not only on the historical events from Polish history but also on the contribution of Poles to the United States mediated through tableaux floats, speeches, traditional and classical dances and exercises of Sokol that became a sports association of many Slavic nations.[108]

Other European countries that emerged as new entities from the war also found it crucial to perform their identity to the exhibition audiences. The Yugoslav Day, held on July 29, 1934, saw "girls in national costume" holding national dances at various sites of the fair.[109] Not without a parallel to the Czechoslovak efforts to be represented as a single nation, the Yugoslavs promoted their new polity that united South Slavs into a newly formed kingdom after the First

108 *Official Guide: Book of the Fair. Chicago: Century of Progress* (Administration Building: A Century of Progress, Chicago, 1933), 112. *Dzień polski na wystawie wszechświatowej niedziela dnia 26go sierpnia 1934 roku. Polish Day at A Century of Progress, Sunday August 26th* (n.p., 1934), 9.

109 *Official Guide: Book of the Fair. Chicago: Century of Progress*, 112.

World War. The Yugoslav Day was planned as a strong proclamation of this new and joint identity of the state, composed of different ethnic groups and nationalities. Here, for instance, the Serbian Juniors performed a circle dance, the Croatian Singing Society Zora presented various songs and the Slovene School Band staged folk dances and a mock wedding.[110]

Drawing out the contrast of the new state and the old order, the guide to the day made an expressive statement about the historical inaccuracies of the first Chicago world's fair in 1893, in which the Austrian Day featured a medley of national costumes presented as Austrian and Hungarian. They were, in fact, comprised of people from Slavic provinces, such as Croatia, Slavonia, Dalmatia, Slovenia, Bosnia-Herzegovina, Bukovina, etc., who represented the "many races, all earnestly desiring self-determination and individual freedom."[111] In contrast to 1893, the Yugoslav Day in 1934 attempted to give recognition to the individual ethnic groups, or at least to the largest ones.

Days dedicated to particular states, based on their new national and state identities, took place alongside other special days that were not necessarily linked to a single political entity. This was, for example, the case of the Scandinavian Day that featured Norway, Sweden, Denmark *and* Finland. And while national pavilions were not granted to groups that did not form a state, these groups could stage their own days. Chicago therefore also hosted the Welsh day and the Jewish day, the latter including a pageant called "The Romance of a People."[112] Similarly, the "Negro Day" on August 12, 1933 (more on this in Race chapter) was presented as an occasion for African Americans to get together and participate in the evening pageant titled "The Epic of a Race" with African dances, scenes of cotton picking slavery and emancipation, as well as spirituals sang by a large-chorus.[113] Yet this particular event ended up a fiasco; it was orchestrated by the fair officials and not members of the African American community who largely ignored it.[114]

Most elaborate yet was the Czechoslovak Day in 1933, held at the Soldier's Field stadium near the exhibition grounds on August 20. It can also be considered a pertinent example of what such Czechoslovak days included and stood for; they were a mixture of nostalgia, pride and ambition of the local diaspora.

110 *Yugoslav Day,* July 29, 1934, Lagoon Theatre. A Century of Progress International Exposition, Chicago (n.p., 1934), 18.
111 John A. Zvetina, "Yugoslav Participation in the Columbian Exposition of 1893," in *Yugoslav Day*, 4.
112 *Official Guide: Book of the Fair* (Chicago: A Century of Progress Administration Building, 1933), 113.
113 "Negros Plan Many Events for Day at Fair Tomorrow," August 11, 1933, scrapbook, CPI.
114 Ganz, *The 1933 Chicago*, 114.

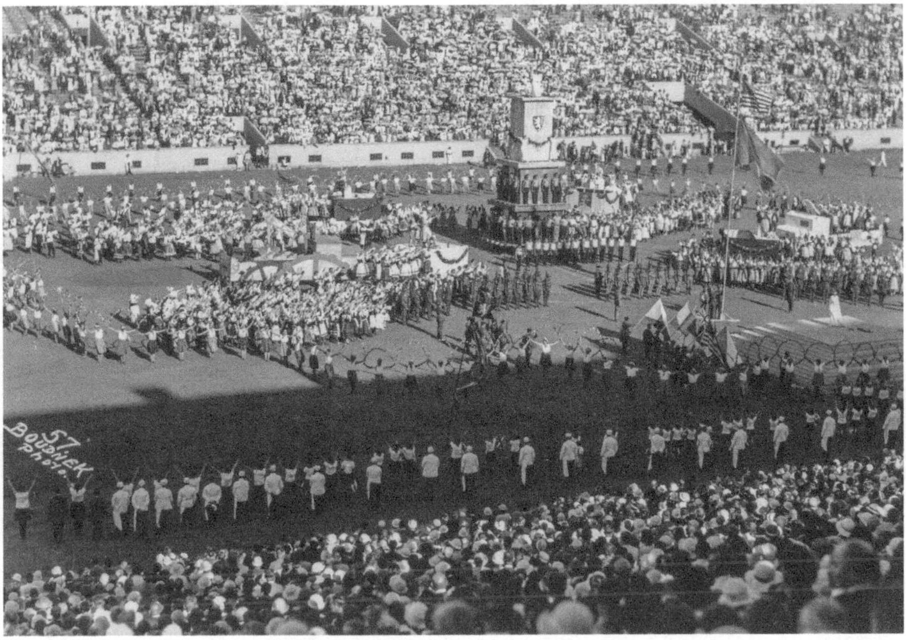

Fig. 52. Pageant "Oh my Country." Century of Progress, Chicago, 1933.

For many, the Czechoslovak Day became "the manifestation of joy" that tried to rekindle the success of the Bohemian day in 1893.[115] In 1933, the manifestation was planned as a "massive expression of [the] united branch overseas," showing the unity of the Czechs and Slovaks to the American audiences.[116] The participants were recruited from émigré communities with small numbers coming to join the festivities from Czechoslovakia. A gathering of 1800 Sokol gymnasts and a tableau vivant under the title "Oh, my country" consisting of some 4000 participants were some of the highlights [fig. 52]. In total, around six thousand people took part in the Day and marched or drove to the Soldier Fields with tableaux on cars of fourteen patriotic societies and companies, followed by children from the freethinkers schools, the legionnaires, representatives of the Red cross and many others.[117] There were reportedly also 120 horse riders in a re-enactment of the folk annual event Ride of the Kings, regional dance performances, and Czech war veterans, who were most probably the Czech and Slovak

115 "Československý den na Výstavě pokroku," *Orgán Československých spolků v Americe Č.S.A.* XLI, no.7 (July, 1933): 255.
116 "Československý den," 255.
117 "Československý den," *Spravedlnost*, August 25, 1933, 3.

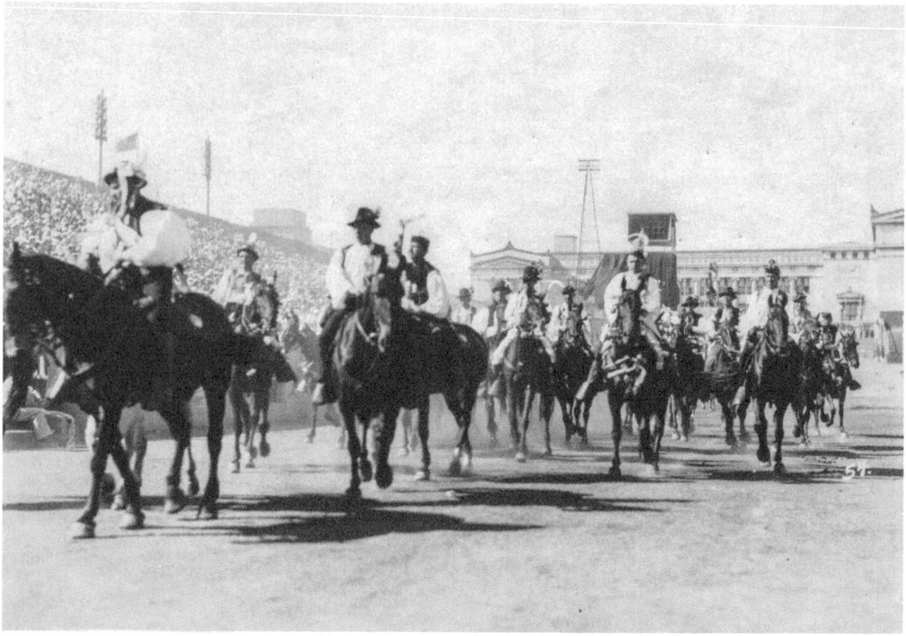

Fig.53. Horse parade at Czechoslovak national day.

legionnaires[118] [fig. 53]. A football match between the clubs Sparta of Chicago and Slavia of Cleveland took place in the afternoon. The day ended with a reception in the Hall of Sciences which included "a program of folklore with native singing and dancing."[119]

In terms of practical organization, the Czechoslovak Day was planned out by the Executive Committee of the Czechoslovak Associations in America, which consisted of five main board members and thirteen deputies and administrators, all men, replicating the gender composition of the organizing committees in Czechoslovakia.[120] The main organizers were the Czechoslovak minister (ambassador) to the United States Ferdinand Veverka, the president of the Czech Senate František Soukup and J. A. Červenka, the president of the Czechoslovak exhibition committee representing all Czech national societies in the USA.[121] While the Day was presented as Czechoslovak and indeed, many Slo-

118 "Huge Crowds at Fair See Czech Parade," Century of Progress Scrapbook, July 17–August 8, 1933, Box 2, folder 1, CPL.

119 "Huge Crowds at Fair."

120 "Orgán bratrstva ČSA," *Orgán československých spolků v Americe ČSA*, vol XLI, July 1933, 275.

121 "Huge Crowds at Fair;" "'Náš' den na výstavě Století Pokroku dnem dokonalého triumfu. Úchvatné scény ve Stadiu Vojenské Pole," *Svornost*, August 21, 1933, 1–2.

vak clubs and organizations participated in it, the Czech element and emphasis prevailed, just like in the national pavilions. Moreover, none of the extensive reports on the Czechoslovak day mentioned participation of the American Ruthenians or representatives from Sub-Carpathian Ruthenia. It therefore seems that Sub-Carpathian Ruthenians in the USA were not engaged with the Czechoslovak pavilion or its activities. It may be linked to their disenchantment with the Czechoslovak politics, because Ruthenia was supposed to have autonomy by 1933. Many Ruthenians also might not have felt any historical or cultural affiliation with the new state.

At the same time, the combination of aspects of folk culture with music and sporting traditions yet again communicated that Czechoslovakia was not a brand-new country, as could have been perceived. "...the form of government is new, but the country is old," claimed the promotional pamphlet which put emphasis on the country's historicity and traditions.[122] The spectacular pageant "Oh, my country" showcased this to the general public, presenting "the struggle of ... the once free people" against tyrants and "the brutal hands of its Teutonic and Magyar neighbors" throughout the centuries.[123] The spectacle was concluded with the literal and metaphorical rise of The Spirit of Slavs and the ultimate birth of the united nation after the war. A tribute to America and to the helping hand of Woodrow Wilson closed the story. From the description, the event—devised by Jarek Jelínek, the chief of the Czechoslovak Sokol in America with the help of local Czechs and Slovaks, is uncannily reminiscent of Alfons Mucha's grandiose painting project, The Slav Epic, which on twenty sizeable canvases follows a similar evolutionary trajectory of the nation. It features the same Herderian narrative of a struggle against a tyrant and completes with a victorious independence backed by allied powers under the arms of a supreme Slav.

Conclusion

National days at fairs were great occasions to showcase live performances of songs, theatre, or dances deemed representative of a particular nation. The alleged authenticity was, indeed, constructed and selective. Czechoslovak days did not include any performances of minorities, however substantial, and in-

122 "Czechoslovaks and the United States," *The World's Fair Memorial.*

123 "'Náš' den na výstavě Století Pokroku dnem dokonalého triumfu. Úchvatné scény ve Stadiu Vojenské Pole," *Svornost*, August 21, 1933, 1–2.

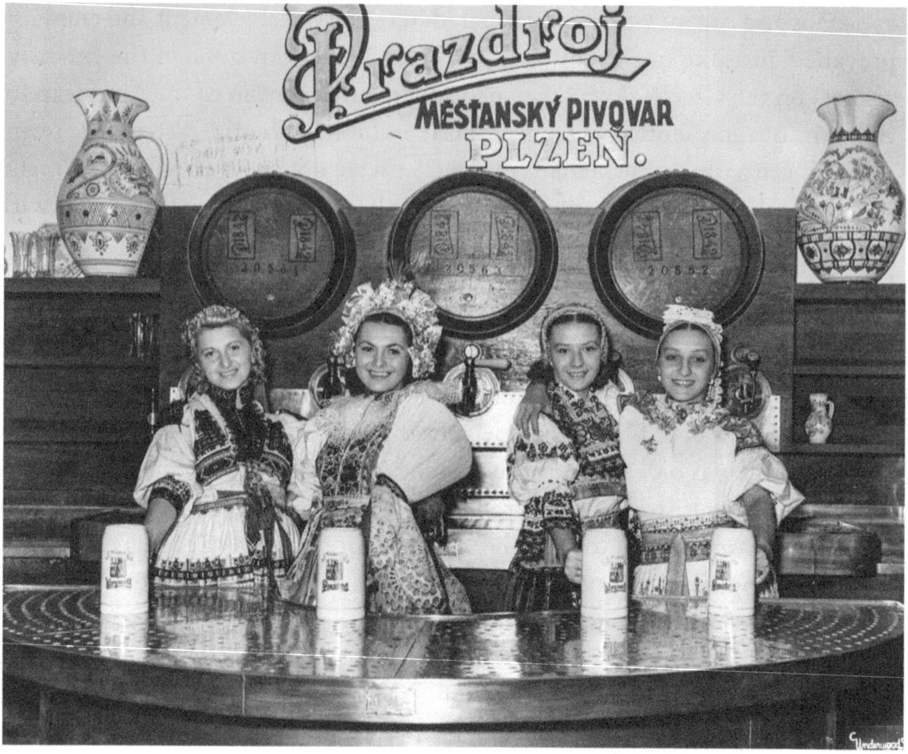

Fig. 54. Women selling beer in Pilsner restaurant.

stead employed the reductive concept of Czechoslovakism. This, nevertheless, was not exclusive to Czechoslovakia as diversity was not something to be commonly broadcasted or explained within the limited space and narrative of national presentations.

National pavilions with the accompanying events like national days were not only constructed by government officials and their proxies but also by those that were not as visible. The more minor players were often marginalized in major society but contributed significantly to Czechoslovak displays at world's fairs. The diaspora was key in supplying the content, ideas and some workforce for the pavilions in the USA. While not fully integrated in the Czechoslovak narrative in Chicago, it significantly contributed to the displays in 1939 when it replaced the disappeared state in San Francisco and New York. Yet the diaspora itself diversified along the lines of ethnic, gender, religious, political and generational divides with groups and members more or less visible or vocal. It did not fully copy the composition of the state in Europe because it consisted of voices that were more embedded in American culture: for instance, younger generations felt integrated with American society, women contributed their exhibits and workforce, and Slovak influence was much stronger [fig. 54].

Together with the marginalized groups in the state itself, the involvement of all these actors and their intersections further complicates the unanimous image Czechoslovakia planned to promote. These many "others" did not immediately fit the dominant narrative of a modern, progress-driven and ethnically homogenous nation. Their absence and marginalization in national pavilions, most noticeable in the case of women and ethnic minorities, nevertheless largely reflect the inner workings of the Czechoslovak state and its society.

Chapter Six

RACE, SCIENCE, AND ETHNICITY

MOLDING THE IDEAL CZECHOSLOVAK

I have so far examined how emphasizing different aspects of culture, society and business contributed to the construction of place a that the Czechoslovak state occupied, both physically and mentally, in the world's fairs. As a newcomer on the political scene, the state had to find a meaningful spot next to the world powers and their colonies as well as other nation-states. I have argued that the construction of the Czechoslovak nation and the state was based on the presentation of Czechs and Slovaks as a single group albeit with important ethnic or cultural distinctions. I have pointed out the hierarchy in what roles the Czechs and the Slovaks (let alone other groups) played within this narrative, whether it was in, for example, art and design or the representation of businesses in the national pavilions. These somewhat crude dividing mechanisms often casually bypassed the inclusions of minorities at the expense of the dominant Czech culture and economy.

The hierarchies in the national pavilions were symptoms of more general perceptions of who the Czechoslovak people were. They also reflected wider trends in understanding how nations and ethnic groups were formed. In the interwar period, many scholars of history, anthropology or culture, as well as politicians, philosophers and writers explained ethnic hierarchies within Czechoslovakia in lectures, articles, illustrations and exhibitions. They represented official attitudes to how racial groups, using contemporary terminology, and human development more generally were classified and systematized.

This chapter concentrates on discussions about the nature of the Czechoslovak people and the nation for the purposes of exhibitions as much as for the more universal location of the Czechoslovaks in the global structures. Throughout the interwar period, a range of individuals tried to describe not only who the nation and the people were but also how they can be improved. World's fairs and large exhibitions were an ideal outlet for such practices as they were themselves driven by the idea of progress, however abstract or utopian. As Robert W. Rydell reminded us, "at the fairs, the idea of technological and national progress became laced with scientific racism" from quite early on.[1] He associated exhibits related to scientific racism with the efforts of exhibitors and organizers to convey a vision of specific progress, linked to ever improving and prospering society.[2] While Rydell has mainly discussed the American world's fairs from the point of view of American society, "scientific racism" and related beliefs have been widespread inside and outside other world's fairs.[3]

In this chapter, I focus on the proposals for the improvement of humanity that were informed by social Darwinism, Mendelianism and eugenics known particularly from the United Kingdom and the USA and I show how these ideas were linked to the constructions of the "Czechoslovaks" inside and outside exhibitions. Social Darwinism argued for the survival of the fittest not only in terms of biology, but also culture, politics, or economics. Mendelianism, a set of theories named after the Moravian monk Gregor Johann Mendel (1822–1884), explained how certain traits were passed from parents to offspring, while eugenics has been considered "a biological theory of human improvement grounded almost exclusively in ideas of race and class."[4]

As a term, eugenics was first introduced by the British scientist Francis Galton in 1883 to define the study of the hereditary differences of mental, moral and physical traits of individuals, classes and races and second, measures of social control to ensure the general improvement of the species.[5] It came to stand for a diversity of things and ideas, adopted into many political ideologies, most notably Nazism, fascism and nationalism. The links between eugenics and the Nazi racial hygiene program have been explored well but as Marius Turda has pointed out, eugenics needs to be scrutinized as a concept closely tied to modernism and the construction of modern identities.[6] The place of eugenics in world's fairs, driven by the idea of progress and improvement, therefore does

1 Robert W. Rydell, *All the World's a Fair: Visions of Empire at American International Expositions, 1876–1916* (Chicago: Chicago University Press, 1985), 23.
2 E.g., Robert Rydell, "Fitter Families for Future Firesides: Eugenics Exhibitions between the Wars," *World of Fairs: The Century-of-Progress Expositions* (Chicago: University of Chicago Press, 1993), 38–60; Rydell, *All the World's a Fair,* 220–233.
3 E.g., Tenorio-Trillo, "Natural History and Sanitation in the Modern Nation," *Mexico,* 142–157.
4 Marius Turda, *Modernism and Eugenics* (Basingstoke: Palgrave Macmillan, 2010), 1.
5 Green, "Veins of Resemblance," 8. Galton, *Inquiries into Human Faculty and its Development* (London – New York: J.M. Dent and E.P. Dutton, 1907).
6 Turda, "Introduction," *Modernism and Eugenics,* 1–2.

not surprise and is informed by the hierarchical distinctions between various ethnic groups established in colonial displays and ethnographic shows that also applied narratives of selectiveness and racial betterment.

Eugenic theories have generally been intertwined with racist ideologies and the belief in the superiority of one group over another; and this was one of the characteristic features for the display of people in these environments. Therefore, as Rydell suggested, "world's fairs provide a partial but crucial explanation for the interpenetration and popularization of evolutionary ideas about race and progress."[7] Considering eugenics in the context of nationalism in modern central and eastern Europe, Turda has further noted that eugenics also served as a cultural and social philosophy of identities, which—and this is what I emphasize here—had a grave impact on exhibitions and world's fairs.[8] Eugenics was therefore an important aspect of both world's fairs and nation-building. Czechoslovakia yet again serves as a pertinent example of how a smaller nation adopted and promoted such ideologies for its own purposes and advancement. The way the internal, or rather national, divisions of the inhabitants of Czechoslovakia were perceived and presented at world's fairs also reveals the attitudes towards more general development of humankind and understanding of race and ethnicity.

Establishing hierarchies

Creating comparisons between pavilions, goods, the arts or different groups of people was a common feature of exhibitions of various sizes.[9] Colonial and ethnographic exhibitions were most known for establishing hierarchies between people, combining political and commercial motivations with entertaining displays of picturesqueness and cultural diversity. From mid-19th century onwards,

---

7 Rydell, *All the World's a Fair*, 24.

8 Turda, *Modernism and Eugenics*, 1; and Robert W. Rydell, *World of Fairs. The Century-of-Progress Expositions* (Chicago: University of Chicago Press, 1993), 38–58.

9 Zeynep Çelik, *Displaying the Orient: Architecture of Islam at Nineteenth-Century World's Fairs* (Berkeley: University of California Press, 1992); Sadiah Qureshi, *Peoples on Parade: Exhibitions, Empire and Anthropology in Nineteenth-Century Britain* (Chicago: Chicago University Press, 2011); Nicolas Bancel et al. eds. *Zoos humains. Autemps des exhibitions humaines* (Paris: La Découverte, 2002); Dana S. Hale, *Races on Display: French Representations of Colonized Peoples, 1886–1940* (Bloomington: Indiana University Press, 2008); Matthew F. Bokovoy, *The San Diego World's Fairs and Southwestern Memory, 1880–1940* (Albuquerque: University of New Mexico Press, 2005); Nancy J. Parezo and Don D. Fowler, *Anthropology Goes to the Fair: The 1904 Louisiana Purchase Exposition* (Lincoln: University of Nebraska Press, 2007); James Gilbert, *Whose Fair? Experience, Memory and the History of the Great St. Louis Exposition* (Chicago: Chicago University Press, 2009).

large colonial powers would develop their presentations into elaborate showcases of goods and culture in the broadest sense. In Europe, colonial expositions or exhibitions with colonial elements established a hierarchy between Europeans and Europe's colonial subjects, motivated by demonstrating progress in human development, economic and political power, and the superiority of western culture in general. As Czechoslovakia would enter these established classifications as a newcomer, it clearly associated itself with the western paradigm. During the First World War, exiled politicians had already established the future state's key political allies when they negotiated the future composition of Central Europe and developed their version of democracy, modernity and attitudes to minorities. Alliances were forged not only in politics but also in the fields of the visual arts and science, all crucial areas when constructing a vision of a nation and when constructing a national pavilion.

In the interwar period, various writers and scholars tried to define the basic traits of the Czechoslovaks and explain (read: justify) their existence, composition and future direction. I have briefly discussed the concept of Czechoslovakism and its adoption in the national pavilions earlier in relation to the place Slovaks and Ruthenians were granted within the project. The historical and biological relationship to other minorities in the state had to also be defined and to this degree, a crucial role was played by anthropology.

Many scholars have examined the close links between anthropology and world's fairs and focused on the displacement and mistreatment of native inhabitants, as well as their occasional rebellious and subversive behavior in these environments.[10] In the context of interwar Czechoslovakia and Central Europe, recent studies have highlighted the uncomfortable involvement of anthropology (and ethnography) in shaping racial policies and racist thinking.[11] Some attention has been paid to the effect this had on specific exhibitions, especially ethnographic displays that either focused on the selective concept of the na-

10 David R. M. Beck, *Unfair Labor? American Indians and the 1893 World's Columbian Exposition in Chicago* (Lincoln: University of Nebraska, 2019); Emily C. Burns, "World's Fairs as Sites of Orientalism East and West," 173–195, in *Near East to Far West. Fictions of French and American Colonialism*, ed. Jennifer R. Hennerman (Denver: Denver Museum of Art, 2023); Robert W. Rydell, John E. Finding, and Kimberly Pelle, *Fair America: World's Fairs in the United States* (Washington, DC: Smithsonian Institution Press, 2000).

11 Demski and Czarnecka, eds., *Staged Otherness*; Brandon, *The Perils of Race-Thinking*; Milan Ducháček et al., *Za rovnocennost evropských plemen. Československá antropologie tváří v tvář rasismu a nacismu* (Prague: Ústav pro soudobé dějiny AV ČR, v. v. i.: Prague: NLN, 2023); Filip Herza, *Imaginace jinakosti. Pražské přehlídky lidských kuriozit v 19. a 20. století* (Prague: Scriptorium, 2020), Markéta Křížová, "Curators, Objects and the Indigenous Agency," *Reviews in Anthropology* 50, no. 1–2 (2021): 22–40, doi: 10.1080/00938157.2021.1962036.

tion, or the so-called *human zoos* that essentialized ethnic and racial differences between people.[12]

Hierarchically classifying and displaying groups of people was a concern of Czech ethnography and anthropology already from the end of the 19th century too. It was aimed at visualizing ideas of their origin, physical differences and diverse living conditions. The main purpose of such practices was to provide definitions and historical justification of the Czech, respectively Czechoslovak people. One of the major events for constructing Czech national and ethnic identity was the Czechoslavic Ethnographic Exhibition held in Prague in 1895. It focused on rural culture and the peasants mainly of the Czech lands, with displays of recreations of rural dwellings, figurines of peasants and live performances of folk customs. The Czechs were presented here as the "repository of identity and seat of patriotism,"[13] while Czech national culture was framed as traditional and historic, free from influences of other ethnic groups in the Czech-speaking lands.[14]

One of the scholars who took an active part in the Exhibition was the Prague-based anthropologist Jindřich Matiegka (1862–1941). The first Czech Professor of Physical Anthropology at Charles-Ferdinand University in Prague played a vital role in establishing ethnic categories of people on the territory of Czechoslovakia. The origins of his classifications date back to the Exhibition in 1895 where he promoted the existence of a joint ethnic group, the Czechoslavs, an invented ethnicity which gave the title to the Exhibition. The term Czechoslavs was chosen to proclaim the unity of all Czech and other Slavic inhabitants against the notion of the Deutschböhmen, which described the German inhabitants in the region. Together with the ethnographer Lubor Niederle (1865–1944), Matiegka displayed tables, graphs and maps of extensive measurements of various individuals and their physical state.[15] The measurements focused on their weight, height, sizes of skulls etc. and it was not uncommon to deduce other features and traits from these physical measurements.

---

12 Apart from Demski and Czarnecka, *Staged Otherness*; Herza, *Imaginace jinakosti,* also Filipová, "Peasants on Display," Corinne Geering, "'Is This Not Just Nationalism?': Disentangling the Threads of Folk Costumes in the History of Central and Eastern Europe," *Nationalities Papers* (2021), doi 10.1017/nps.2021.21.

13 Derek Sayer, *Coasts of Bohemia: A Czech History* (Princeton, NJ: Princeton University Press, 1998), 119; also Catherine Albrecht, "Pride in Production: The Jubilee Exhibition of 1891 and Economic Competition between Czechs and Germans in Bohemia." *Austrian History Yearbook* 24 (1993): 101–118.

14 Filipová, "Peasants on Display."

15 F. A. Šubert, "Národopisná výstava československá," *Osvěta, listy pro rozhled v umění, vědě a politice* 25 (1895): 34.

Many of these findings had specific aims, and the establishment of distinctions between ethnic groups was one of the most crucial. Matiegka tried to determine with his research of the brain weight of different "races," for instance, that the Czech brain was heavier than that of Romanians, Slovaks, Germans, or Poles. These today questionable measurements led him to conclude that "while we cannot overestimate the importance of the brain weight for intelligence, we need to acknowledge that the organ which serves mental activities corresponds in development with the high abilities and mental capabilities of the Czech people."[16] Matiegka therefore proposed certain superiority of the Czechs based on their physical and mental capacities.

After independent Czechoslovakia was founded, Matiegka continued his research of the physical qualities of local inhabitants in the service of the new political establishment. He also presented and published his findings in Czechoslovakia as well as abroad.[17] His foreign publications helped to spread internationally his ideas about what and who the Czechoslovak people was. He published the article "The origin and the beginnings of the Czechoslovak people," for example, in the *Annual report of the Board of Regents of the Smithsonian Institution* in 1921 where he explained the history, ethnic composition (or physical characteristics) of the Czechoslovak people.[18]

Now that the Czechoslavs became Czechoslovaks, anthropologists, together with scholars from other disciplines, provided evidence for existence and standing of the Czechoslovaks to international as well as local audiences. It was clear that a nation, especially a new one, could be described and defined in many ways and using various arguments and proofs. As Matiegka's approach demonstrates, detecting the similarity of the physical traits (especially people's height and pigmentation), identified by anthropologists and ethnogenetists, was a suitable tool for many scholars and politicians to determine a common base of the people or nation as the two were often used as synonyms for each other.[19]

It was not only shared commonalities that were important for the composition of ethnic groups but also external influences. In the texts written for and about the new state, Matiegka acknowledged the beneficial impact of historic

16 Jindřich Matiegka, *Vznik a počátky národa československého* (Prague: Vilímek, 1917).

17 Matiegka, *Vznik a počátky*. Matiegka, *Tělesné vlastnosti českého lidu* (Prague: Vilímek, 1917), Matiegka, "Tělesná povaha dnešního lidu československého," in *Československá vlastivěda II. Člověk* (Prague: Sfinx, 1933), 193–240.

18 Matiegka, "The origin and the beginnings of the Czechoslovak people," *Annual Report of the Board of Regents of the Smithsonian Institution 1919* (Washington: Government Printing Office, 1921), 471–486.

19 Matiegka, "Ethnogenie českého národa na základě archaeologickém a anthropologickém," *Památky archaeologické* XXIX, no. 2 (1917): 109–123.

mixing between races. Expanding on his research of the mental capabilities of Czechs, he concluded that the Czech nation originated from "various racial and national elements including Slavic, Gallic and Germanic influences. Their 'mixing and crossing' strengthened the physical capabilities and [...] mental abilities of the Czechs."[20] Matiegka argued that the most advanced nations were also the most complex in their racial and ethnic composition and that this "produced a strong culture."[21] He recognized diversity as a necessary element in the composition of the new nation.

Matiegka's views were not dissimilar to his colleague Aleš Hrdlička (1869–1943), a highly influential anthropologist of Czech origin based in the USA from 1881 whose anthropological theories, approaches, and activities played a vital role in Czechoslovakia. Hrdlička is, until today, a recognized figure in Czech and American anthropology whose often unethical practices have only slowly been critically re-assessed by recent research.[22] Hrdlička built an important position as the curator of physical anthropology of the U.S. National Museum, today part of the Smithsonian Museums, and became advisor to President Franklin D. Roosevelt on migration (and to a large extent racial) issues during the Second World War. He also founded the influential *American Journal of Physical Anthropology* and was a frequent participant at international scientific congresses, including those on eugenics which he himself defined as the "science of improving the human stock."[23] Such orientation of interwar anthropology and ethnography was universal and adopted in variations in many national contexts, including Czechoslovakia.

Hrdlička conducted field research in Northern and Southern America, Africa and Asia either on his own or via his colleagues, some from Czechoslovakia.[24] He was also implicated in questionable practices at the turn of the century world's fairs and was intimately involved in the Louisiana Purchase Exhibition in St. Louis in 1904, as well as the Panama California Exhibition in San Diego in 1915. At San Diego, for instance, Hrdlička co-curated the natural history exhibit Science of Man which showcased the racial division of people according

20 Matiegka, *Vznik a počátky*, 53.
21 Matiegka, *Vznik a počátky*, 53.
22 Brandon, *The Perils*, Markéta Křížová, "Aleš Hrdlička a transatlantický rozměr českého antropologického a eugenického myšlení první poloviny 20. století," 42–75, in *Za rovnocennost evropských plemen Československ. antropologie tváří v tvář* (Prague: NLN – ÚSD, 2023).
23 Rydell, *All the World's a Fair*, 223.
24 Herza, "Ne/rovnocennost evropských plemen a cesty k autoritativním formám vládnutí: Jindřich Matiegka, Jiří Malý a Vojtěch Suk v meziválečných diskusích o rasách, národech a eugenice," 159–185, in *Za rovnocennost.*

to technological development, classifying them into stages between savages and civilized peoples.[25] It used visualizations of the different types in sculpture and casts and was accompanied by "The Races of Man" chart in which Hrdlička included a chart of categorized racial types. And before that, at the St. Louis World's Fair, he infamously retrieved brains of native Filipino persons who died there of pneumonia for his further research at the Smithsonian, causing considerable controversy in recent years.[26]

What are the Czechoslovaks

As an anthropologist and Czech native, Hrdlička maintained close relationships with colleagues in Czechoslovakia, helped to shape the development of the discipline of anthropology there and financed the establishment of the Czech journal *Anthropologie*, published between 1923 and 1941.[27] Hrdlička is often recognized as a promoter of the theory that all humans had a common origin and he applied it to explaining the existence and composition of the Czechoslovak people.

He also retained close contacts with the Czech and Slovak émigré communities in the USA and appeared in connection with the Czechoslovak pavilion at the Chicago world's fair in 1933. The pamphlet *World's Fair Memorial of the Czechoslovak Group*, put together by the local diaspora, which I have discussed in the previous chapter, included not only a description of the national pavilion but also a brief explanation of various aspects of the history and culture of the new country and its people.[28] Both the brochure and Hrdlička's contribution to it im-

---

25 Matthew Bokovoy, "San Diego 1915–16," in *Encyclopedia of World's Fairs*, 225. Bokovoy, *The San Diego World's Fairs and the Southwestern Memory, 1880–1940* (Albuquerque: The University of New Mexico Press, 2005), 90–93; Aleš Hrdlička, *Descriptive Catalog of the Section of Physical Anthropology, Panama-California Exhibition* (National Views, 1915).

26 Michael L. Krenn, ed., *Race and U.S. Foreign Policy from 1900 Through World War II* (New York – London: Garland Pub., 1998), 273, note 22. Also A. L. Lawshe to Edwards, 28 March 1905, RBIA RG 350, General Classified Files, no. 10699-3. Nicole Dungca, Claire Healey and Andrew Ba Tran, "The Smithsonian's 'Bone Doctor' Scavenged Thousands of Body Parts," *The Washington Post* 15 August 2023, https://www.washingtonpost.com/history/interactive/2023/ales-hrdlicka-smithsonian-brains-racism. The story of one of the subjects from the St. Louis fair was turned into a graphic novel Searching for Maura by Claire Healy, Nicole Dungca and Ren Galeno at https://www.washingtonpost.com/history/interactive/2023/maura-philippines-smithsonian-brain-collection.

27 The full title was *Anthropologie, časopis věnovaný fysické anthropologii, nauce o plemenech, demografii, eugenice a tělesné výchově se zvláštním zřetelem k Slovanům* (Anthropology, a journal on physical anthropology, learnings about races, demography, eugenics and physical education with special interest in the Slavs).

28 *World's Fair Memorial of the Czechoslovak Group (Czechs and Slovaks)* (Chicago: Czechoslovak Group, 1933).

ply that world's fairs and the related publications were indeed formative for verbalizing and visualizing the basic features of new nations like Czechoslovakia.

Hrdlička supplied an article on the physical and mental characteristics of "Czechoslovaks," which outlined the historic circumstances of the ethnic formations of the group and offers an insight into how the invented ethnicity was framed for external audiences.[29] The anthropologist overviewed here the geography of the newly formed state, its history and the different ethnic groups as well as the various traits of the inhabitants. Discussing the effects of the various historical struggles on the current racial composition, the invasions of the territory from Germans and Magyars led, in his view, to so-called *admixtures*.

According to Hrdlička, this mixing between racial groups contributed to the different cultural development of the individual parts of the nation and brought certain cultural advantages: "the Czechs particularly suffered, were much admixed, but through force of circumstances also rose culturally."[30] The Slovaks, on the other hand, were repressed and isolated from the west by Magyars, Hrdlička held.[31] The Slovak seclusion had restricted their cultural advance but in direct proportion to this, the Slovaks (and to an extent Moravians) had better preserved their folk art and their individuality, while the Czechs became more cosmopolitan.[32] The positive outcome of the Slovak isolation was for Hrdlička that the Slovaks and most Moravians "remained purer than the Czechs."[33] The assumed purity of the Slovak culture and arts could therefore be contrasted with the more sophisticated cultural progress of the Czechs.

Hrdlička's observations, backed by his position at the Smithsonian, therefore played an important role in communicating who the Czechoslovaks were to the world and the world's fair. As an anthropologist, he tried to demonstrate that diversity within the people can be traced in different body types and behavioral traits in a way reminiscent of Johann Gottfried Herder's characteristics of the Slavs and their typical features.[34] The notions of what constituted typical characteristics of Slavs, Germans and other groups were so influential for national

29 Aleš Hrdlička, "What are the Czechoslovaks," in *World's Fair Memorial*, 22–24. Hrdlička used "Czechoslovaks," the "Czechoslovak people" and the "Czechoslovak tribes" as an umbrella term common at the time.

30 Hrdlička, "What are the Czechoslovaks," 23.

31 Hrdlička, "What are the Czechoslovaks," 23.

32 Hrdlička, "What are the Czechoslovaks," 23.

33 Hrdlička, "What are the Czechoslovaks," 23.

34 Johann G. von Herder, *Ideen zur Philosophie der Geschichte der Menschheit*, 4 vol. (Riga and Leipzig: J. F. Hartknoch, 1784–1791).

revivalists across Central Europe.[35] In Hrdlička's view, the most preserved (read: ancient) types could be found in Moravia and Slovakia and had:

> good stature, strong, well-proportioned body, face more rounded than oval, physiognomy frank, smiling, intelligent and attractive, hair and eyes ranging from light to medium brown, absence of prognathism. Their principal mental characteristics are cordiality, sensitiveness, idealism, valour, with love of family, music, dance, and of everything good and beautiful. Also, considerable individualism, ingrained love of the soil and all that goes with it, of order and cleanliness.[36]

Such commentary from a scholar who engaged in establishing classifications of people based on anthropometric data should be read as his attempt at combining the quantifiable and measurable physiognomy with unmeasurable cultural and mental characteristics of specific ethnic groups. Hrdlička put together the latter approach, more typical of Romantic Nationalism of the previous century, with more rigorous methods in order to explain the hierarchies between the various ethnicities in Czechoslovakia as well as the historicity of the people in rural areas. These features were then indirectly used in presentations of Czechoslovakia. As I have shown earlier, the image of the state was partly embedded in folk culture and it was the sensitiveness, cordiality, and love of dance and music of the Moravians and Slovaks that materialized in the use of folk costumes and customs at world's fairs. The answers to Hrdlička's question what the Czechoslovaks were therefore helped to justify the divide in the presentation of the people from the eastern and western parts of the country.

Eugenics at exhibitions and congresses

Creating hierarchies of peoples within one (albeit invented) ethnic group was a fairly common practice, which in many cases was applied to determining racial differences. Hrdlička and his fellow anthropologists, too, worked on race and racial divisions. Hrdlička, for instance, was a member of the Committee on the Negro, established in 1926 by the American Association of Physical Anthropol-

35 For example Jan Kollár, "Dobré vlastnosti národu slovanského," in *Obrození národa, Svědectví a dokumenty*, edited by Jan Novotný, 191–205 (Prague: Melantrich, 1979); Karel Havlíček Borovský, "Slovan a Čech," *Pražské noviny*, February 15 – March 12, 1846.

36 Hrdlička, "What are the Czechoslovaks," 24.

ogy and the National Research Council in the USA. The committee conducted anthropometric research on black adults and babies, comparing them to young apes.[37] On top of the unethical side of such approach, data received from such simplified measurements were often misconstrued and misused to advance racist claims about supremacy of white people.[38]

One of the other researchers in the research group, Earnest A. Hooton, later published his findings in "What is an American?," in which he searched for common physical and other traits of Americans.[39] The study was not dissimilar to Hrdlička's enquiry about the Czechoslovaks and provides an insight into how race was subjected to the allegedly scientific classification that was so often visualized in world's fairs and various publications. The two authors shared not only a similar title but also line of thinking.

Hooton's American was descended from a European white settler, whose characteristics can be deduced on the basis of data and measurements. Hooton distinguished four different types, the eldest being called "old American" already by Hrdlička in 1925.[40] Hooton acknowledged the existence of Afro-Americans as well as "Real Americans," whom he identified as "Indians." They, however, "of course [did] not count" in his consideration of Americanism, without giving the reason why.[41] Yet Hooton also concluded that there were no anthropological grounds to create a selection between the different racial, ethnic or national groups. The purpose of the research was to "segregate and to eliminate the unfit, worthless, degenerate and antisocial portion of each racial and ethnic strain."[42] He, nevertheless, added that "candidates for such biological extinction would not be selected on the basis of Aryan or Semitic descent, blond hair or black skin, but solely on the score of their individual physical, mental and

37 A. E. Samaan, *From a "Race of Masters" to a "Master Race": 1948 to 1848* (published through www.createspace.com, 2013), 256; Aleš Hrdlička, "The Full-Blood American Negro," *The American Journal of Physical Anthropology* 12, no. 1 (1928): 15–33.

38 For a critical revision of scientific racism and anthropology, see for example Nicholas Thomas, *Colonialism's Culture. Anthropology, Travel and Government* (Cambridge: Polity Press, 1994); James Clifford, *The Predicament of Culture: Twentieth-Century Ethnography, Literature and Art* (Cambridge, MA: Harvard University Press, 1988); Daniel J. Kevles, *The Name of Eugenics: Genetics and the Uses of Human Heredity* (Berkeley, CA: University of California Press, 1986); Ta-Nehisi Coates, *Between the World and Me* (New York: Random House, 2015); Steve Garner, *Racisms: An Introduction* (Los Angeles: Sage Publications, 2010).

39 Earnest A. Hooton, "What is an American," *The American Journal of Physical Anthropology* 22, no. 1 (1936): 1–26. In fact, Hrdlička also offered his views on the American characteristics in his articles and lectures, see for example Katherine Scarborough, "Heart of America is Sound, Says Scientist," *The Sun*, 21 November 1926, 1.

40 Hrdlička, The Old Americans, Baltimore, 1925.

41 Hooton, "What is an American," 4.

42 Hooton, "What is an American," 26.

moral bankruptcy."[43] While Hooton, Hrdlička and many of their fellow anthropologists in the USA avoided applying theories that would openly endorse racial hygiene, they created other problematic criteria between groups of people based on their physical and mental capacities.

Some of the research for the study on Americans was conducted at the Century of Progress Exhibition in 1933–34, where anthropometric data was taken from about 6,000 visitors.[44] Many American world's fairs as much as the colonial expositions in Europe were considered a paradise for anthropologists because they brought together various peoples from distant parts of the world to one place who could be studied and categorized. And it also was at the Century of Progress in Chicago that eugenics, considered at the time the science of human improvement by better breeding, was applied for "classification for the basic sciences" for the first time at a world's fair.[45]

The relationship between eugenics and exhibiting had been established at the beginning of the 20th century when smaller eugenics exhibits cropped up at various places across the USA and Europe. They were frequently attended by Hrdlička and eventually his Czech colleagues. The First International Congress of Eugenics which was held in London in 1912, included a small eugenics display which mostly consisted of charts.[46] Subsequent congresses, like the Second International Congress of Eugenics in 1921 that took place in New York, included increasingly larger exhibitions with various casts, specimens and photographs. In 1921, A. H. Schultz of the Carnegie Institution displayed plaster casts of Caucasians and negro foetuses which were to illustrate the difference between the two races.[47] Hrdlička played an important role in the content of these exhibits and arranged here seven cases with his findings, aimed at proving that race was a biological category. The exhibits consisted of, for example, American Indian skulls that alleged the persistence of Neanderthaloid forms and "other primitive features" or "Old" (white) Americans that included so-called "degenerates ... producing unsocial offspring."[48] This exhibit was used repeatedly and would eventually appear—in part—at the Century of Progress too.

43 Hooton, "What is an American," 26.
44 Hooton, "What is an American," 6.
45 Charles Davenport, *Eugenics: The Science of Human Improvement by Better Breeding* (New York: Henry Holt, 1910), Harry L. Laughlin, "The Eugenics exhibit at Chicago," *The Journal of Heredity* 26, no. 4 (1935), 155.
46 *Catalogue of the exhibition, July 24th to July 30th, 1912 / First International Eugenics Congress*, London, University of London, South Kensington International Eugenics Congress 1912 (London: Charles Knight and co, 1912).
47 Rydell, *All the World's a Fair*, 47.
48 Rydell, *World of Fairs*, 47.

Hrdlička also participated in the Third International Congress of Eugenics that took place at the American Museum of Natural History in New York in 1932. This was also the first international meeting attended by Czech eugenicists in the role of contributors and exhibitors who had not taken part previously for financial reasons. The Czech biologist Vladislav Růžička (1870–1934), for instance, was appointed one of the vice-presidents of the Congress while the ethnographer František Pospíšil of the Moravian Museum in Brno and Vojtěch Suk, an anthropologist from Masaryk University in the same city, contributed to the exhibit at the Congress. The former, for instance, supplied "Photographs showing Asiatic traits in American Indians," based on his field research, while the latter added charts and models of teeth decay and blood groups in "white and negro races" as well as a section on vacations and health.[49] Comparison of what was seen as different stages of development of human races was used to establish order and classification between them. Czech scholars therefore engaged in the eugenic methods, which included visualizing difference and emphasizing desired characteristics racial hierarchies between groups of people.

Eugenics in Czechoslovakia

The active involvement of Czechs[50] in the Eugenics congresses and the views of anthropologists like Hrdlička and Matiegka on race and ethnicity indicate that eugenics in the interwar period was practiced in various forms worldwide as well as in Czechoslovakia. Its wide acceptance and presence thus resonate with the narratives about the Czechoslovak people presented in national pavilions which included the envisaged forward direction of the nation. Czech scientists and eugenicists had been in close contact with their colleagues in the United Kingdom, France and the USA, and some studied theories of racial hygiene promoted in Germany.[51] The scope of beliefs was also wide but many scholars and scientists at the beginning of the 20th century were quick to disassociate their

49 Henry Farnham Perkins and Harry Hamilton Laughlin, *A Decade of Progress in Eugenics; Scientific Papers of The Third International Congress of Eugenics, Held at American Museum of Natural History, New York, August 21–23, 1932* (Baltimore, The Williams & Wilkins co, 1934), 490, 499.

50 I refer to *Czech* eugenicists here rather than Czech and Slovak, because for most of the interwar period, Czechs were more dominant in this field. Eugenics in Slovakia is often discussed more in connection with the Slovak State during the Second World War. Anton Hruboň, "Creating the Paradigm of 'New Nation.' Eugenic Thinking and the Culture of Racial-Hygiene in the Slovak State," *Fascism* 10, no. 2 (2021): 275–297, https://doi.org/10.1163/22116257-bja10032.

51 Turda, "Introduction," *The History of East-Central European Eugenics, 1900–1945* (London – New York: Bloomsbury, 2015), xi–xii.

aims for applying eugenics from those of the German racial hygiene policy. If not outright racist, the German approach was seen as undemocratic and aristocratic, i.e., it was far too selective.[52]

The most important impulses were taken from the United Kingdom and the USA, especially from scholars like Francis Galton and the British statistician Karl Pearson whose first preoccupations were mainly in heredity and inheritance.[53] Another strong influence came from Mendelian genetics and focused on establishing patterns in heredity. As a "science," eugenics was institutionalized quite early on in the Czech environment; the Czech Eugenics Society was established already in 1915 with the neurologist Ladislav Haškovec (1866–1944) elected the first chairman.[54] Membership in the Society included a range of scientists as well as prominent figures, for instance, Edvard Beneš, the interwar foreign minister, prime minister, and eventually Czechoslovak president between 1935 and 1938.[55]

The Czechoslovak Eugenics Society, as it was renamed after 1918, played an active role in interwar institutional and medical practice of the new state. For example, it appealed to the government and President Masaryk to incorporate eugenics in the health program of the new state and give it official backing.[56] This included a call to support popular instruction on eugenics through public discussions, or theatrical and cinematographic performances. Exhibitions and a planned establishment of a Museum of Hygiene were also central, alongside Hrdlička's attempt to set up the Museum of Man in Prague and his involvement in founding *Anthropologie*. To evidence how prominent eugenics was in the state structures of the time, a few more bodies can be mentioned. Both the Czechoslovak Institute of National Eugenics and the Eugenics Committee at the Masaryk Academy of Labor, an academic institution for technicians and engineers, were established quite soon after 1918 and tried to put forward various legislation aimed at improving the health of and in society.[57]

Eugenics was therefore intimately bound with the Czechoslovak state. Soon after the end of the First World War and the establishment of the new state, Czechoslovak eugenicists, for example, the sociologist and philosopher Břetislav

52 Turda, *Blood and Homeland*, 155.
53 Turda, *Blood and Homeland*, 147.
54 Michal V. Šimůnek, "Overview," in *The History of East-Central European Eugenics*, ed. Turda, 131–132.
55 Turda, *Blood and Homeland*, 152.
56 Turda, *Blood and Homeland*, 153.
57 Ludmila Cuřínová, "Ústav pro národní eugeniku," 151–156, in *Technokracie v Českých zemích*, eds. Jan Janko and Emílie Těšínská (Prague: Akademie věd, 1999); Hruboň, "Creating the Paradigm," 280.

Foustka (1862–1947), presented eugenics as in the service of the healthy, plebeian and young nation.[58] Haškovec, too, promoted a similar view when already in 1921, he proclaimed,

> I remain convinced that through well-conducted and developed eugenic efforts we may strengthen and increase in power the life of our nation, both internally and externally. Eugenics is concerned with the health, not only of the individual but also of the entire nation.[59]

Presented as a western and democratic science, eugenics in Czechoslovakia could thus be employed to construct (and display) the new nation of healthy Czechs and Czechoslovaks.

At the same time, the findings of its advocates could be used in the name of the new political orientation of the state and as a reason to break off with the Austrian and Hungarian aristocracy that had ruled the regions until 1918. In this regard, the biologist and promoter of Mendelianism, Jaroslav Křížnecký (1896–1964), for instance, claimed he detected cases of inbreeding amongst the Habsburgs that led, in his view, to their degeneration.[60] Selective breeding by eugenics—Foustka suggested—could instead create a "new noble aristocratism," and a "new democratism for a new mankind."[61] The old aristocracy of the world that disappeared with the Habsburg rule would be replaced by a new class of healthy and fit families in the democratic system.

The question of class was, indeed, an important one as the new Czechoslovak nation was, at least in theory, built on liberal values of a just class system. Such ideas associated with eugenics therefore in many ways fitted into the pragmatic, democratic orientation of the state, or rather its political, cultural and academic protagonists. Moreover, many Czech scientists tried to develop their own national eugenics which considered the biological uniqueness of the nation as well as the influence of the external environment. Environmental determinism and the potential to shape the environment to improve humanity was

58 Foustka, "Etika a eugenika," 121–129, in *Pamětní spis ku oslavě stých narozenin J. G. Mendela, ed. Artur Brožek* (Prague: Borový, 1925), translated by Anna Pilátová, "Ethics and Eugenics," in *The History of East-Central Eugenics*, ed. Turda, 153–155.

59 Haškovec, quoted in Michal Šimůnek, "Eugenics, Social Genetics and Racial Hygiene: Plans for the Scientific Regulation of Human Heredity in the Czech Lands, 1900–1925," in *Blood and Homeland*, ed. Turda, 150, note 27. 145–166.

60 Jaroslav Křížnecký: "Zákonná úprava příbuzenskch sňatků s hlediska biologicko-lékařského," *Právník* 57 (1918): 325–31, in Šimůnek, "Eugenics," 133.

61 Foustka, "Etika a eugenika," 125.

a key issue which became prominent in the displays in Czechoslovak national pavilions, as I will mention a bit later.[62]

However, I also need to point out that eugenics was not accepted in Czechoslovakia unilaterally and it found many critics. G. K. Chesterton's critique *Eugenics and Other Evils* on the dangers of eugenic theories from 1922 was translated into Czech in 1928.[63] Already in 1918, the philosopher and biologist Emanuel Rádl (1873–1942) wrote *Rassové theorie a národ* (Racial Theories and the Nation) as a warning against racial politics in Germany and its relationship to modern-day German nationalism.[64] In the mid-1930s, many others disassociated themselves from eugenics interpreted as racial hygiene and from focusing on the superiority of specific ethnic groups based on biological differences. The psychologist and biologist Josef Meisner, for instance, published the book *Rasismus hrozí kultuře* (Racism Threatens Culture) in 1934, in which he advised against the misapplication of eugenics. The scepticism, however, was not a rejection of eugenics. Rather, it was a turn away from the German understanding of eugenics as supremacy of one race over others to a more careful attention to the selectiveness of desired traits and to a focus on hereditary qualities which had the potential to improve the nation.

Race and ethnicity displays

Visualization of the ideas that first took place at international congresses and smaller exhibitions, limited primarily to specialists, quickly transferred to the widely accessible world's fairs. I will leave aside the context of colonial exhibitions like the British Empire Exhibition in London in 1924–25 or the Colonial Exhibition in Paris in 1931 because Czechoslovakia did not take an official part in them. It did participate in The Century of Progress in 1933, though, where eugenic ideas infiltrated many instances of the fair. A keen proclamation of the organizers' belief in the ability of science to enable progress, to offer a better future and to overcome global depression, meant the fair was to "demonstrate the indebtedness of industry to pure science, the beneficent results of invention

62 E.g., Artur Brožek et al., *Rovnocennost evropských plemen a cesty k jejich ušlechťování* (Prague: Akademie věd, 1934). Cf. also Milan Ducháček et al., *Za rovnocennost.*

63 G. K. Chesterton, *Eugenika a jiná zla* (Prague: Ladislav Kuncíř, 1928).

64 Emanuel Rádl, *Rassové theorie a národ* (Prague: Kočí, 1918).

and applied science and the dependence of society upon all of these."[65] For example, the organizers of the third eugenics Congress in New York attempted to sell their exhibition of 267 items to the world's fair in Chicago, although only a small portion appeared there in the end.[66]

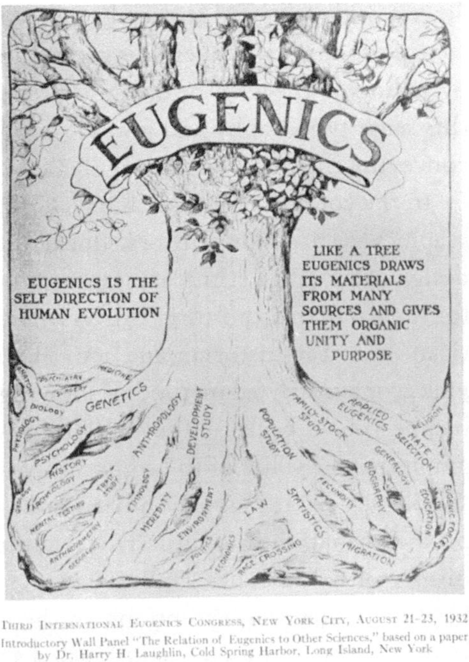

Fig. 55. "The Relation of Eugenics to Other Sciences" from A decade of progress in Eugenics. Scientific papers of the Third International Congress of Eugenics.

The Eugenics Exhibit in Chicago was placed in the Hall of Science, the largest pavilion of the fair and fit with the general tone of the depiction of progress at the exhibition. The Exhibit consisted of four large wall panels and charts showing the aims of eugenics for the "average intelligent world's fair visitor."[67] The first panel showed a graphic rendition of a tree of eugenics with roots in disciplines like genetics, environment, or anthropology [fig. 55]. The accompanying description outlined the role of eugenics "in improvement in the breed of man." Panel 2 included a chart explaining changes in "families, communities, races and nations" through generations caused by migration or birth rate.[68] The remaining two panels displayed family trees of two strikingly different groups: the "old, extensive and degenerate Ishmael tribe" from Indiana was contrasted with "a superior family" of the Roosevelts. They were to show the importance of hereditary qualities of the different family *stocks* as well as of the environment, meaning the physical, social, economic, educational and spiritual surroundings.

Although the Eugenics Exhibit was not extensive, by using infographics and charts, the displayed information was made official and legitimized as science.

65 *The Basic Sciences and Medicine. A Century of Progress International Exposition Chicago 1933* (Chicago, 1932), 3.
66 Rydell, *World of Fairs*, 55.
67 Laughlin, "The Eugenics Exhibit," 156.
68 Laughlin, "The Eugenics Exhibit," 158.

Since world's fairs had already become known as the world's universities and effective educational venues, this kind of information could be taken at face value by visitors.[69] The way the data was presented to visitors was also crucial and in this sense typography and graphic design played a crucial role in successfully conveying the intended message.

In the context of world's fairs, communication of complex information by graphics has had a long history. One of the earliest examples of the use of graphic design to visualize ethnic issues was produced by W.E.B. DuBois in his exhibit for the Parisian Exposition Universelle in 1900. DuBois (1868–1963) was an African American historian and activist who was tasked to come up with a tangible, contextualized method of showing why the African diaspora in America had been held back.[70] "The Exhibit of American Negros" consisted of photographs from the life of the community in the USA, accompanied by visualized sociological data in graphs and diagrams with vibrant colors, which turned complex information into visually appealing and comprehensive data sets. In the history of world's fairs, DuBois's images of the emancipation of the American blacks are, however, exceptions. In the USA as well as in Europe, world's fairs were very much initiated, constructed and populated by white Europeans and the Anglo-Americans who put themselves in the position of driving the progress, industrialization and modernity on display.[71]

Thirty years later, white supremacy in many respects still ruled world's fairs and its surroundings. In Chicago, the large Field Museum, for instance, located at the northern end of the grounds and focused on natural history, opened a new exhibit in 1933 entitled *The Races of Mankind*. It consisted of a series of 104 bronze sculptures of different "racial types" from around the world were created by the sculptress Malvina Hoffman (1885–1966), who was commissioned by Stanley Field (1875–1964), a Chicago businessman and banker, and the president of the Field Museum at the time.[72]

69 Rydell, *All the World's a Fair*, 23, referring to, e.g., "The World's Columbian Exposition," *World's Columbian Illustrated* 1 (February 1891): 2.

70 Britt Rusert and Whitney Battle-Baptiste, *W. E. B. DuBois' Data Portraits* (Princeton Architectural Press, 2018).

71 Rusert and Whitney-Baptiste, *W. E. B. DuBois' Data Portraits*, 19.

72 Linda Nochlin, "Malvina Hoffman: A Life in Sculpture," *Arts Magazine* (Nov.1984): 106–110; P. H. Decoteau, "Malvina Hoffman and the 'Races of Mankind,'" *Art Journal* (Fall 1989/Winter 1990): 7–12; Jeff Rosen, "Of Monsters and Fossils: The Making of Racial Difference in Malvina Hoffman's Hall of the Races of Mankind," *History and Anthropology* 12, no.2 (2001): 101–158; Gregory Foster-Rice, "The Visuality of Race: 'The Old Americans,' 'The New Negro' and American Art, c. 1925" PhD dissertation, The Northwestern University (2003); Linda Kim, "Malvina Hoffman's Races of Mankind and the Materiality of Race in Early Twentieth-Century Sculpture and Photography," PhD dissertation, University of California, Berkeley (2006); Marianne Kinkel, *Races of Mankind: The*

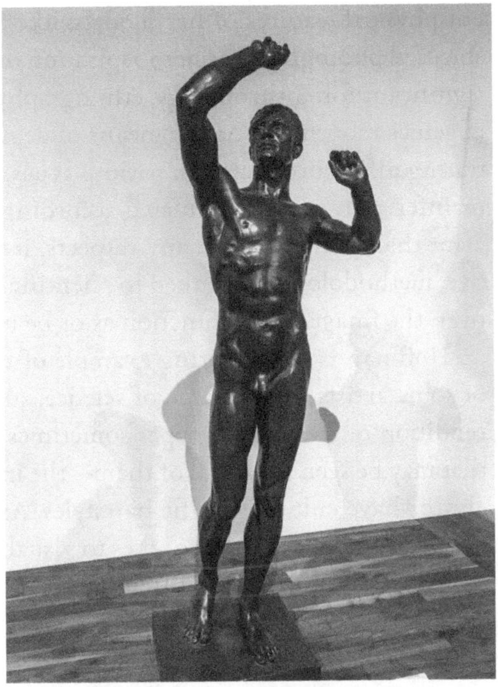

Fig.56. Malvina Hoffman, Tony Sansone, A Man from New York (The Nordic Type).

The title could be seen as a nod to the 1915 exhibit, the Science of Man, at San Diego world's fair where Hrdlička's chart "The Races of Man" was included. In 1933, Hoffman contributed 106 facial casts of people's types to the San Diego exhibit.[73] Her sculptures of full-size bodies and busts made very clear divisions of racial categories split into three main groups: the white, the negro and the mongoloid "racial stock." They included examples of a Native American, Japanese man, Sudanese or Inuit woman as well as the so-called Nordic type. The latter was an example of a "man of the white stock," using the terminology of the time. The model for the Nordic type as ideal representative of the white race was an Italian professional bodybuilder and dancer living in Brooklyn.[74] [fig. 56] Based on a classical statue, the young, muscular man was naked with his arms raised in a gesture that would be replicated in many different body representations around the fair that I will discuss shortly.

As it formed a part of the visualization of races at the Century of Progress, the sculptural group requires a bit more consideration. The original aim of "The Races of Mankind" was to show human progress with the white race as the pinnacle, supplemented by the wish to capture the "primitive man," disappearing due to progress, in a permanent medium of bronze sculpture.[75] Hoffman created the sculptures based on her research in the field and collaboration with anthropologists. This included taking anthropomorphic measurements of differ-

Sculptures of Malvina Hoffman (Urbana, IL: University of Illinois Press, 2011); Rebecca Peabody, "Race and Literary Sculpture in Malvina Hoffman's 'Heads and Tales,'" *Getty Research Journal*, no.5 (2013): 119–132.

73 Bokovoy, *San Diego*, 96.

74 Teslow, *Constructing Race*, 78.

75 Laufer, "Hall of the Races of Mankind," 1931, p. 3 quoted in Teslow, *Constructing Race*, 84.

ent physical features of her models, like height or skull size and occasionally, she used photographs as her inspiration or additional source. Photography was significant for anthropology, ethnography and fine art as it was crucial for the eugenics movement. As the means of depicting human physiognomy and illustrating differences between various types, it could serve as proof of superiority or inferiority and manipulated accordingly.[76] The medium was also more efficient than sculpture in many respects. It was believed as more accurate, objective, methodologically linked to scientific naturalism, and transportable. Moreover, the images could function as or be mistaken for the reality itself.[77]

Hoffman is thus a fitting example of the way racism appeared in the work of some artists in the name of science, such as anthropology or eugenics. Her rendition of the various types sometimes showed them with iconic attributes that may be seen as typical of them—the individuals would hold tools, wear jewelry or have culture-specific hairstyles. As such, they used both anthropological and ethnographic approaches to visualizing racial types in a combination of the physical features with often stereotyped cultural and social manifestations.

Visualizing race and humanity

Payment of attention to the human body, its features, traits and functions was another common denominator of the Century of Progress fair, albeit not publicly advertised. It deserves further attention because anthropology, social sciences, medicine and genetics exhibits visualized race and racial differences in the name of science, while the visual arts contributed to such practice by figurative depictions around the grounds. It was especially sculptures that depicted the idealized human body in its classical form; they appeared on pylons of the Hall of Social Science, the courts of the Hall of Science, and as reliefs on buildings like the Radio and Communications or Electrical Buildings. The naked figures here were stylized to recall Art Deco's interest in historic artistic precedents and their whiteness was detectable not only in the material but also in their physical features; they represented the ideal people.

More futuristic was a large sculptural group in white bronze at the entrance to the Hall of Science by another sculptress Louise Lentz Woodruff (1893–1966). De-

76 David Green, "Veins of Resemblance: Photography and Eugenics" in *Photography/Politics: Two*, eds. Patricia Holland, Jo Spence and Simon Watney (London: Comedia, 1986), 9–21.

77 Green, "Veins of Resemblance," 4.

scribed as "modern in tone," the fountain was referred to as "Science Leading Youth" or "Science Advancing Mankind" [fig. 57]. It depicted a larger-than-life, robotlike figure that bent forward, leading two people. The woman and man have Egyptian clothing and hairstyles, suggesting the origins of science in the ancient world. They were walking forward led by science represented by the technological and futuristic robotic marvel. The advancement of humanity, its forward-looking progress, was brought by a robot in a male form.[78]

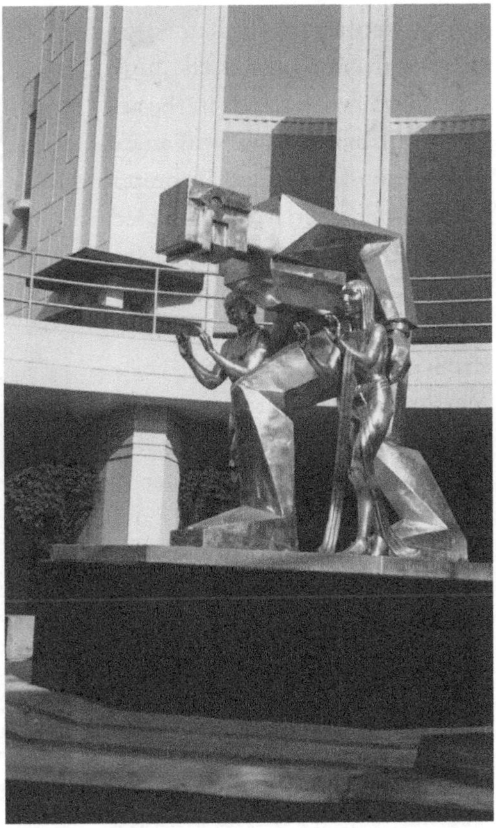

Fig.57. Louise Lentz Woodruff, *Science Leading the Youth*, 1933.

Another popular scientific visualization at the fair that conformed to the emphasis on improvement of the human and humanity was the "Transparent Man" in the Hall of Science. The six feet tall glass figure of a man with raised arms was made by "German experts,"[79] and supplied by the Deutsches Hygiene Museum in Dresden.[80] In terms of formal appearance, the figure evoked similar visual renditions of the idealized human body found at the time. Woodruff's sculpture, for instance, also featured an upward movement of the raised arms in an act of worshipping the sun. Malvina Hoffman's Nordic type to an

78 Cheryl Ganz reflects American identification of science and technology with masculinity in Cheryl Ganz, "Science Advancing Mankind" Technology and Culture, Oct., 2000, Vol. 41, No. 4 (Oct., 2000): 783–787.

79 "The Story of Medicine," *Official Guide: Book of the Fair, 1933* (Chicago: A Century of Progress Administration Building, 1933), 39.

80 Eric McLeary and Elizabeth Toon, "'Here Man learns about Himself': Visual Education and the Rise and Fall of the American Museum of Health," *American Journal of Public Health* (July 2012): 27–36, doi.org/10.2105/AJPH.2011.300560; cf. also Klaus Vogel, "The Transparent Man—Some Comments on the History of a Symbol," 31–61, in *Manifesting Medicine: Bodies and Machines*, eds. by Robert Bud, Bernard Finn, and Helmuth Trischler (Amsterdam: Harwood Academic Publishers, 1999).

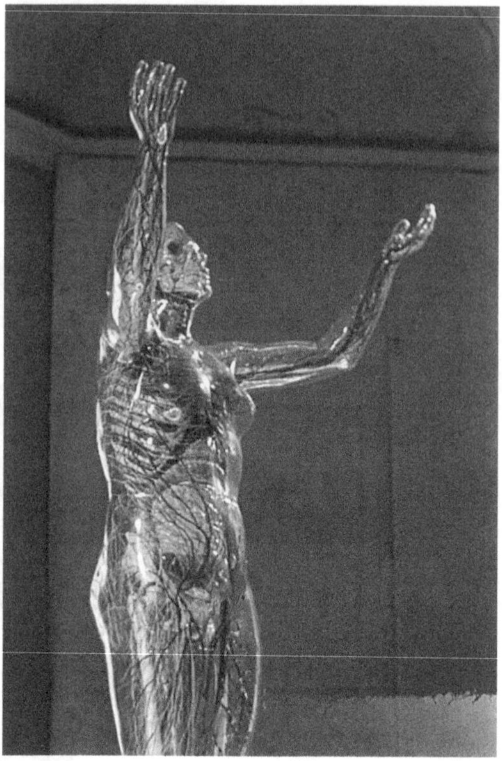

Fig.58. The Transparent Woman in the Museum of Hygiene, Dresden, 1947.

extent replicated the same posture that emphasized the muscular body with hands in the air.

"Transparent Man" was an explanation of the physical composition and genetics that affected individuals. It demonstrated the processes happening inside the body and how "cells combined from parents unite to create a new person."[81] The figure was accompanied by photographs, charts, moving models and show mechanisms of speech and thought and was located among exhibits from other countries, like France, Germany, Italy, and Austria.

For its ability to visually communicate the complex biological processes in the human body, "Transparent Man" was used in further displays which included eugenics exhibits. Other copies were made in Germany, one in 1936 for the Stockholm exhibition on mothers and children which became a popular exhibit a year later in Paris.[82] In 1936, "Transparent Woman" was created for the Museum of Science at the Rockefeller Center which toured the USA more as an advertising accompaniment to sell corsets[83] [fig. 58]. And another copy of the "Transparent Man" started touring the USA in 1934 as part of the *Eugenics in the New Germany* exhibition.[84] It was also put on display at the world's fair in New York in 1939 as part of the American Museum of Health exhibit in the so-called Hall of Man. While Germany was not officially represented at the World of Tomorrow, it still had a presence

81 "Biology Section Tells the Story of Man," *Scrapbook*, CPL.
82 Jeffrey T. Schnapp, "Crystalline Bodies: Fragments of a Cultural History of Glass," *West 86th A Journal of Decorative Arts, Design, History and Material Culture* 20, no. 2 (2013): 189, https://doi.org/10.1086/674728.
83 Schnapp, "Crystalline Bodies," 190.
84 Schnapp, "Crystalline Bodies," 189.

through such scientific contributions. This visual health education enshrined the aestheticized, normalized body in "a new type of dramatic textbook that appeals to both children and adults."[85]

The ideal man, the ideal Czechoslovak

The advancement of humanity, proclaimed at fairs, could be achieved by improvements in humanity's smaller units like nations, ethnic groups or individuals. As I have been trying to show, there are many overlaps between the nature of world's fairs, eugenic theories and nation-building. Creating a better nation by selecting and emphasizing wanted features is not far from showcasing the nation in world's fairs and exhibitions in its idealized form where the best aspects are accentuated, and weaknesses suppressed. And such practices were not limited to nations or groups of people, they extended to their individual components. The idealized man and the human body also commonly appeared in various forms in the Czechoslovak pavilions that further demonstrate how closely Czechoslovak exhibits were entangled in the international emphasis on classification of people.

Already in 1928, the national exhibition in Brno, that I discussed in Chapter Two, included the Pavilion of Man and Mankind in which the evolution of humanity through scientific discoveries by archaeologists and anthropologists was presented. The Brno-based organizational committee of the pavilion included previously mentioned palaeontologist Karel Absolon (1877–1960) and the anthropologists Niederle, Matiegka and Suk. Matiegka put together an exhibit about the composition and development of the inhabitants of current Czechoslovakia. They featured maps, hair samples, photographs and skulls, documenting the physical development, variations in their mental development and racial composition of the people.[86] They were complemented by skulls of native Americans and plaster casts of "Eskimos," which were Suk's contribution.[87] They were supposed to demonstrate that the features of so-called contemporary prim-

85 Schnapp, "Crystalline Bodies,"189.

86 Vladimír Úlehla, Jaroslav B. Svrček and František V. Vaníček, "Oddíl A. Člověk a příroda živá," *Výstava soudobé kultury v Československu, Brno, 1928*, 97; Petr Kostrhun, "90 let muzea Anthropos: Proměny muzea v období první Československé republiky," in *Transformations of the Anthropos Pavilion in Time (1928–2018)*, eds. by Petr Kostrhun, Barbora Půtová and Zdeňka Nerudová (Brno: Moravian Museum, 2018), 26.

87 Úlehla, Svrček and Vaníček, "Oddíl A.," 98.

itive nations could be compared to those of the diluvian predecessors of modern man in central Europe.[88]

In the context of the Exhibition of Czechoslovak Culture, the display was meant to help visitors understand the development of the inhabitants of modern-day Czechoslovakia. At the beginning of the 20th century, collections of "Mankind" were institutionalised in disciplines like anthropology and ethnography, and displayed in museums and at exhibitions.[89] In Brno, the Man and Mankind pavilion appeared in the Man and Nature section of the Exhibition which also covered exhibits on health, hygiene or medical care and health resorts. These created a context which could have an impact on the state of the Czechoslovak people and their well-being.

Man and Mankind was also the pavilion where a reconstituted mammoth, mentioned in Chapter Four, was placed. The prehistoric animal very well corresponded to the image of a pragmatic state which believed in evolution.[90] Its reconstitution was sponsored by Baťa and the company's financial involvement in the animal reconstruction can thus be seen not only as a philanthropic gesture, but also an expression of Baťa's philosophy. The mammoth, a symbol of evolution, corresponded with the vision of an empirically-oriented state presented at the exhibition, and also with the way Baťa ran their company with the emphasis on evolution (and even improvement) of man.[91] In this connection, the company Baťa, represented by its two directors, Tomáš and Jan Antonín, illustrates how private enterprises can relate to the official orientation of the state, as well as to worldviews, be it pragmatism, evolution or even eugenics.

In Chapter Four, I have also stressed the intimate links the company forged with the state in and outside of the world's fairs. Further parallels can be found between Baťa's approach to life and work on the one hand and the affinity to eugenics detectable in the Czechoslovak and other pavilions on the other. The attitude of Baťa to its workers, where the company looked after and oversaw every aspect of their life, was also geared towards creating a new man.[92] This man (using the period terminology which privileged the male gender) would be assisted by the machine that "rescued him from slavery."[93] This proposition

88 Letter of Karel Absolon addressed to the Moravian Land Committee from 19 November 1928. Archive of Anthropos Institute, ECC 1928, File November, quoted in Kostrhun, "90 let muzea," 39.
89 Clifford, *The Predicament of Culture*, 12–13.
90 Úlehla, Svrček and Vaníček, "Oddíl A.," 96.
91 Úlehla, Svrček and Vaníček, "Oddíl A.," 96.
92 *Výběr a výchova průmyslového člověka, instrukční příručka* (Zlín: Baťa, osobní oddělení, 1938), 15.
93 Coudenhove-Kalergi, quoted in *Výběr a výchova*, 12.

could, to a large extent, be visualized as the robot-like figure of Science leading the youth that appeared in Chicago.

This philosophy of Baťa was also compiled in the instruction guidebook *Výběr a výchova průmyslového člověka* (The Selection and Education of an Industrial Man) published in 1938, which comprized ideas of Tomáš and Jan Antonín Baťa and enthusiastic essay on the machine by Count Richard Coudenhove-Kalergi, an Austrian nobleman with Czechoslovak citizenship.[94] It also included the views of Emile Malespine on occupational health and of Alexis Carrel on positive eugenics. Carrel was a French doctor whose book Man, the Unknown was published in 1935. The Baťa booklet quoted this text and provided suggestions how to distinguish between people's roles. While women should return to their original natural role of not only bearing children but also educating them, men should perform such tasks they are most suitable for. Individuality should be encouraged instead of equality. And although the booklet avoided explicit mention of selective breeding that Carrel suggested elsewhere, the selectiveness in the focus on fit individuals with desirable qualities was certainly there.

The purpose of the instruction booklet was to give guidance to the Baťa managers on how to "create" the best employees. A new man, the employee of the modern age, could be improved by education and it was best to start as soon as possible. Nurseries, schools, training institutes that Baťa built conformed to these beliefs. Constructing and maintaining these facilities was an important topic that was also visualized for promotional purposes. Images of hospitals, maternity wards, nurseries and schools were used in international and national displays to broadcast the company's caring approach. Yet as Zachary Doleshal has noted, this was a part of the social engineering project of the management who put emphasis on appearance of the company[95] [fig. 59].

The Baťa vision of its workers further shows the company's understanding of hierarchy in society. It was around the Brno exhibition in 1928 that President Masaryk visited Zlín. He was greeted by and taken around by Tomáš Baťa who in turn acknowledged that the democratic state provided a thriving environment for the company. Addressing the president, Baťa asserted,

> The investment that our people made in you brought an immense benefit—our independence, which made free citizens from us and gave us the right

94 *Výběr a výchova.*
95 Doleshal, *In the Kingdom*, 133–134.

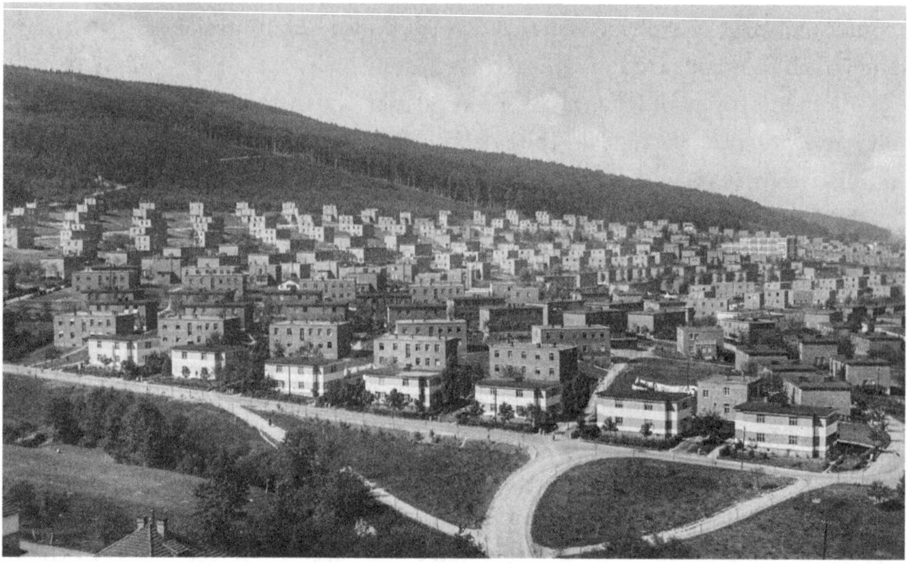

Fig.59. Zlín, houses for Baťa workers.

to run our business in our own way. This right binds us with the obligation of conducting business to the benefit of each citizen, for a better life of all.[96]

What is of note here is the language of trade and business related to the evolutionary progress of the Czechoslovak people that Baťa used for expressing his sentiments about the president and the state.

Such thinking was typical of the company. After all, in the mid-1930s, Jan Antonín Baťa proclaimed that "our success lies in the Czechoslovak man... in that we managed to free him from the position of servitude, proletariat and being a scared coward... We made him into a man of the world... " with high self-esteem.[97] Along the lines of business and people management based on Fordism and Taylorism, the Czechoslovak man evolved into a self-sufficient individual without a proletarian revolution. Under the same ideology, Baťa banned creation of trade unions in the company while the Communist party did notoriously badly in Zlín elections.

The President, too, played an important role in the evolutionary process of the Czechoslovak state. As I have noted already in Chapter One, Masaryk's like-

96 President republiky, dr. T.G. Masaryk na návštěvě ve Zlíně v Baťových závodech v červnu 1928, film, *The National Film Archive*, https://www.youtube.com/watch?v=db1ajG9Uiew.

97 J. A. Baťa, Brazilské stopy, https://www.youtube.com/watch?v=uDOxYghcT4A. Also quoted in Výběr a výchova, 23.

ness became an indispensable part of Czechoslovak pavilions and exhibitions and appeared in a great many locations across the country and abroad. Masaryk's age also made him a great role model for young people, and he was popularly known as "tatíček" in Czech, which translated to "little father."[98] The image conjured up by this appellation was one of a protector of the people, closer in age to a grandfather than a father. Masaryk depicted as the mature man was an exception in the image of Czechoslovak people conjured up at national pavilions, which stressed youth and vitality.

Anthropologists and other scholarly contributors helped to create a sense of the existence of healthy people, or suggested ways this could be achieved, through graphs, charts, historic artefacts and findings displayed at exhibitions. Yet the most visually effective was the representation of the Czechoslovak people in photographs and works of art that inevitably contributed to creating an image of the nation as healthy and ethnically homogenous. Large-scale photographs became frequent additions to the pavilions and showed life in the cities and the countryside, people at work and leisure, modern facilities and modes of entertainment. In this regard, Baťa's views of Zlín, which featured frequently in the national and company pavilions, conformed to this vision of healthy, happy people and emphasized youth, education, work and health.

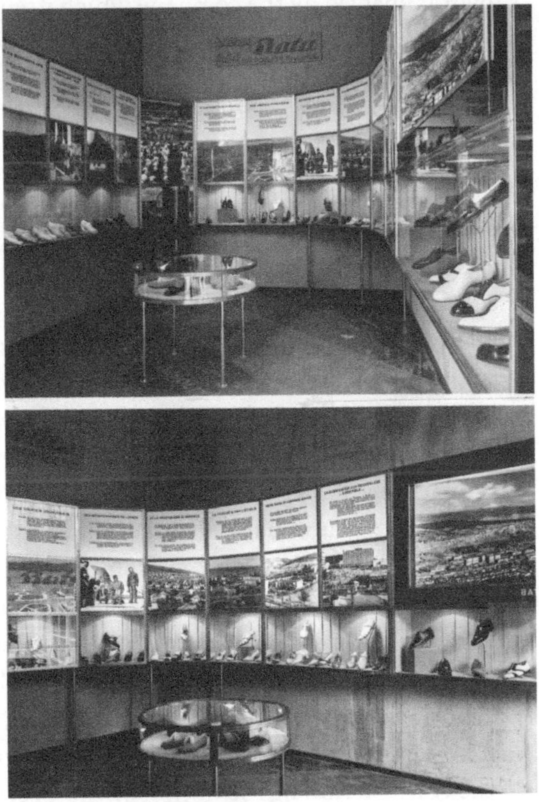

Fig. 60. View of the Baťa pavilion, Brussels, 1935.

Such an approach can be illustrated by the Baťa section at

98 Cf. Gustav Jaroš-Gamma, *Náš tatíček Masaryk* (Prague: Gamma 1918–25), Arnošt Caha, *Tatíček Masaryk Osvoboditel* (Brno: A. Caha, 1920), Lída Merlínová, *Tatíček Masaryk* (Prague: Šolc and Šimáček, 1934).

the Brussels exposition in 1935. It contained a display of shoes in glass cabinets around the room which bore panels above them with large-scale photographs and explanatory descriptions [fig. 60]. Altogether they demonstrated the progress of the people, the city and the company from the humble origins of tedious shoemaking in primitive conditions to the modernity of metropolitan life. The panels showed in succession the "immense factories," "rapid means of connection" and "the psychology of happiness," "progress through schooling," "men's sana in corpore sano (A sound mind in a sound body)," and "the glorification of a marvellous whole." Most of these inscriptions were placed above photographs of modern architecture and urban sprawl of Zlín, with one panel showing a mass gathering of workers on May Day. The evolution of the people was shown as enabled by the company which itself flourished towards improvement by modernization and technology. And as I will show in the concluding chapter, visualization of such a trajectory appeared in other works related to Baťa.

Hygiene and health

Outside of the Baťa space, exhibits promoting Czechoslovak spas, sanatoria and medical facilities became a common addition to national pavilions at international exhibitions. Their aim was not only to boost tourism but also to send out a message about the way the state looked after the nation's health. Architects, designers, politicians, businessmen as well as eugenicists were concerned with projects of "better housing, practical household furnishing, healthy nutrition, clothing."[99] Hygiene became one of the topics widely debated in this connection and often visualized at national and international exhibitions and museums or publicized by film or promotional books and articles.

Health and hygiene exhibitions provide further insight into the topic of classification of people and objects. While they were held on a smaller scale than world's fairs, they were significant events in Central Europe, in which interwar Czechoslovakia participated. For that reason, I will briefly discuss them to further argue that what was presented as scientific approaches contributed greatly to the construction of the notion of the Czechoslovak people.

The first large exhibition on the topic was held in Dresden in 1911. The International Hygiene Exhibition (Internationale Hygiene-Ausstellung) included "a

99 A. Grác, "Mezinárodní hygienická výstava v Drážďanech," *Zlín. Sdělení zaměstnancům firmy Baťa* 13, August 2, 1930, 3.

continuous historical narrative around hygiene as central to the civilizing process" and many of its exhibits formed the basis of the Hygiene Museum, mentioned earlier.[100] Such displays drew on the earlier examples of orientalized and racialized exhibits at large expositions and world's fairs of for example Paris, St. Louis or Chicago before and after the war. The achievements on show included scientific advancement of the "western world" in medicine, sanitation, and hygiene, which became particularly relevant in the period immediately following the First World War when Spanish flu, bad living conditions and shortages of medical supplies had to be dealt with.

In Austria, for instance, large waves of refugees from across the Habsburg monarchy arrived in the capital during the war in need of shelter and decent standards of living. Such changes in population numbers, concentration and composition, as well as the impact they had on cities, were addressed by various exhibitions in the interwar period. One of the earliest ones, the Hygiene Exhibition, took place in Vienna in 1925 with sections of "The New Household," "Public Welfare and Hygiene" and "Reproduction, Inheritance, Racial Hygiene." A much larger event was organized in Germany the following year. The Great Düsseldorf Exhibition of Healthcare, Social Welfare and Physical Exercise (Die Große Ausstellung Düsseldorf 1926 für Gesundheitspflege, soziale Fürsorge und Leibesübungen, also known as GeSoLei) aimed at promoting the idea that new capable people can be formed by education.[101] Drawing in some 7.5 million visitors, it featured sections on personal health and hygiene, sports, working conditions and heredity.

The follow up to the 1911 Dresden exhibition was the 1930 Second International Hygiene Exhibition in the same city at the Museum where the "Transparent Man" was shown as the main attraction.[102] Other temporary pavilions there included displays of healthcare, racial hygiene, children, women in the family and at work, fables and science, nutrition, housing, mental health and hygiene, and hospitals.[103] A separate building, the Place of Nations, contained

100 Claudia Stein, "Organising the History of Hygiene at the *Internationale Hygiene-Ausstellung* in Dresden in 1911," NTM Zeitschrift für Geschichte der Wissenschaften, Technik und Medizin 21, 355–387 (2013). https://doi.org/10.1007/s00048-014-0109-5.

101 The author of most of the buildings was the chief architect of the fair Wilhelm Kreis, with contributions from Max Taut and Peter Behrens, for example.

102 The image of a man with raised hands was adopted by Nazi propaganda after 1933. Cf. for example Klaus Vogel, "The Transparent Man — Some comments on the history of a symbol," In *Manifesting Medicine*, ed. by Robert Bud (London: Science Museum, 2004), 36–38.

103 A. Grác, "Mezinárodní hygienická výstava," *Zlín, Sdělení zaměstnanců firmy Baťa* 13, no. 30, 2 August 1930, 3; Z, "Mezinárodní hygienická výstava," *Lidové noviny* 38, May 25, 1930, 7.

contributions from other states, such as France, Great Britain, Turkey, Austria, Japan, Mexico, and Czechoslovakia.[104]

The Czechoslovak section was put together by the architect Rudolf Kvěch, who worked at the Ministry of Health, and was assisted by Ladislav Sutnar and a few other designers.[105] It showed examples of the Czechoslovak healthcare system, communal hygiene, spas, and various care institutions through photographs, diagrams, and models.[106] A panorama of the High Tatras was found impressive while the central piece was the habitual bust of Masaryk by Štursa, paralleling the setup of Czechoslovak pavilions in world's fairs.[107] There was also an eugenics exhibit, showing the family trees of the Rieger, Palacký, Pejša. Unfortunately, no further information about what it looked like is known to establish what parallels it had with the display of family trees a few years later in Chicago and whether it created the same juxtaposition of elite and "common" families.

What can be established is the eugenics framing of the display. František Ladislav Rieger (1818–1903) was a prominent nationalist politician of the second half of the 19th century, while František Palacký was a well-known Czech historian of the 19th century. However, Pejša is not a household name, and it refers to the extensive Pejša "tribe" from southern Bohemia that, by 1929, consisted of some 150 families. Their social developments and biological and mental characteristics were researched by Czech anthropologists and geneticists, especially the professor of heredity and genetics Artur Brožek (1882–1934), author of many texts on eugenics.[108] An active participant in eugenics conferences and knowledgeable of the contemporary debates in the USA and the UK, Brožek established the Eugenics Office at Prague's lunatic asylum, but in the 1930s condemned Nazi racial science.[109] He, as was characteristic of the Czech approach at the time, nevertheless remained convinced that positive eugenics and social

104 George C. Dunham, "The International Hygiene Exhibition," *American Journal of Public Health and the Nation's Health* XXI no. 1 (1930): 3. (1–10).

105 Photograph of the Czechoslovak pavilion is available in the Deutsche Fotothek at https://www.deutschefotothek.de/documents/obj/81023410/df_hauptkatalog_0359025.

106 "Hygienická výstava v Drážďanech," *Rozkvět* 23 (1930): 13.

107 "Československo na mezinárodní hygienické výstavě v Drážďanech," *Nová doba* 36, May 15, 1930, 3.

108 Artur Brožek, "Correlation between the Mating Surrounding Conditions in Man. The Tribe of the Pejsas, Sedlcan-branch," *Institut international d'anthropologie, 11e session Prague*, Prague 14–21 Septembre 1924 (Paris: Librairie E. Nourry, 1926); Artur Brožek, *Zušlechtění lidstva (Eugenika)* (Prague: F. Topič, 1922); Artur Brožek, *Ukázka studia vloh a vlivů okolí v rodokmenech* (Prague: A. Brožek, 1927).

109 Michal V. Šimůnek, "Artur (Arthur) Brožek," in *The History of East Central European Eugenics*, ed. Marius Turda, 147–148.

hygiene should be practiced to improve nations.[110] Condoning racism did not necessary mean condoning another kind of selectiveness.

These few examples speak of the intimate engagement of the Czech scientists, representing the Czechoslovak state, with the international movement promoting hygiene and health. The medium of exhibitions proved again to be an excellent one. Apart from these specialized events, Czech achievements in these areas, education, visions as well as the more tangible health facilities, medical institutions and spas featured prominently in the Czechoslovak national displays at interwar world's fairs. Already in Rio de Janeiro in 1922, the Czechoslovak pavilion contained a section on spas. Its inclusion here had an economic rationale; featuring spas and health retreats aimed at finding new clients at a time when many German tourists were economically restrained by a weak Mark. There was, however, an underlying message of such displays which emphasized the improvement of the body. Subsequent fairs would also include sections on such facilities directed both outwards towards potential customers and inwards explaining the state's activities in social care and various medical branches. They would also frequently focus on tackling socially pressing topics such as alcoholism, which was a frequent concern of eugenicist medical professionals.

Conclusion

Concerns with health of the body and society or nation were universal. The various improvements and solutions suggested at exhibitions and world's fairs aimed at creating the ideal (or rather, idealized) member of mankind. As the last example I can metion one of the most popular exhibits in the second year of the New York World's Fair, which was the "Typical American Family" display.[111] In an essay contest which had preceded it, Americans were asked to write their explanations of why they deemed themselves typical. The winners from each state were promised a free trip to the fair in a new Ford car, as the company co-sponsored the competition.[112] The selection process by the world's fair author-

110 Artur Brožek, "Biologický pojem rasy," in *Rovnocennost evropských plemen a cesta k jejich zušlechťování*, ed. Karel Weigner (Prague: Česká akademie věd a umění, 1934), 19–32. Cf. Milan Ducháček and Michaela Lenčéšová, "Aleš Hrdlička a transatlantický rozměr českého antropologického a eugenického myšlení první poloviny 20. století," in *Za rovnocennost*, 6–41.

111 Rydell, "Fitter Families," 56. Deborah B. Shepherd, "The 1939–1940 New York World's Fair: Typical American Families Build Tomorrow," 2011, https://api.semanticscholar.org/CorpusID:111490811, accessed February 20, 2024.

112 Rydell, "Fitter Families," 56.

ities included assessing the racial origins of the families and their composition, favoring nuclear and Caucasian families. All the winners were white Americans born in the USA to European descendants, the kind that Hrdlička and Hooton would call old Americans. Their selection promoted a very specific image of what a typical American family was and the exhibit helped to reinforce the hierarchies and supremacies promoted at world's fairs.

Many countries that participated in exhibitions adopted the practice of systematization and classification in their displays as well as in their understanding of their place in the world order that they broadcasted in the fairs. In New York, Czechoslovakia wanted to locate itself above the non-democratic countries tormented by revolutions, contra-revolutions and inflations and show it belonged among powers with established democratic order, stable economy and advanced social system.

Exhibitors and organizers at exhibitions of any size had the power to simplify and legitimize complex ideas through displays, transmitting them quickly to large audiences. Presenting what was deemed as scientific facts about racial or ethnic composition and development was a common feature of many fairs, whether they appeared explicitly as eugenics exhibits or more subtly as comparative material and artwork depicting ideal bodies. Czechoslovakia—through its scientists, companies, or governmental representatives—also engaged in the classification of humanity in its displays and presented its views of superiority and distinctions among the different ethnic groups in the state. These various agents took the opportunity of participation in world's fairs and other exhibitions to advance ideologies informed by eugenics and promote their visions of a better, healthier society, companies or states. While eugenics in Czechoslovakia and its exhibition activities was not related to racial hygiene and open discrimination of specific ethnic groups, it still operated with selectiveness of desired biological features. These inevitably helped to construct hierarchical reading of the nation and explain the complex and unequal composition of its people.

CONCLUSION

National pavilions at world's fairs, including the Czechoslovak ones, were not condensed versions of the world around them; they were designed to create new worlds that interpreted the past, selected the present and constructed the future. Organizers, exhibitors and designers ordered complex phenomena and information in an accessible manner and refrained from controversial or difficult subjects. A final example illustrates these processes—a large painted window by the Czech artist Cyril Bouda titled *The Hymn to Work*, that was a prominent part of the façade of the Czechoslovak pavilion in New York. Commissioned by Baťa, it appeared on the first floor of the curved side of the building and was one of the few large objects that were delivered to the USA before the transportation of exhibits was halted by the German authorities. The window was a distinguishing and indispensable feature of the pavilion through which visitors could look out to the outside world or glance into the interior [fig. 61].

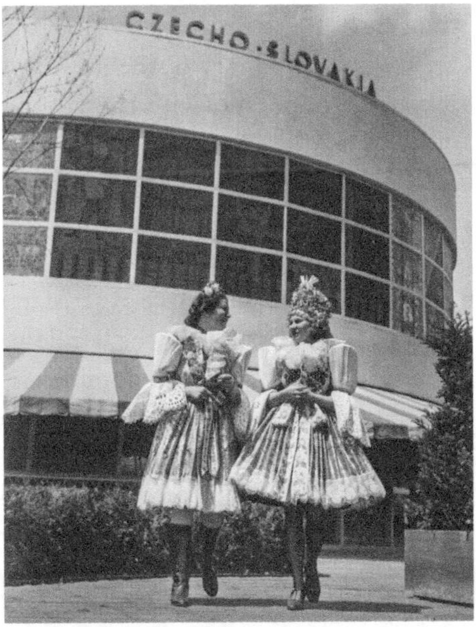

Fig.61. Czechoslovakia Participation - Czechs in traditional dress - In front of building, 1939.

Bouda's subject matter, captured in narrative-driven, representational visual language, was a celebration of work and people depicted as an evo-

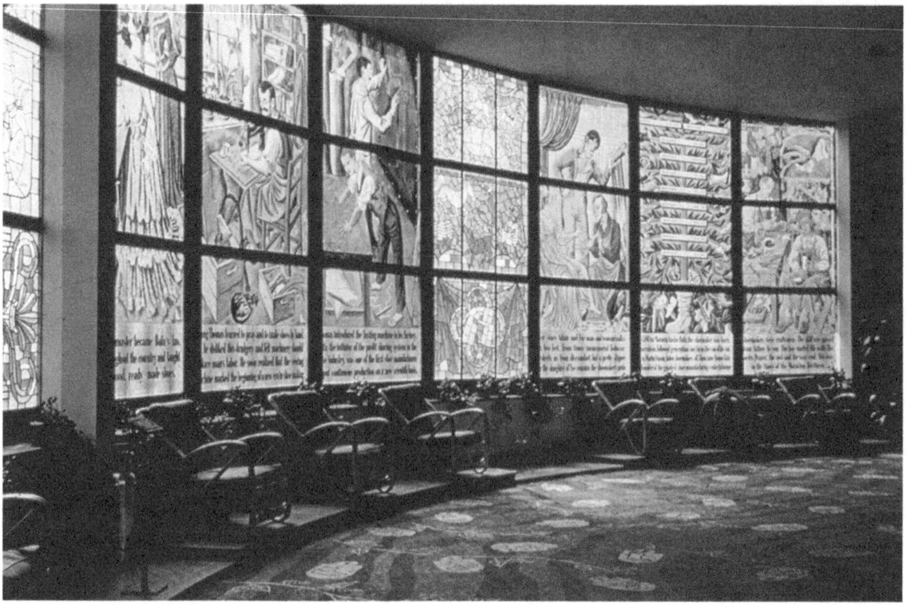

Fig.62. World's Fair. Czech mural I., New York, 1939.

lution, which was enabled by industrialization and modernization. It can serve as a swan song not only of the representation of the state but also of its ambitions during the short span of its interwar existence. At the same time, it can be seen as a summary of the most significant aspects of the Czechoslovak displays read in parallel with the visions of the Baťa company and the trajectory of the state [fig. 62].

The immense set of windows was 2 meters high with a total length of 35 meters.[1] The 21 panes of "The Hymn to Work" showed "how a humble, little man managed to reach heights, thanks to his craftsmanship, and how he managed to build for himself a place of satisfaction and high living standards around his work."[2] In the first place, the window captured the story of Tomáš and Jan Antonín Baťa who came from humble origins to organize the work and life of new generations of young people. The historic sequence was divided into five "chapters" of mostly three, and in the case of the central motif five, images each, interspersed with black and white stained-glass sections. The main protagonists were the two Baťa half-brothers who dominated the central images and the evolution

1 Ing. Černovský, *Výstavnictví*, October 1938, National Archive Baťa, SV-NY, 1937-1940.

2 "Cestovní zpráva – Černovský, New York: Referát amerických novin o otevření čs. pavilonu," May 31, 1939, sheet no. 75, June 10, 1939, in Korespondence ředitelny fy Baťa s USA CZ z čs. pavilonu v New Yorku – Světová výstava. Baťa r. 1939 I/4 i.no. 412. MRA in Brno, branch Zlín.

of shoemaking and working life they represented. Their presence in the depiction is as dominant as that of the statue of President Masaryk that adorned the interiors of every Czechoslovak pavilion in the interwar period. Given the ambition—albeit failed—of Jan Antonín Baťa to devise the entire ideology of the Czechoslovak display in New York, it is not surprising that the two managers were elevated to such a significant place.

The allegorical journey of the Baťa company from the harsh beginnings towards a large shoe empire starts with the historical precedent of a cobbler family in the first window. Here, a father is passing his skills onto his son who sits on a stool laboriously making a shoe by hand. This hardship is followed by a depiction of the Baťa family tree growing out of a boot with the date 1580, pointing to the hundreds of years of shoemaking legacy. The various branches of the family tree bear no fruit but shoes which are modernised by each successive generation. The tree could be seen in parallel to the pedigree panels of the Eugenics Exhibit in Chicago in 1933, which I discussed in the previous chapter. The depictions of families and the tree of eugenics in the Exhibit suggested a similar hierarchical and symbolic visualization of information alongside the developmental trajectory.

The subsequent set of images in the window reinforces the narrative of improvement by modernization. The next window shows a nineteenth-century boudoir where a shoemaker tries a shoe on a young woman and is supervised by her husband or father. This was the time when a single person was responsible for the procurement of material, shoe production, customer care and sales (using modern-day terminology). This progresses in the next set of images where the benefits of modernization are delivered by the invention of the sewing machine and the split of the different tasks. Cobblery turns into shoe manufacture, and viewers are presented with how shoes are designed, and the hide prepared in the first window of this set. The second one depicts how shoes are still made at home at this time and then passed for sale in a shop shown in the third image. Here, two women are inspecting a wide selection of footwear and are helped by a shop assistant. The division of labor sped up the manufacture, provided better consumer choice, increased sales, and lowered prices [fig. 63].

The central set of windows gives the stage to the two Baťas who flank the panorama of Zlín. The newly- built city is represented by factories, the company headquarters, the employee housing and the so-called Gahura's prospect which consists of a green belt leading towards the Tomáš Baťa memorial building and to the forests beyond the city. Tomáš Baťa, who had passed in 1932, is commemorated here as the city's builder and the company's founder. He is shown stand-

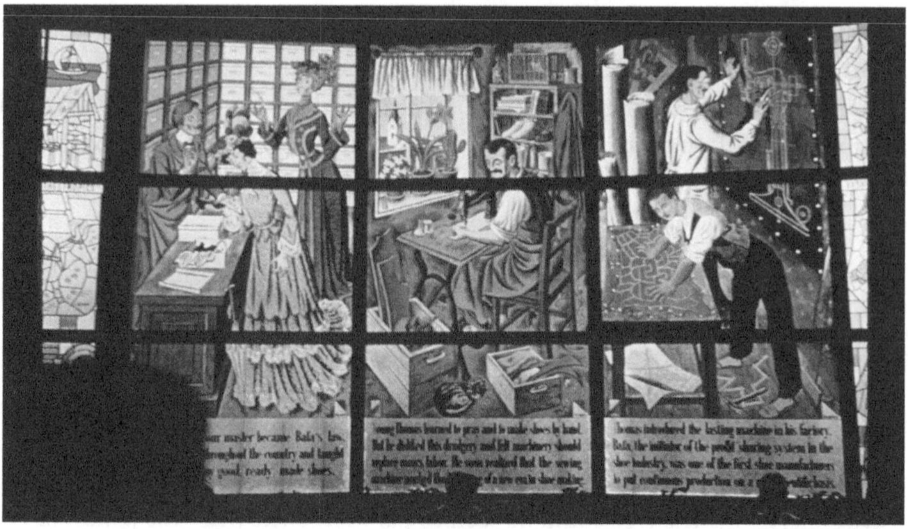

Fig. 63. World's Fair. Window panels with illustrations and text on the profitsharing system in the shoe industry III.

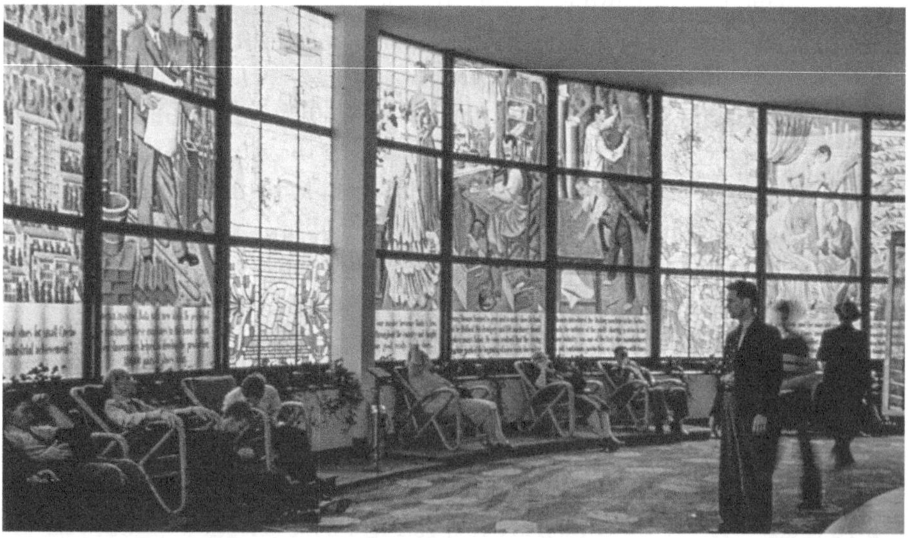

Fig. 64. World's Fair. Window panels with illustrations and text on the profitsharing system in the shoe industry I.

ing on a construction site with cranes and Czechoslovak flags in the background, holding plans and pointing to some aspect of "his" city as it is being built. His counterpart Jan Antonín is seen aboard a ship that leaves America, symbolized by skyscrapers behind him. He holds a notebook in which he probably keeps notes from his visit to the USA and is surrounded by four maritime flags which inconspicuously read the letters Z, L, I, and N. The link between America and Zlín is spelt out clearly because Baťa's ideas of manufacturing, welfare and architecture principles were greatly inspired by the USA. America was also significant for the Czechoslovak state as the cradle of democracy and capitalism as well as one of the refuges for representatives of the forming government during the war [fig. 64].

The next set of windows continues with the present life that has been created for the workers in Zlín and inspired overseas: education and sport on the one hand and factory work and customer care on the other. The final series of this evolution looks into the future of the workers and the company. On the side windows, May Day is celebrated by young, smiling employees featuring a placard with a graphic rendition of the company's growth "Always further, always higher" and upbeat slogans like "The world needs shoes." The middle pane shows a nuclear family in front of a typical Zlín house, with a grandmother helping with the children and a grandfather attending to the garden. Inscriptions under each of the windows further comment on the scenes and here we see a "happy family" looked after by Baťa. The text in the surviving photographs is hard to read, but generally, it helps to explain the achievements and progress the company brought, by freeing people from hard labor and giving them better, modern lives [fig. 65].

The advancement of the company and shoemaking can also be read alongside New York fair's motto "The World of Tomorrow," which, despite the looming war in Europe, insisted on promoting a concept of a better global future. Baťa's vision for the company, society and the state fulfilled this idea and visual-

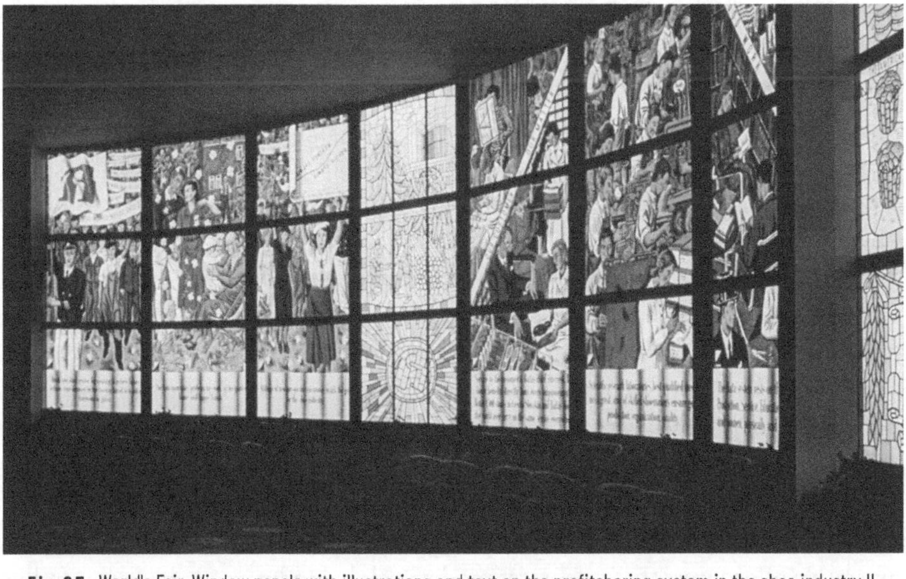

Fig.65. World's Fair. Window panels with illustrations and text on the profitsharing system in the shoe industry II.

ized it in the windows. Moreover, the story could also be seen as an analogy of the trajectory Czechoslovakia followed as a state. Many politicians, historians and novelists portrayed the period of the Habsburg rule as an oppressive one during which the Czech and Slovak people survived in primitive conditions. According to this narrative, the nation, nevertheless, had developed through hard work and modernization into a prosperous and confident state. The emphasis on progress and evolution becomes another feature shared by the company and Czechoslovakia, both giving importance to young, healthy people that compose the firm and the state respectively.

As a final point, the painted window also puts in plain view the key role of the two main male protagonists of the company. They are portrayed here, through their central position, as the guardians of the company, the workers and the pavilion. Such elevated place could be reminiscent to the role of the President of the state in the Czechoslovak pavilions. Masaryk, and to a lesser extent his successor Beneš, were depicted as the symbolic protectors in the sculptures placed in central locations of the national pavilions.

Modernity and modernism

The window can be taken as a metaphor for the state and its efforts at presenting itself at world's fairs in several areas, foremost in its attitude to modernity and modernism. While the curved line of the building and the use of large glass areas suggest an innovative approach, Bouda's rendition of the individual scenes in the Baťa window is less experimental. The comprehensible narrative of the series is comparable to the composition, space and figuration of other artists and designers that often featured in the Czechoslovak pavilions. František Kysela, for example, whose narrative-driven tapestries of the crafts featured as one of the highlights of the Czechoslovak pavilion in Paris in 1925 was a regular contributor in most interwar exhibitions. He represented both the School of Decorative Arts and the importance of crafts under the auspices of the state. Bouda's similarly didactic presentation of the Baťa story also shares traces of the visual approach taken from Renaissance painting. This suggests that he (and his sponsor) opted for a tamer version of modernism that emphasized the clarity of the message and links to historic precedents rather than the abstraction and experimentation of more radical avant-garde artists. However, there was one main difference between Bouda's and Kysela's approach and it was in the evolutionary narrative of progress that celebrates modern, global industry in the Baťa window.

The search for an appropriate language to represent Czechoslovakia underlays many of the choices the official organizers made with each world's fair. Works like "The Hymn to Work" built on a tradition found in the crafts, whether it was glass painting, tapestry or shoemaking. Familiarity with established forms of visual expression on the one hand and showcasing a progression leading to modernism was embedded in the duality between national authenticity and universal modernity of the national presentations in the interwar period. The most progressive embrace of modernism could be detected in how the exterior and interior architecture of the national pavilions developed. From the late 1920s, architects and designers used building materials and technologies like large glass areas to tap into contemporary construction techniques and to respond to current trends in pavilion architecture. Interiors developed from museum-like sequences of exhibition rooms with display vitrines and wall hangings to sophisticated and engaging routes that presented visitors with visually impressive objects and large-scale photographs.

Designers like Ladislav Sutnar and Antonín Heythum worked with a visual flow of information that led the visitor through a continuous space, taking into account the psychology of such experience. Or at least, this is how they envisioned their exhibition interiors on paper. In reality, the Czechoslovak pavilions were filled and sometimes overfilled with objects to satisfy various traders, associations and individuals. This was most obvious at the Century of Progress in Chicago in 1933 where the national pavilion was crammed with stalls selling "trinkets" and in the International Exhibition in Paris of 1937, where the Czechoslovak pavilion, originally planned as a flow with a logical sequence of exhibits and breaks, was crowded with exhibits of various sizes and of various companies.

The visual arts and especially the displays of fine art within national pavilions and in art halls had the primary function of representing the state. They were therefore also subjected to a search for appropriate modernism. Personal connections played a significant role here and the various members of the exhibition elites, like the exhibition commissioner Ladislav Turnovský, the pavilion supervisor Bohuslav Soumar or the ministerial official V. V. Štech, could easily influence the choice of participants and exhibits. As a result, The School of Decorative Arts and The Association of the Czechoslovak Werkbund, linked to Štech, featured in many exhibitions and stood for art and design that was comprehensible and, in the case of designers, had strong commercial potential for export. This selectiveness also meant that many omissions were made. While the avant-garde with more experimental work was often excluded, regional artists and associations had no space either.

Classifications and hierarchies

Art for display in national pavilions and its relation to the modernist framework also had an unofficial representative function when it came to regions outside of the Czech part of the state (but not the minorities). Slovakia and Ruthenia did not generally have any significant place in the Czechoslovak pavilions, except when presented as regions with touristic sites and where folk cultures had been preserved. While folk art could largely be seen as the antithesis of modernism, the dichotomy in the case of world's fairs was not so straightforward. On the one hand, folk art, as adopted for commercial use by the company Detva, for example, could be modernized and updated for contemporary consumers. It would also be commonly incorporated into Czechoslovak exhibits to increase the attractiveness of performances, restaurants or hostesses.

On the other hand, folk art of the eastern regions signified their permanent quality and purity embedded in the belief that the culture there was more insular and devoid of external influences. This was the view of contemporary anthropologists like Matiegka and Hrdlička who helped to construct the more "scientific" narrative of the Czechoslovak people and explain its existence based on physical features and ethnographic patterns of behavior. Classifications of people, objects and knowledge was one of the crucial aspects of national displays at world's fairs. Creating and visualizing hierarchies and trajectories in graphs, charts, collages and photographs gave some sense to the complex phenomena that filled the world as well as the world's fairs. Within the space of the fairs, such visualizations also presented information as scientific facts, justifying them as infallible [fig. 66].

Such displays were planned to turn the chaos of the outside world, both past and present, into a more comprehensible order.[3] This was clear in the Czechoslovak entries that presented the state as the home of a single Czechoslovak nation with discernible hierarchies between its two main components, the Czechs and the Slovaks using the simplified formula of the industrialized (and seemingly civilized) west opposed to the rural (and allegedly backwards) east. Such hierarchy had several functions but one of them was to demonstrate Czech supremacy and justify the existence of the joint state. Autonomy of Slovakia and Sub-Carpathian Ruthenia agreed in the negotiations at the birth of the state was never achieved and images of the primitivized and under-industrialized, albeit picturesque, countryside helped to explain why. Such an approach was

3 Zygmunt Bauman, *Liquid Modernity* (Cambridge: Polity Press, 2012), 138.

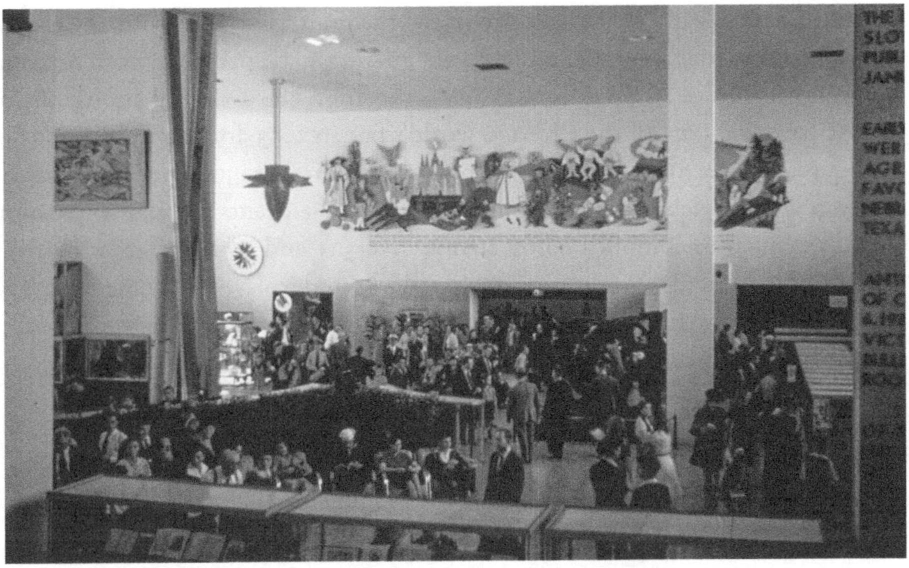

Fig. 66. World's Fair. Interior of Czech Building.

not explicitly or consciously racist but bore similarities with colonial attitudes justifying colonizing missions. It contained dangerous messages about the differentiation among ethnicities within Czechoslovakia and their relationship to each other, prioritized the Czechs and placed them in the role of civilization bearers and guardians of Slovaks and Ruthenians.

Such selectiveness helped to construct a picture of what typified Czechoslovakia and Czechoslovaks. Very much facilitated by eugenic ideology, the displayed graphs, charts, photographs and sculptures position the Czechs in relation to other ethnic and national groups in Central Europe and in the world. Ethnic diversity in Czechoslovakia was almost non-existent in the narrative of Czechoslovakism put on display at national pavilions. The exceptions were the Germans and Jews, represented via their companies like the Vítkovice Steel Works and Moser glass, or vocational schools from the border regions of Bohemia. Other groups like the Hungarians, Poles, or Roma were mostly absent, accentuating the dominance of Czech businesses, arts and culture. A similar image was created outside the pavilions in phenomena that accompanied the main exhibits, most prominently in the Czechoslovak restaurants and national days. The restaurants were important places that would serve visitors with the flavors of Czechoslovak culture and cuisine via the food and beverages offered here. The way they were served by waitresses in folk costumes, or in their redesigned versions, accompanied by regional music, only strengthened the native experience.

National days at world's fairs promoted the homogenous picture of Czechoslovak culture further with music, dance and dramatic performances, often in national costumes. They were massive gatherings, consisting of thousands of participants. As part of the Chicago world's fair in 1933, for example, 1,800 Sokol gymnasts performed their routines while a further 4,000 people created a tableau vivant celebrating the home country. These events were an expression of uniformity, uncomplicated by ethnic diversity and tensions growing over the interwar decade. Modernity was similarly underplayed in these events in favor of what was promoted as a traditional, authentic culture with references to the historicity of the people.

Agency

The local diaspora was often involved in running the pavilion restaurants, delivering the program of the national days and contributing to the shape of the displays. It was especially the American Czechs and Slovaks who offered their own understanding of what the Czechoslovak nation was at the world's fairs in Philadelphia, Chicago, New York and San Francisco. They stood outside the Prague-based exhibition elites in charge of the main aspects of the exhibits and included a range of contributors from casual workers like builders and waitresses to performers and journalists. These individuals, nevertheless, were not a single group with a unanimous view. In the USA, the diaspora consisted of such diverse voices as those of Slovak autonomists, former legionnaires, or women's associations, all trying to make their mark in the representation.

The real agency these groups and individuals had was, nevertheless, limited. While the diaspora became formative for the content of the displays in the American world's fairs in the last interwar years, its impact on the overall ideology of the pavilions was limited. Their input was the result of external political and economic circumstances like the loss of statehood rather than an outcome of a conscious plan of the pavilion organizers for inclusivity. The presence and involvement of women in the exhibition activities linked to the official Czechoslovak presentation were similarly curtailed. On only a few occasions, women appeared as exhibiting fine artists and designers or appeared in the roles of assistants to architects and designers. More often, though, women occupied anonymous service positions in the restaurants and shops and as attendants in the pavilions. They featured in parades, performances and official events but their agency was minimal and unrecognized. The male organizers could only envisage their roles in increasing the attractiveness of the pavilion and sales. The main input into the content, structure and delivery of the national pavilions remained with the exhibition elites, a group women were excluded from.

Trajectories and developments

The displays in national pavilions were put together to show a desirable interpretation of the outside reality, yet they did not function in a vacuum and responded to the political and economic situation. The Czechoslovak organizers reacted especially to the global financial circumstances when considering whether to participate in the fairs. While the earlier pavilions in Rio de Janeiro in 1922 and Paris in 1925 were influenced by discussions of the main focus of the exhibits, the cost of the subsequent participation in Philadelphia, Barcelona and Chicago in the late 1920s and early 1930s, affected by the global economic downturn, became one of the major issues. However, the fact that the government ultimately always opted to invest money and energy in the national pavilion or section, despite the financial difficulties, speaks of the importance Czechoslovakia placed on participating in these events.

Throughout the interwar period, a set of regular exhibitors and features formed the basis of the national presentations. Companies like Baťa, Vítkovice and Pilsner provided not only the content but also materials and commodities which could be used for the construction and offset the overall cost of the participation. In cases like beer, they could also be consumed while the exhibition was open. The commercial side was habitually complemented by the visual arts which took the form of fine art and design displays with a representative function. Several references had their fixed place in the national pavilions, and they included the bust of President Masaryk and large-scale paintings of Prague and the Tatra mountains.

Crumbling façade

The case of Czechoslovakia at world's fairs therefore reflects the general attitudes of the world's fairs towards minorities, typical of other, often larger political entities. The pavilion might have been built from below by casual workers, members of the diaspora, and visitors, but it was Czechoslovakia, through its exhibition elites, that carved itself a space in the ranks of major exhibitors and the world order established at the fairs. Like other great powers, it contributed to the classification of humanity in its displays, presenting its own views of superiority and distinction among different groups of people and objects. Baťa's "Hymn to Work" in New York provided such a visualization of a world, that could be easily classified and represented in windowpanes. In a world threatened by the coming war, the colorful vision in glass as well as the entire pavilion became a nostalgic reminder of the interwar ambition of the Czechoslovak state.

Compared to other European, especially Central European countries, Czechoslovakia indeed gave the impression of an island of exceptional democracy. While Poland, Austria, Hungary and Germany grappled with autocratic regimes and internal conflicts and did not invest the same energies or ideologies into world's fairs, the external image of Czechoslovakia was one of democracy, stability, and modernity. The national pavilions at world's fairs helped to sustain such a notion with the modern structures and objects in them and the drive towards progress and improvement. Behind this façade, however, the reality was different and internal tensions and alliances, ethnic and gender stratification of society and the belief in Czech supremacy in the state penetrated the displays. Despite the concerted effort to present a unified image of the "island of democracy," the project of the democratic state and its pavilions ultimately crumbled. One of the reasons for this failure was the fact that the Czechoslovak identity was built and visualized on flawed and simplified attributes which suppressed the multi-ethnic aspect of the state. The different actors behind the Czechoslovak displays at world's fairs ultimately presented a limited, wishful portrayal of reality through the inclusions, exclusions and proposed hierarchies and classifications in these highly ambitious exhibition projects.

Bibliography

Archival sources

Abbreviations:

AACE NTM – Archive of Architecture and Civil Engineering, National Technical Museum, Prague
ACASA – Archive of the Czech and Slovak Association, University of Chicago
APRO – Archive of the President of the Republic Office, Prague
CPL - Chicago Public Library
DETVA - Detva, československý ľudový umelecký priemysel, účastinná spoločnosť v Bratislave (1919) 1927–1949 (1957)
GRI – Getty Research Institute
MFA – Ministry of Foreign Affairs, Prague
MRA – Moravian Regional Archive
NACR – National Archive of the Czech Republic, Prague
SNA – Slovak National Archive, Bratislava

"Article 9," Podmínky soutěže, 20041215/07 f.140 Kejř (mezinárodní výstavy), AACE NTM, unpag.

"Biology Section Tells the Story of Man," Century of Progress Scrapbooks, 26 June – July 22, 1933, box 1, folder 3, CPL.

"Cestovní zpráva – Černovský, New York: Referát amerických novin o otevření čs. pavilonu," 31 May 1939, sheet no. 75, June 10, 1939, in Korespondence ředitelny fy Baťa s USA CZ z čs. pavilonu v New Yorku – Světová výstava. Baťa r. 1939 I/4 i.no. 412, MRA in Brno, branch Zlín.

"Čs. úspěch na výstavě v Bruselu," *Lidové noviny,* June 29, 1935. Newspaper Excerpts Archive, Oh8, MFA Archive.

"Čsl. pavilon na mezinárodní výstavě v Bruselu," *Národní listy,* May 29, 1935, Newspaper Excerpts archive, Oh8, MFA Archive.

"Čsl. účast na světové výstavě v Bruselu 1935," Ministr obchodu pro předsednictvo ministerské rady, August 30, 1934, no. 944, file 115, D6983/34, Výstavy v Belgii, APRO.

"Czech Fair Centre is Now an Orphan," *New York Times,* March 17, 1939, Personal papers, 1897-1976, Ladislav Sutnar Papers, 1897-1976, GRI, Los Angeles, Accession no. 2013.M.6, bx18.

"Dojmy českého reportéra ze světové výstavy v Bruselu," *Národní listy*, May 12, 1935, Newspaper Excerpts Archive, Oh8, MFA Archive.
"Exhibition in Rio de Janeiro," The Ministry of Education, January 23, 1922, 22II Rio de Janeiro Jubilejní výstava v Rio de Janeiro no. 35490/22, Výstava v Rio de Janeiro, no. 7037, APRO.
"From the Ministry of Trade to the Presidium of the Ministerial Committee," March 25, 1922, 5, 22II Rio de Janeiro Jubilejní výstava v Rio de Janeiro no. 35490/22, Výstava v Rio de Janeiro, no. 35490/22, APRO.
"General consulate of the Czechoslovak Republic in Chicago to the Ministry of Foreign Affairs," May 25, 1932, ACASA University of Chicago, Petr Zenkl Papers, f2, Box 55.
"Ideový návrh exponátů ministerstva sociální péče pro Nový York 1939," 1. Výstavy v cizině - Amerika - Světová 1939, Section III, Part 5, MFA Archive.
"List of Austrian, Hungarian, Finish and Norwegian Exhibits Consisting of Paintings, Graphics, Bronzes, Porcelains, Sculpture, Embroideries and Rugs Originating Abroad Through J. Nilsen Laurvik and on Exhibition at the Palace of Fine Arts by the San Francisco Art Association," Panama Pacific International Exposition, Records 1911-1929, carton 167, BANC MSS C-A 190, U.C. Berkley Bancroft Library, Manuscript Collection.
"Mezinárodní výstava umění a technika v moderním životě v Paříži v roce 1937 - účast ČSR," inv. no. 1033, file 134, D, APRO.
"Ministerstvo obchodu, průmyslu a živností," June 9, 1939, Ministerstvo obchodu SV-NY, 1939-1940, NACR.
"Ministerstvo školství a osvěty předsednictvu ministerské rady," February 4, 1925, Mezinárodní výstava dekorativního umění v Paříži 1925, D1441/25, APRO.
"Náčrt programu čsl. oddělení mezinárodní výstavy v Paříži," *Mezinárodní výstava Umění a technika v moderním životě v Paříži 1937*, MFA Archives, Division III, Section 2.
"Naše propagace cizineckého ruchu získala v Paříži velkou cenu," *Národní střed*, November 27, 1937, Newspaper Excerpts archive, Oh8, MFA Archive.
"Nástin organisace československých exposic na výstavách v cizině," 1927, Organisace Československé účasti na výstavách v cizině, kD2519/7, APRO.
"Native delicacies to be served in Czecho-Slovakia's outdoor beer garden and restaurant at the New York World's Fair 1939," Release prepared for E. F. Roosevelt, Acting Director. Foreign Government Participation. Series I.A. Administration, I. Central Files. New York World's Fair 1939 and 1940 Incorporated Records, 1935-1945. New York Public Library holdings.
"Návrh na utvoření stálé meziministerské komise pro organisaci státní účasti na výstavách v cizině," July 26, 1927, Organisace Československé účasti na výstavách v cizině, kD2519/7, APRO.
"Negros Plan Many Events for Day at Fair Tomorrow," August 11, 1933, Scrapbooks 1933 August 6 - 1933 August 25, Box 2, Folder 2, Century of Progress Collection, box 5, CPL.
"Pilsner Urquell auf der Brüsseler Ausstellung," *Prager Presse*, June 1, 1935, Newspaper Excerpts archive, Oh8, MFA Archive.
"Pp. Karfík, Gahura, Vojta Baťa, Ševčík, Grác, Kadlec, Novosad, Rojt, Meisl." October 24, 1936, letter on Paris 1937, Baťa, X. prodejní oddělení, inv.no. 208, MRA in Brno, branch Zlín.
"Procházky po výstavě," *Právo lidu*, October 19, 1937, Newspaper Excerpts archive, Oh8, MFA Archive.
"Projektová dokumentace," 8, f. 140, Kejř, AACE NTM.
"Úřad pro zahraniční obchod v Praze - Protokol," The Office for Foreign Trade, February 4, 1922 22II Rio de Janeir, Výstava v Rio de Janeiro, no. 13765/22, APRO.

"Veřejná soutěž na ideové náčrty na čsl. státní pavilon pro mezinárodní výstavu Umění a technika v moderním životě" v Paříži 1937, 20041215/07 f.140 Kejř, AACE NTM.
"Výročná zpráva DETVY československý ľudový priemysel, úč.spol. v Bratislave za správný rok od 1.septembera 1930 do 31.augusta 1931," 1932, I.B.8, DETVA, SNA.
"Výročná zpráva DETVY československý ľudový priemysel, úč.spol. v Bratislave za správný rok od 1.septembera 1927 do 31.augusta 1928," 1929, DETVA, SNA.
"Výročná zpráva DETVY československý ľudový priemysel, úč.spol. v Bratislave za správný rok od 1.septembera 1928 do 31.augusta 1929," 1930, DETVA, SNA.
"Výročná zpráva DETVY československý ľudový priemysel, úč.spol. v Bratislave za správný rok 1934," 1935, DETVA, SNA.
"Výstava v Paříži 1937, protest Alsasanů, aby jejich svéráz sloužil jako atrakce," letter from The Consulate of the Czechoslovak Republic in Strasburg to the Ministry of Foreign Affairs and the Ministry of Trade of 5 December 1936, section III, part V, MFA Archive.
"Výstava v Rio de Janeiro. Jaroslav Bohdalecký, pracovní síla pro pavilon ČSR," Presidium pro zahraniční obchod. *Rio de Janeiro* no. 232 1921-22, no. 10128, NACR
"Vysvětlivky k průvodní zprávě projektu československého pavilonu na Výstavě umění a techniky v moderním životě v Paříži v roce 1937 k soukromému použití poroty," f.140, Kejř, 5, AACE NTM.
"Wagner Sings for Czeko-Slovakians," Century of Progress Scrapbooks, June 26 - July 22, 1933, box 1, folder 3, CPL.
"Z výstavy v Paříži: na třetím místě," *Večerní České slovo,* November 26, 1937, Newspapers Excerpts archive, oh8, MFA Archive.
"Zpráva o zvelebovací a propagační činnosti ústavu "DETVY" v Bratislavě v období 1924/25 a 1925/26," I.B.1. DETVA, SNA.
"Zpráva o zvelebovací a propagační činnosti ústavu Detvy v Bratislavě v období 1924/1925 a 1925/1926," inv. no. 2, I.B.1, DETVA, SNA.
"Zpráva odborné komise pro výtvarnou práci z prvé schůze dne 11.1.1937," Výstavy v cizině. Francie, Paříž 1937. sec. III, part V, no. 6328, MFA Archives.
A letter from Colonel Finch of September 28, 1933, Century of Progress Government Correspondence, no.2, box 54, Czechoslovakia-Denmark, Special Collections, University of Illinois at Chicago.
Advertisement, Century of Progress Scrapbook, June 3, 1933 - June 29, 1933, Century of Progress Collection, Box 1, folder 2, CPL.
Baťa, J. A. "Jak by měl vypadat čsl. pavilon v New Yorku," May 26, 1938, X. prodejní oddělení, Inv. no. 208, f. 26, doc. 5493, MRA in Brno, branch Zlín.
Čapek, Thomas Jr. *Czechoslovak Immigration.* Classroom material. Service Bureau for Intercultural Education (New York), CPL.
Černovský. "Cestovní zpráva. Referát amerických novin o otevření čsl. pavilonu 31.V.1939," June 10, 1939, Korespondence ředitelny fy Baťa s USA, I/4 Ředitelna, inv.no. 412 fond Baťa 1939, MRA in Brno, branch Zlín.
Černovský. *Výstavnictví*, October 1938, Baťa, SV-NY, 1937-1940, NACR.
Chicago on Parade, 1933, Century of Progress Collection, box 5.13, CPL.
Dušan Kubány to Central committee of Century of Progress, Government Correspondence, no.2, box 54 Century of Progress, University of Illinois at Chicago_
Heinz 57 Exhibit Manual Brochure. New York, 1939. III. Golden Gate International Exposition, Fairs and Expositions Collection, 1893-1967. 216-X-863. Environmental Design Archives, University of California Berkeley.

Kerkhoff, J. D. "Czech Pavilion, Flags at Half-Staff, Opens," *New York Evening Journal*, May 31, 1939, Personal papers, 1897-1976, Ladislav Sutnar papers, 1897-1976, GRI, Los Angeles, Accession no. 2013.M.6, bx18.
La seccion de Checoeslovaquia, *La Veu di Catalnuya, Las noticias*, July 16, 1929, Newspapers Excerpts archive, oh8, MFA Archive.
Ladislav Sutnar Papers, Series II. Ladislav Sutnar papers, 1897-1976. (Accession no. 2013.M.6), 1897-1976. GRI Los Angeles.
Letter from Biere Urquell Pilsner, Paris, to Legation Tchecoslovaque from May 30, 1936, no. 6328, 1937, division III, section 2, MFA Archive.
Letter from J. C. Holmes to Oliver M. Sayler, of April 12, 1939, New York World's Fair 1939 and 1940 Incorporated records 1935-1945, Series I, P03.Foreign, b. 306 f. 2, New York Public Library.
Letter from Jiřina Václavek to her grandfather in Czechoslovakia of February 1, 1934, courtesy of Peggy Hazard.
Letter from Josef Černovský to George Janeček, 26 April 1939, New York World's Fair 1939 and 1940 Incorporated records 1935-1945, entry no. 492, New York Public Library.
Letter from Kejř to the editors of *Přítomnost* of November 12, 1937, f.140 Kejř, AACE NTM.
Letter from Konzulát Československé republiky v Singapore to Ministerstvo zahraničních věcí ČSR o vývozu československého piva do Singapuru, May 29, 1935. Division III, 1918-1939, no. 599. MFA Archive.
Letter from Pešánek to the General commissioner's office of the Czechoslovak exposition, August 11, 1937, AACE NTM.
Letter from T. G. Masaryk to the mayor of Philadelphia of May 29, 1926, D2982/26, APRO.
Letter from the Czechoslovak American Chamber of Commerce to the Publicity Department of August 28, 1934, Century of Progress Government Correspondence, no. 2, box 54, Special Collections, University of Illinois at Chicago.
Letter from the Ministry of Trade to the Board of the Ministerial Committee re World's Fair in Chicago, October 31, 1932, DM2251, APRO.
Letter to the Ministry of Foreign Affairs, May 11, 1936, Mezinárodní výstava v Paříži 1937, kD5640/36, MFA Archive.
Manuscripts Collections. Panama Pacific International Exposition, Records 1911-1929, BANC MSS C-A 190, carton 167, U.C. Berkley Bancroft Library.
Pechar, Jaroslav. "Mezinárodní výstava v Barceloně," *Právo lidu*, April 19, 1929, Newspaper Excerpts Archive, Oh8, MFA Archive.
Špaček, Stanislav. "XV. Zpráva. V. Konečná zpráva o instalaci československého pavilonu na výstavě ve Philadelphii," September 16, 1926, Světová výstava ve Filadelfii, D5531/26, 2, APRO.
Styblík. "Československo v Bruselu," *Večerní Lidové noviny*, July 20, 1935, Newspaper Excerpts archive, Oh8. MFA Archive.
Sucharda, Jan. "Zpráva o Mezinárodní výstavě umění a techniky v Paříži," Mezinárodní výstava umění a technika v moderním životě v Paříži v r. 1937 – Účast ČSR, inv. No. 1033, f. 134, k D1977/38. APRO.
Sucharda, Jan. "Zpráva o stavu přípravných prací na výstavním pavilonu v New Yorku," April 19, 1938, Světová výstava v New Yorku 1939, inv. no. 1264, box 199. APRO.
Sutnar, Ladislav. "The Bauhaus, as seen by its neighbors South of the Border," Manuscript, 1971, Ladislav Sutnar Papers, Box 22, Biographical Information, Miscellaneous Correspondence, Exhibitions. Cooper Hewitt Museum archive.
Sutnar, Ladislav. *The Czechoslovak Pavilion*, pamphlet, Ladislav Sutnar Papers, Accession no. 2013.M.6, Exhibition design, 1939-1976, Box 2013.M.6.bx16, folder 1. GRI.

Sutnar, Ladislav. "Curriculum vitae draft," Ladislav Sutnar papers, 1897-1976, GRI, Los Angeles, Accession no. 2013.M.6, box 17 f. 6.
Sutnar's notes for The Czechoslovak Pavilion pamphlet, Ladislav Sutnar Papers, Accession no. 2013.M.6, Exhibition design, 1939-1976, Box 2013.M.6.bx16, folder 1. GRI.
Telegram from Josef Černovský to Vavrečka and Hlavnička of April 28, 1939. MRA in Brno, branch Zlín.
Telegram from Malota to Černovský of April 21, 1939, Korespondence ředitelny fy Baťa s USA CZ z Čs. pavilonu v New Yorku - Světová výstava. I/4 Ředitelna inv.no. 412, MRA in Brno, branch Zlín.
The Woman's World's Fair, Coliseum - Chicago - May - 1927. The Czechoslovak Folk-Art Exhibit, The University of Chicago Library.
Vs. "Viníci ostudy s čs. Pavilonem na pařížské výstavě musí býti potrestáni," *Národní listy*, June 23, 1937, Newspapers Excerpts Archive, MFA, oh8.
Zmrhal, J. J. "Masaryk v Americe," in *TGM 1850-1950 sborník*, ed. Otakar Odložník (Chicago: Národní jednota čsl. protestantů v Americe a v Kanadě. 1950), CPL.

Secondary sources

("Review: May-October 1850"). https://www.marxists.org/archive/marx/works/1850/11/01.htm.
"Československé oddělení na mezinárodním knižním veletrhu ve Florencii 1928. Podle zprávy výstavního komisaře Prause," *Zprávy Svazu československého díla* 9, no. 8 (1927–28): 60–64.
"Československo bude zase svobodno!" *Věstník Česko-Americké jednoty* XXVI (July 1939): 4.
"Československo na mezinárodní hygienické výstavě v Drážďanech," *Nová doba* 36, 15 May 1930, 3.
"Československý den na Výstavě pokroku," *Orgán Československých spolků v Americe Č.S.A.* XLI, no.7, July 1933, 255 and 261.
"Československý den," *Spravedlnost*, August 25, 1933, 3.
"Československý státní pavilon na světové výstavě," *Umění* 10 (1936/7): 246–247.
"Contemporary Slovakian Art," *New York Times*, November 13, 1938, 185.
"Cream of Contemporary Europe's Art Exhibited at Golden Gate Fair," *The Art Digest. The News and Opinions of the Art World*, 15 March 1939, 37.
"Čs. pavilon na svetovej výstave znázorňuje republiku aká bola," *Souvenir Programme of the Eleventh Slovak National Day* (Chicago: Mally Press, 1939).
"Čsl. legionář píše,'*Věstník českoslov.obce legionářské v Americe*, March 1934, 1.
"Czechoslovakia," *Contemporary Art. Official Catalog, Department of Fine Arts* (San Francisco: Golden Gate International Exposition, 1939), 14.
"Czechoslovaks and the United States," *Czechs and Slovaks. World's Fair Memorial of the Czechoslovak Group* (Chicago, 1933).
"Declaration of Common Aims of the Independent Mid-European Nations," http://www.carpatho-rusyn.org/fame/proc.htm.
"Delegácia a jej poslanie," *Zájazd Delegácie Matice Slovenskej do Ameriky v r. 1935-1936* (Chicago: Výbor pre Zájazd Matice Slovenskej v Chicagu, 1936).
"Fair Facts," Officially Licenced New York World' Fair Comics 1940 (New York: Detective Comics, 1940).
"Glänzendes Abschneiden der Tschekoslovakei in Paris," *Prager Presse*, November 26, 1937, 5.
"Huge Crowds at Fair See Czech Parade," Century of Progress Scrapbook, July 17–August 8, 1933, Box 2, folder 1, Chicago Public Library.

"Hygienická výstava v Drážďanech," *Rozkvět* 23 (1930): 13.
"Inaguració de la Secció Txechoslovaca," *La publicitat* July 16, 1929.
"Jako skvělá pohádka z oceli a sklad bude zářit skleněný čs. pavilon na světové výstavě v Paříži," *Venkov* 32, February 7, 1937, 8.
"Koho representuje čs. pavilion na svetovej výstave v Paríži?" *Slovenský denník*, August 21, 1937, 5.
"Košíkářství na výstavě 'Století pokroku,'" *Spravedlnost*, June 16, 1933, 10.
"Mezinárodní uznání našim vystavovatelům," *Lidové noviny* 45, no. 595, November 26, 1937, 5.
"My na světové výstavě v Paříži," *Venkov* 32, no. 197, August 22,1937, 5.
"Na výstavě se lidi už nepřijímají do práce," *Svornost*, May 20, 1933, 2.
"Na výstavišti," *Svornost*, June 16, 1933, 3.
"Náš pavilon na pařížské výstavě," *Venkov* 32, no. 187, August 11, 1937, 8.
"'Náš' den na výstavě Století Pokroku dnem dokonalého triumfu. Úchvatné scény ve Stadiu Vojenské Pole," *Svornost*, August 21, 1933, 1-2.
"Orgán bratrstva ČSA," *Orgán československých spolků v Americe ČSA*, vol. XLI, July 1933, 275.
"Our history," *Bureau International des Expositions*, https://www.bie-paris.org/site/en/about-the-bie/our-history. Accessed on July 2, 2023.
"Paris," in Brigitte Schroeder-Gudehus and Anne Rasmussen, *Les Fastes du Progrès. Le guide des Expositions universelles 1851-1992* (Paris: Flammarion, 1992), 192-200.
"Pavillon national," *Catalogue officiel de la Section Tchéchoslovaque. Exposition internationale des arts décoratifs et industriels modernes Paris 1925*, ed. Vilém Dvořák (no publisher, 1925), 13.
"Préambule," 5-6, *Tchecoslovaquie à l'Exposition Internationale des Artes et Techniques dans la Vie Moderne Paris 1937.*
"Prof. František Kysela," *Výtvarné snahy* 6, no. 2 (1925): 48.
"Průvodní zpráva k projektu arch. Jaromíra Krejcara a spolupracovníků," *Stavba* XIII (1936-37): 175-181.
"Rašínova měnová reforma," *Historie ČNB*, https://www.historie.cnb.cz/cs/menova_politika/1_vznik_a_stabilizace_samostatne_ceskoslovenske_meny/rasinova_menova_reforma.html.
"Skutok, ktorý sme veru neočekávali," *Náš svet,* 4, no. 25, 1933, 2.
"Slováci vystúpili z výboru světovej výstavy," *Náš svet*, January 6, 1933, 3.
"Slováci, zastúpení na Columbickej výstave 1893. Súčastnení na Českom dni 12. augusta 1893. "Z pamätí Boženy Kosmanovej, rod. Mallý," *The Fifth Slovak National Day,* 1933, 77.
"Účast Československa na mezinárodní výstavě v Paříži r. 1937," *Volné směry* 1, no. 32 (1936): 298-299.
"Výzva," *Pásmo* 7-8, 1924-25, 1.
"Zlín na světové výstavě v Paříži," *Zlín* VII, no. 10, March 8, 1937, 1.
A Week at the Fair, Illustrating the Exhibits and Wonders of the World's Columbian Exposition. Rand, McNally & Co., 1893.
Abbattista, Guido. *Moving Bodies, Displaying Nations. National Cultures, Race and Gender in World Expositions 19th to 21st Century.* EUT Edizioni Università di Trieste, Trieste, 2014.
Act 122/1920 Coll. article 129 of the Constitution, on the principles of the language right in the Czechoslovak Republic, passed on February 29, 1920 and valid as of March 6, 1920, published in the Code no. 26/1920, 268.
Albert, Samuel D. "The National for Itself: The 1896 Hungarian Millennium and the 1906 Romanian National General Exhibition," in *Cultures of International Exhibitions 1840-1940. Great Exhibitions in the Margins*, ed. Marta Filipová, 113-136. Farnham: Ashgate Publishing, 2015.

Albrecht, Catherine. "Pride in Production: The Jubilee Exhibition of 1891 and Economic Competition between Czechs and Germans in Bohemia." *Austrian History Yearbook* 24 (1993): 101–118.

Anderson, A. "The Paris Exhibition and Some of its Buildings," *Architectural Review* 7 (1900): 29–37.

Arpáš, Róbert and Matej Hanula. "Postoje hlavných slovenských politických prúdov k čechoslovakizmu v medzivojnovom období." In *Čecho/slovakismus*, eds. Adam Hudek, Michal Kopeček and Jan Mervart, 182–201. Prague: NLN, 2019.

Auerhahn, Jan. "Pokus o demografii zahraničních Čechů a Slováků." In *Československá vlastivěda II. Národopis*, ed. Jiří Horák, Karel Chotek and Jindřich Matiegka. Prague: Sfinx 1936.

Austin, E. L. and Odell Hauser. *The Sesqui-Centennial International Exposition*. New York: Arno Press, 1976.

Aynsley, Jeremy. "Pressa Cologne, 1928: Exhibitions and Publication Design in the Weimar Period." *Design Issues* 10 (1994): 52–76.

Badger, Elizabeth. "The World's Fare: Food and Culture at American World Fairs from 1893–1939," MA Thesis, Western Washington University, 2012.

Baker, Roland. *Official Guidebook: New York World's Fair 1939 "Building the World of Tomorrow."* New York, Exposition Publications, 1939.

Bancel, Nicolas et al, eds. *Zoos humains. Autemps des exhibitions humaines*. Paris: La Découverte, 2002.

Barter, Judith, and Jennifer Downes. *Shaping the Modern: American Decorative Arts at The Art Institute of Chicago, 1917–65*. Chicago: Art Institute of Chicago, 2001.

Baťa, Jan Antonín. "Brazilské stopy," https://www.youtube.com/watch?v=uDOxYghcT4A.

Baťa, Jan Antonín. "Dva pavilony Světové výstavy v Paříži." *Zlín* 4, October 1937, 1.

Baťa, Jan Antonín. *Budujme stát pro 40,000,000 lidí*. Zlín: Tisk, 1937.

Baudrillard, Jean. *Simulacra and Simulation*. New York: Semiotext(e), 1983.

Bauer, Zdeněk. *Jak vytvořit atraktivní obchodní název firmy, služby, produktu, značky: tvorba názvu - obchodního jména krok za krokem*. Prague: Zdeněk Bauer, 2014.

Bauman, Zygmunt. *Liquid Modernity*. Cambridge: Polity Press, 2012.

Bauman, Zygmunt. *Modernity and Ambivalence*. Cambridge: Polity Press, 1991.

Bauman, Zygmunt. *Modernity and the Holocaust*. Cambridge: Polity Press, 1989.

Bayer, Herbert. "Aspects of Design of Exhibitions and Museums." *Curator* 4, no. 3 (1961): 257–288.

Bayer, Herbert. "Fundamentals of Exhibition Design." *Production Manager* (1939/40): 19.

Beck, David R. M. *Unfair Labor? American Indians and the 1893 World's Columbian Exposition in Chicago*. Lincoln: University of Nebraska, 2019.

Benedict, Burton. *The Anthropology of World's Fairs*. Berkeley: University of California Press, 1983.

Benešová, Marie. *Pavel Janá*. Prague: Nakladatelství československých výtvarných umělců, 1959.

Benjamin, Walter. "Paris, Capital of the Nineteenth Century (1939)." In *Reflections*, edited by Peter Demetz. New York: Harcourt, Brace, Jovanovich, 1979.

Benjamin, Walter. *The Arcades Project*. Harvard University Press, 1999.

Bennett, Tony. "The Exhibitionary Complex." *new formations* 4, no. 1 (1998): 73–102.

Benš, Adolf. "Mezinárodní výstava v Paříži dekorativního a průmyslového moderního umění v Paříži 1925." *Stavitel* VI, no. 8–9 (1925): 109–118.

Birnholz, Alan C. "El Lissitzky and the Spectator: From Passivity to Participation." In *The Avant-garde in Russia 1910–1930. New Perspectives*, edited by Jeanne D'Andrea and Stephen West. Los Angeles: LACMA-Cambridge and MA: MIT Press, 1980.

Bitgood, Stephen. "Museum Fatigue: A Critical Review." *Visitor Studies* 12, no. 2 (2009): 93–111.

Bitgood, Stephen. "When is "museum fatigue" not fatigue?" *Curator: The Museum Journal* 52 (2009): 193–202.

Blanchard, Pascal et al, eds. *Human Zoos: Science and Spectacle in the Age of Colonial Empires*. Liverpool: Liverpool University Press 2008.

Boisseau, Tracey Jean and Abigail M. Markwyn, eds. *Gendering the Fair: Histories of Women and Gender at World's Fairs*. Urbana: University of Illinois Press, 2010.

Boisseau, Tracey Jean. "Once Again in Chicago: Revisioning Women as Workers at the Chicago Woman's World's Fairs of 1925–1928." *Women's History Review* 18, no. 2 (2009): 265–291.

Bokovoy, Matthew F. *The San Diego World's Fairs and Southwestern Memory, 1880–1940*. Albuquerque: University of New Mexico Press, 2005.

Bokovoy, Matthew. "San Diego 1915–16." In *Encyclopedia of World's Fairs and Expositions*, edited by John E. Findling and Kimberley D. Pelle, 222–228, Jefferson N.C.: McFarland, 2008.

Bokovoy, Matthew. *The San Diego World's Fairs and the Southwestern Memory, 1880–1940*. Albuquerque: The University of New Mexico Press, 2005.

Bontempi, Bruno Jr., and Noah W. Sobe. "Exposição Internacional do Centenário da Independência." In *Encyclopedia of World's Fairs and Expositions*, edited by John E. Findling and Kimberly D. Pelle, 228–230, Jefferson, N.C.: McFarland, 2008.

Boone, M. Elizabeth. "*The Spanish Element in Our Nationality,*" *Spain and America at the World's Fairs and Centennial Celebrations, 1876–1915*. University Park, PA: Penn State University Press, 2021.

Borovský, Karel Havlíček. "Slovan a Čech," *Pražské noviny*, February 15 – March 12, 1846.

Bořutová, Dana. *Architekt Dušan Samuel Jurkovič*. Bratislava: Slovart, 2009.

Boussahba-Bravard, Myriam and Rebecca Rogers, eds. *Women in International and Universal Exhibitions, 1876–1937*. London: Routledge, 2018.

Brandon, Mark A. *The Perils of Race-Thinking: A Portrait of Aleš Hrdlička*. Budapest–New York: Central European University Press, 2023.

Brožek Artur. *Ukázka studia vloh a vlivů okolí v rodokmenech*. Prague: A. Brožek, 1927.

Brožek, Artur et al. *Rovnocennost evropských plemen a cesty k jejich ušlechťování*. Prague: Akademie věd, 1934.

Brožek, Artur. "Correlation between the Mating Surrounding Conditions in Man. The Tribe of the Pejsas, Sedlcan-branch," *Institut international d'anthropologie, 11e session Prague, Prague 14–21 Septembre 1924*. Paris: Librairie E. Nourry, 1926

Brožek, Artur. *Zušlechtění lidstva (Eugenika)*. Prague: F. Topič, 1922.

Bugge, Peter. "A Nation Allied with History. Czech Ideas of Democracy, 1890–1948." In *Democracy in Modern Europe: A Conceptual History*, edited by Jussi Kurunmäki, Jeppe Nevers and Henk te Velde, 208–230. New York; Oxford: Berghahn Books, 2018.

Burešová, Jana. *Proměny společenského postavení českých žen v první polovině 20. století*. Olomouc: Univerzita Palackého v Olomouci, 2001.

Burns, Emily C. "World's Fairs as Sites of Orientalism East and West." In *Near East to Far West. Fictions of French and American Colonialism*, edited by Jennifer R. Hennerman, 173–195. Denver: Denver Museum of Art, 2023.

Burrows, Anna. "The San Francisco Golden Gate Exhibition 1939–1940." The Digital Collection, University of Maryland libraries, December 5, 2023, https://digital.lib.umd.edu/worldsfairs/result/id/umd:1010.

Č. "Československo na světové výstavě v Bruselu 1935," *Sklářské rozhledy* 12, no. 1 (1935): 12–13.
Caha, Arnošt. *Tatíček Masaryk Osvoboditel.* Brno: A. Caha, 1920.
Cameron, Marcy. "The Dual Identity of the Czechoslovak Pavilion at the 1933 Chicago World's Fair." A Master Thesis at the School of Architecture, The University of Virginia, 2012.
Čapek, Josef. "Na rozcestí." *Bytová kultura* I, no. 6–8 (1924–25): 115–116.
Čapek, Karel. "Hle, Československo." *Srdce Evropy* I, unpag. (1936). Reprinted in *Od* člověka k člověku III. Prague: Československý spisovatel, 2018.
Čapek, Karel. "La visage de la Tchécoslovaquie." In *Tchecoslovaquie à l'Exposition Internationale des Arts et Techniques dans la Vie Moderne Paris 1937*, edited by B. Soumar, 15–21. Prague: Melantrich, 1937.
Čapek, Karel. *Talks with T.G. Masaryk*, translated by M. H. Heim and D. Round. North Haven: Catbird Press, 1995.
Čapková, Kateřina. *Czechs, Germans, Jews?: National Identity and the Jews of Bohemia.* New York: Berghahn Books, 2015.
Carpenter, Patricia F. and Paul Totah, eds. *The San Francisco Fair. Treasure Island 1939–1940.* San Francisco: Scottwall Associates, 1989.
Catalogue of a Century of Progress Exhibition of Painting and Sculpture. Chicago: The Art institute of Chicago 1933.
Catalogue of the exhibition, July 24th to July 30th, 1912 / First International Eugenics Congress, London, University of London, South Kensington International Eugenics Congress 1912. London: Charles Knight and co, 1912.
Catalogue officiel de la section Tchécoslovaque. Exposition internationale des arts décoratifs et industriels modernes Paris 1925. Prague: Comité d'Exposition, 1925.
Çelik, Zeynep. *Displaying the Orient: Architecture of Islam at Nineteenth-Century World's Fairs.* Berkeley: University of California Press, 1992.
Chandler, Arthur and Philip Whalen. "Paris 1937." In *Encyclopedia of World's Fairs and Expositions*, edited by John E. Findling and Kimberley D. Pelle, 290–296. Jefferson N.C.: McFarland, 2008.
Chesterton, G. K. *Eugenika a jiná zla.* Prague: Ladislav Kuncíř, 1928.
Chrobák, Tomáš and Doubravka Olšáková. *Ernest Denis.* Prague: Eva – Milan Nevole, 2003.
Čipera, Dominik. *Ve službách práce a lidu.* Zlín: Tisk, 1944.
Clifford, James. *The Predicament of Culture: Twentieth-Century Ethnography, Literature and Art.* Cambridge, MA: Harvard University Press, 1988.
Cogniat, Raymond. "Moderní československé umění." Život 16, no. 1 (1937): 15.
Cornwall, Mark and R. J. W. Evans, eds. *Czechoslovakia in a Nationalist and Fascist Europe 1918–1948.* Oxford: Oxford University Press, 2007.
Cronin, Elizabeth. *Heimat Photography in Austria. A Politicized Vision of Peasants and Skiers.* Vienna: Photoinstitut Bonartes: Albertina and Salzburg: Fotohof Edition, 2015.
Crough, Olivia. "El Lissitzky's Screening Rooms." In *Screens. From Materiality to Spectatorship – A Historical and Theoretical Reassessment*, edited by Dominique Chateau and José Moure, 223–235. Amsterdam University Press, 2016.
Čtyroký, V[áclav]. "Československé sklářství v Bruselu." *Sklářské rozhledy* XII, no. 6-7 (1935): 82–85.
Čtyroký, V[áclav]. "Naše sklářství v novém státě." *Sklářské rozhledy* XV, no. 1 (1938): 168.
Čtyroký, Václav. "Naše sklářství v Paříži." *Národní politika* (supplement), September 25, 1937, 16.
Cude, Michael. "The Imagined Exiles: Slovak-Americans and the Slovak Question during the First Czechoslovak Republic." *Studia Historica Gedanensia* V (2014): 287–305.
Čulen, Konštantín. *Slováci v Amerike.* Turčianský Sv. Martin: Matica slovenská, 1938.

Cuřínová, Ludmila. "Ústav pro národní eugeniku." In *Technokracie v Českých zemích*, edited by Jan Janko and Emílie Těšínská, 151–156. Prague: Akademie věd, 1999.

Czerner, Olgierd and Hieronim Listowski, eds. *The Polish Avant-garde: Architecture and Town Planning, 1918–1933*. Paris: Editions du Moniteur, 1981.

Darras, V. "Praktický průvodce po pařížské výstavě umění a techniky v moderním životě." *L'Indépendance Tchécoslovaque*, no. 11 (1937): 3–4.

Davenport, Charles. *Eugenics: The Science of Human Improvement by Better Breeding*. New York: Henry Holt, 1910.

Davey, Gareth. "What is museum fatigue?" *Visitor Studies Today* 8, no. 3 (2005): 17–21.

Davis, John R. "Marginal Exhibition? The All-German Exhibition in Berlin, 1844." In *Cultures of International Exhibitions, 1840–1940. Great Exhibitions in the Margins*, edited by Marta Filipová, 69–90. Farnham: Ashgate Publishing, 2015.

de Quirielle, Pierre. "Une exposition d'art populaire tchécoslovaque." *Journal des débats politiques et littéraires*, April 28, 1920, 1, https://gallica.bnf.fr/ark:/12148/bpt6k4893140.

Decoteau, H. "Malvina Hoffman and the 'Races of Mankind.' *Art Journal* (Fall1989/Winter1990): 7–12.

Demski, Dagnosław, and Dominika Czarnecka, eds. *Staged Otherness: Ethnic Shows in Central and Eastern Europe 1850–1939*. Budapest–New York: Central European University Press, 2022.

Detva, československý ľudový umelecký priemysel, úč.spol. Bratislava: Detva, 1929.

Dluhosch, Eric and Rostislav Švácha, eds. *Karel Teige / 1900-1951: L'enfant terrible of the Czech modernist avant-garde*. Cambridge, MA: MIT Press, 1999.

Doleshal, Zachary Austin. "Life and Death in the Kingdom of Shoes: Zlín, Bat'a, and Czechoslovakia, 1923–1941." A Dissertation at the University of Texas at Austin, 2012.

Doleshal, Zachary Austin. *In the Kingdom of Shoes: Bata, Zlín, Globalization, 1894–1945*. Toronto: University of Toronto Press, 2021.

Ducháček, Milan et al. *Za rovnocennost evropských plemen. Československá antropologie tváří v tvář rasismu a nacismu*. Prague: Ústav pro soudobé dějiny AV ČR, v. v. i.: Prague: NLN, 2023.

Ducháček, Milan. "Čechoslovakismus v prvním poločase ČSR: státotvorný koncept nebo floskule?" In *Čechoslovakismus*, edited by Michal Kopeček, Jan Mervart, Adam Hudek, 149–181. Prague: NLN, 2019.

Dungca, Nicole, Claire Healey and Andrew Ba Tran. "The Smithsonian's 'Bone Doctor' Scavenged Thousands of Body Parts." *The Washington Post*, August 15, 2023, https://www.washingtonpost.com/history/interactive/2023/ales-hrdlicka-smithsonian-brains-racism, accessed February 25, 2024.

Dunham, George C. "The International Hygiene Exhibition." *American Journal of Public Health and the Nation's Health* XXI, no. 1 (1930): 1–10.

Dvořák, Vilém. "Gobeliny umělecko-průmyslové školy v Praze na mezinárodní výstavě dekorativního umění v Paříži 1925." *Výtvarná práce* 4, no. 2–3 (1925): 79–82.

Dzień polski na wystawie wszechświatowej niedziela dnia 26go sierpnia 1934 roku. Polish Day at A Century of Progress, August 26th (n.p., 1934).

Dziewanowski-Stefańczyk, Bartosz. "World's Fairs as Tools of Diplomacy: Interwar Poland." In *World Fairs and the Global Moulding*, edited by Joseph Leerssen and Eric Storm, 300–328. Leiden: Brill, 2022.

Eco, Umberto. "A Theory of Expositions." *Travels in Hyperreality. Essays*, 453–480. San Diego – New York – London: Harcourt, 1990.

Ed. "Bude účast Československa na mezinárodní výstavě v Paříži 1937 ostudou?" *Volné směry* 1, no. 32 (1936): 306.

Ed., "Československé umění na mezinárodní výstavě v Paříži," *Dílo* 29 (1938–39): 50.

Edgar, Emil. *Život a umění. Soudobá architektura*. Prague: Česká grafická unie, 1921.

Exposition internationale d"Art moderne catalogue. Dietrich and co, 1935.

Exposition internationale des arts et des techniques dans la vie moderne, Paris 1937, Catalogue Officiel, vol. II. Paris: M. Déchaux, 1937.

Feinberg, Melissa. *Elusive Equality: Gender, Citizenship and the Limits of Democracy in Czechoslovakia, 1918–1950.* Pittsburg: University of Pittsburg Press, 2006.

Findling, John E. "Barcelona 1929/30." In *Encyclopedia of World's Fairs and Expositions,* edited by John E. Findling and Kimberley D. Pelle, 251–253. Jefferson N.C.: McFarland, 2008.

Findling, John E. "Chicago 1933–1934." In *Encyclopedia of World's Fairs and Expositions,* edited by John E. Findling and Kimberley D. Pelle, 268–277. Jefferson N.C.: McFarland, 2008.

Filipová, Marta, ed. *Cultures of International Exhibitions 1840–1940. Great Exhibitions in the Margins,* London: Ashgate, 2017.

Filipová, Marta. "Colonial Complicities beyond the Empire. Czechoslovakia in between Worlds and World's Fairs." In *Routledge Companion to Art and the Formation of Empire,* edited by Emily C. Burns and Alice M. R. Price. New York: Routledge, 2024.

Filipová, Marta. "Peasants on Display. The Czechoslavic Ethnographic Exhibition of 1895." *Journal of Design History* 24, no. 1 (2011): 15–36.

Filipová, Marta. "The Theatre of Exhibitions. Czechoslovakia at the International Exhibition in Paris, 1937." *The Journal of Design History* 35, no. 2 (2021): 132–150, https://doi.org/10.1093/jdh/epab052.

Filipová, Marta. *Modernity, History and Politics in Czech Art.* New York: Routledge, 2020.

Filipová, Marta. "Czech Glass or Bohemian Crystal? The Nationality of Design in the Czech Context." In *Designing Worlds: National Design Histories in an Age of Globalization,* edited by Kjetil Fallan and Grace Lees-Maffei, 141–155. New York: Berghahn Books, 2016.

Fiss, Karen. *Grand Illusion: Third Reich, the Paris Exposition, and the Cultural Seduction of France.* Chicago: University of Chicago Press, 2009.

Forgács, Éva. "Definitive Space: The Many Utopias of El Lissitzky's *Proun Room.*" In *Situating El Lissitzky: Vitebsk, Berlin, Moscow,* edited by Nancy Perloff and Brian Reed, 47–75. Los Angeles: Getty Publications, 2003.

Formanová, Markéta and Zdeněk Kuchyňka. *Ztracené dioráma: příběh unikátního československého exponátu na Světové výstavě v New Yorku v roce 1939 = The Lost Diorama: The Story of a Unique Czechoslovak Exhibit at the 1939 World's Fair in New York City.* Plzeň: Západočeské muzeum, 2019.

Foster-Rice, Gregory. "The Visuality of Race: 'The Old Americans,' 'The New Negro' and American Art, c.1925." PhD dissertation, The Northwestern University, 2003.

Foucault, Michel. *Discipline and Punish, The Birth of the Prison.* London: Allen Lane, 1977.

Foustka, Břetislav. "Ethics and Eugenics," translated by Anna Pilátová. In *The History of East-Central Eugenics, 1900–1945: Sources and Commentaries,* edited by Marius Turda, 153–155. London and New York: Bloomsbury Academic, 2015.

Frejka, Jiří. "Děláme československou divadelní exposici." *České slovo,* August 22, 1937.

Gaillard, Marc. *Paris. Les Expositions Universelles de 1855 à 1937.* Paris: Les Presses Franciliennes, 2003.

Gál, Vilmos. *Hungary at the World Fairs.* Budapest: Holnap Kiadó, 2010.

Galton, Francis. *Inquiries into Human Faculty and its Development.* London–New York: J.M. Dent and E.P. Dutton, 1907.

Ganz, Cheryl. "Science Advancing Mankind." *Technology and Culture* 41, no. 4 (Oct., 2000): 783–787.

Ganz, Cheryl. *The 1933 Chicago World's Fair: A Century of Progress.* Urbana, Chicago: University of Illinois Press, 2012.

Geddes, Patrick. "The Closing Exhibition – Paris 1900." *The Contemporary Review*, November 1900, 653–668.

Geering, Corinne. "'Is This Not Just Nationalism?': Disentangling the Threads of Folk Costumes in the History of Central and Eastern Europe." *Nationalities Papers* 50, no. 4 (2021): 722–741, doi 10.1017/nps.2021.21.

George, Waldemar. "L'Exposition des Arts Décoratifs et Industriels de 1925: les tendances générales." *L'Amour de l'art* (1925): 283–291.

Geppert, Alexander C. T. *Fleeting Cities: Imperial Expositions in Fin-de-Siècle Europe.* London: Palgrave Macmillan, 2010.

Géraldy, Paul. "L'Architecture vivante." *L'illustration, Special section on L'exposition internationale des Arts décoratifs et industriels modernes*, April 15, 1925, n.p.

Gilbert, James. *Whose Fair? Experience, Memory and the History of the Great St. Louis Exposition.* Chicago: University of Chicago Press, 2009.

Gilman, Benjamin. "Museum fatigue." *Scientific Monthly* 12 (1916): 67–74.

Giustino, Cathleen M. "Industrial Design and the Czechoslovak Pavilion at EXPO '58: Artistic Autonomy, Party Control and Cold War Common Ground." *Journal of Contemporary History* 47, no. 1 (2012): 185–212, https://doi.org/10.1177/0022009411422.

Glassberg, David. "Philadelphia, 1926." In *Encyclopedia of World's Fairs and Expositions*, eds. by John E. Findling and Kimberley D. Pelle, 246–249. Jefferson N.C.: McFarland, 2008.

Golan, Romy. *Modernity and Nostalgia. Art and Politics in France between the Wars.* New Haven and London: Yale University Press, 1995.

Grác, A. "Mezinárodní hygienická výstava v Drážďanech." *Zlín. Sdělení zaměstnancům firmy Baťa* 13, no. 2, August 1930, 3.

Graul, Richard. *Ausstellung europäisches Kunstgewerbe 1927.* Leipzig: Rudolf Schick, 1927.

Green, David. "Veins of Resemblance: Photography and Eugenics." In *Photography/Politics: Two*, edited by Patricia Holland, Jo Spence and Simon Watney, 9–21. London: Comedia, 1986.

Greenhalgh, Paul. *Ephemeral Vistas: The Expositions Universelles, Great Exhibitions, And World's Fairs, 1851–1939.* Manchester, UK – New York: Manchester University Press, 1988.

Griffin, Roger. 'Tunnel Visions and Mysterious Trees: Modernist Projects of National and Racial Regeneration: 1880–1939." In *Blood and Homeland: Eugenics and Racial Nationalism in Central and Southeastern Europe, 1900–1940*, edited by Marius Turda and Paul J. Weindling, 162–189. Budapest–New York: Central European University Press, 2007.

Hájková, Dagmar. *"Naše česká věc," Češi v Americe za první světové války.* Prague: NLN, 2011.

Halada, Jaroslav and Milan Hlavačka, *Světové výstavy: Od Londýna 1851 po Hannover 2000.* Prague: Libri 2000.

Hale, Dana S. *Races on Display: French Representations of Colonized Peoples, 1886–1940.* Bloomington: Indiana University Press, 2008.

Hamlin, Talbot F. "Paris 1937. A Critique." *American Architect* (1937): 25–34.

Hanson, Debra. "Re-presenting the Arab-Islamic World at the Nineteenth-Century World's Fairs." In *Expanding Nationalisms at World's Fairs: Identity, Diversity and Exchange, 1851–1915*, edited by David Raizman and Ethan Robey, 15–32. London: Routledge, 2017.

Heimann, Mary. *Czechoslovakia, The State that Failed.* Yale University Press, 1997.

Herbert, James D. *Paris 1937. Worlds on Exhibition.* Ithaca: Cornell University Press, 2018.

Herder, Johann G. *Ideen zur Philosophie der Geschichte der Menschheit*, 4 vol. Riga and Leipzig: J. F. Hartknoch, 1784–1791.

Heruday-Kiełczewska, Magdalena. "Wystawa Pracy Kobiet na Powszechnej Wystawie Krajowej." *Hrabia Tytus,* September 23, 2018, https://tytus.edu.pl/2018/09/23/wystawa-pracy-kobiet-na-powszechnej-wystawie-krajowej.

Herza, Filip. "Ne/rovnocennost evropských plemen a cesty k autoritativním formám vládnutí: Jindřich Matiegka, Jiří Malý a Vojtěch Suk v meziválečných diskusích o rasách, národech a eugenice." In *Za rovnocennost evropských plemen. Československá antropologie tváří v tvář rasismu a nacismu*, edited by Ducháček, Milan et al., 159–185. Prague: Ústav pro soudobé dějiny AV ČR, v. v. i.: Prague: NLN, 2023.

Herza, Filip. *Imaginace jinakosti. Pražské přehlídky lidských kuriozit v 19. a 20. Století.* Prague: Scriptorium, 2020.

Hitchcock, Henry Russell. "Paris 1937." *Architectural Forum* (September 1937).

Hlavačka, Milan, Jana Orlíková, and Petr Šembera. *Alfons Mucha—Paříž 1900. Pavilon Bosny a Herzegoviny na světové výstavě Alfons Mucha—Paris 1900. The Pavilion of Bosnia and Herzegovina at the World Exhibition.* Prague, Obecní dům, 2002.

Hlaveš, Milan. "Moser." In *Design v českých zemích 1900–2000: instituce moderního designu*, edited by Iva Knobloch and Radim Vondráček, 582. Prague: Academia, 2016.

Hnídková, Vendula. *Národní styl, kultura a politika.* Prague: VŠUO, 2013.

Hoffenberg, Peter H. *An Empire on Display: English, Indian and Australian Exhibitions from the Crystal Palace to the Great War.* Berkeley: University of California Press, 2001.

Holá, Mariana. "Teige, Karel (1900–1951)." *Routledge Encyclopedia of Modernism.* London: Routledge, 2016, doi:10.4324/9781135000356-rem1653-1.

Holý, Ladislav. *The Little Czech and the Great Czech Nation. National Identity and the Post-Communist Transformation of Society.* Cambridge: Cambridge University Press 1996.

Hooton, Earnest A. "What is an American." *The American Journal of Physical Anthropology* 22, no. 1 (1936): 1–26.

Horáková, Monika. "Dominik Čipera - spolupracovník Tomáše Bati." In *Tomáš Baťa: Doba a společnost*, edited by Marek Tomaštík, 97–118. Brno: Viribus Unitis, 2007.

Houze, Rebecca. *Textiles, Fashion, and Design Reform in Austria-Hungary Before the First World War.* London: Routledge, 2017.

Hoyer, Egbert, "Über die heutige Praxis der Ausstellungen." *Zeitschrift* für deutsche *Volkswirtschaft* 1 (1880): 16–23.

Hrabová, Martina. "Více tváří mistra. Le Corbusier a Československo." *Art and Antiques* 2 (2012): 40–49.

Hrdlička Aleš. *Descriptive Catalog of the Section of Physical Anthropology, Panama-California Exhibition.* National Views, 1915.

Hrdlička, Aleš. "The Full-Blood American Negro." *The American Journal of Physical Anthropology* 12, no. 1 (1928): 15–33.

Hrdlička, Aleš. "What are the Czechoslovaks." In *World's Fair Memorial of the Czechoslovak Group (Czechs and Slovaks)*, 22–24. Chicago, 1933.

Hruboň, Anton. "Creating the Paradigm of 'New Nation.' Eugenic Thinking and the Culture of Racial-Hygiene in the Slovak State." *Fascismus* 10, no. 2 (2021): 275–297, https://doi.org/10.1163/22116257-bja10032.

Hruška, J. H. "Výstavka československé Ameriky." *The World's Fair Memorial of the Czechoslovak Group (Czechs and Slovaks)*, 106–107. Chicago, 1933.

Hubatová-Vacková, Lada. "Užití, využití i zneužití folkloru a lidového umění." In *Budování státu: reprezentace Československa v umění, architektuře a designu = Building a State: The Representation of Czechoslovakia in Art, Architecture and Design*, edited by Milena Bartlová, 180–244, Prague: UMPRUM, 2015.

Jaklová, Alena. *Proměny etnické a kulturní identity českých přistěhovalců ve Spojených státech amerických. Interpretace analýz českoamerického periodického tisku 19. a 20. Století.* Prague: ČVUT, 2014.

James, Jack and Earle Weller. *Treasure Island. "The Magic City" 1939–1940. The Story of the Golden Gate International Exposition.* San Francisco: Pisani Printing and Publishing Company, 1941.
Janáková, Iva. "Exhibition Design." In *Ladislav Sutnar. Praha – New York, Design in Action*, edited by Iva Janáková, 324–325. Prague: UPM – Argo, 2003.
Janik, L. I. "Konec výstavy." *Svornost*, November 17, 1933, 4.
Jarábek. "Slovenské zprávy o svetovej výstave." *Svornost*, May 3, 1933, 2.
Jaroš-Gamma, Gustav. *Náš tatíček Masaryk.* Prague: Gamma 1918–25.
Johnson, Philip. *Machine Art. March 6 to April 30, 1934.* New York: Museum of Modern Art, 1934.
Judson, Pieter M. *The Habsburg Empire: A New History.* Cambridge, Massachusetts: The Belknap Press of Harvard University Press, 2016.
Kargon, Robert H. et al. "Modernity a la française. The 1937 Paris Exposition." In *World's Fairs on the Eve of War: Science, Technology and Modernity, 1937–1942*, 7–29. Pittsburgh: University of Pittsburgh Press, 2015.
Kargon, Robert H. et al. "Whose Modernity? Utopia and Commerce at the 1939 New York World's Fair." In *World's Fairs on the Eve of War. Science, Technology and Modernity, 1937–1942*, 57–82. Pittsburgh: Pittsburgh University Press, 2015.
Kargon, Robert H. et al. *World's Fairs on the Eve of War. Science, Technology and Modernity, 1937–1942.* Pittsburgh: Pittsburgh University Press, 2015.
Kárník, Zdeněk. *České země v éře První republiky (1918–1938) II. Československo a české země v krizi a v ohrožení (1930–1935).* Prague: Libri, 2002.
Kárník, Zdeněk. *České země v éře První republiky (1918–1938) III. O přežití a o život (1936–1938).* Prague: Libri, 2003.
Kárník, Zdeněk. *České země v éře První republiky 1918–1938. I. Vznik, budování a zlatá léta republiky (1918–1929).* Prague: Libri, 2000.
Kaufman, Edward N. "The Architectural Museum from World's Fair to Restoration Village." *Assemblage* 9 (1989): 20–39.
KB [Katarína Bajcurová]. "Ľudovít Fulla." In *Budování státu: reprezentace Československa v umění, architektuře a designu = Building a State: The Representation of Czechoslovakia in Art, Architecture and Design*, edited by Milena Bartlová, cat. no. 9/5. Prague: UMPRUM, 2015.
Keels, Thomas H. *Sesqui! Greed, Graft, and the Forgotten World's Fair of 1926.* Philadelphia: Temple University Press, 2017.
Kejha, Josef, Jiří Janouškovec and Vladimír Jurina. *Plzeňský Prazdroj, příběh, který nepřestává inspirovat. Plzeňský Prazdroj – od roku 1842.* Plzeň: Nava, 2012.
Kiesling, Norbert. *Pavel Janák.* Weitra: Bibliothek der Provinz, 2012.
Kim, Linda. "Malvina Hoffman's Races of Mankind and the Materiality of Race in Early Twentieth-Century Sculpture and Photography." PhD dissertation, University of California, Berkeley, 2006
Kinkel, Marianne. *Races of Mankind: The Sculptures of Malvina Hoffman.* Urbana, IL: University of Illinois Press, 2011.
Kisch, Egon Erwin. *Továrna na boty (Socialisovaný Baťa): K přejmenování nár. podniku Baťa na Svit, nár. Podnik.* Gottwaldov: Svit, n.p., 1949.
Klečka, J. "Naše výstava." In *World's Fair Memorial of the Czechoslovak Group (Czechs and Slovaks)*, 99–105. Chicago, 1933.
Klír, Bohumil. "Čsl. legionář píše svému bratrstvu – z Chicaga!" *Věstník československé obce legionářské v Americe*, March 1934, 1.
Klír, Bohumil. "Čsl. Legionář píše svému bratrstvu – z Chicaga!" *Věstník československé obce legionářské v Americe*, November 1933, 2.

Klír, Bohumil. "Čsl. legionář píše svému bratrstvu." *Věstník československé obce legionářské v Americe*, October 1934, 2.

Klír, Bohumil. "Věc tapná a nechutná." *Svornost*, October 6, 1933, 10.

Knobloch, Iva ed. *Žijeme lidsky? Reforma bydlení 1914–1948. Svaz československého díla Housing Reform 1914–1948. The Czechoslovak Werkbund.* Prague: UPM – Slovart, 2023.

Knobloch, Iva. "Krásná jizba a Svaz československého díla jako modernizační hnutí." In *Krásná jizba DP 1927–1948: design pro demokracii*, edited by Lucie Vlčková and Alice Hekrdlová, 69–84. Prague: Uměleckoprůmyslové museum, 2019.

Knop, Karel. "Organisace lidové výroby na Slovensku i Podkarpatské Rusi a dnešní snahy o její umělecké zvelebení." *Výtvarné snahy* 7 (1926): 136–139.

Know, Katelyn K. *Race on Display in the 20th- and 21st Century France.* Liverpool: Liverpool University Press, 2016.

Koch, Alois. "Budoucnost umělecko průmyslových výstav." *Styl* I (1909): 174–180.

Kohut, Leo. "Bauhaus. Ungarn-Tschechoslowakei. Zur Bauhaus-Rezeption in Ost-Europa." In *Bauhaus-Archiv, Museum. Sammlungs-Katalog, (Auswahl), Architektur, Design, Malerei, Graphik, Kunstpädagogik*, edited by Hans Maria Wingler et al. 283–287. Berlin: Gebr. Mann, 1981.

Kollár, Jan. "Dobré vlastnosti národu slovanského." In *Obrození národa, Svědectví a dokumenty*, edited by Jan Novotný, 191–205. Prague: Melantrich, 1979.

Komora, Pavol. *Hospodárske a všeobecné výstavy 1842 – 1940.* Bratislava: Slovenské národné múzeum - Historické múzeum, 2016.

Kopeček, Michal. "Czechoslovak interwar democracy and its critical introspections." *Journal of Modern European History / Zeitschrift* für *moderne europäische Geschichte / Revue d'histoire européenne contemporaine* 17, no. 1 (February 2019): 7–15.

Kosatík, Pavel. *Slovenské století.* Prague: Torst, 2021.

Kostova, Julia. "Spectacles of Modernity. Anxiety and Contradiction at the Interwar Paris Fairs of 1925, 1931 and 1937." PhD Dissertation, Rutgers, The State University of New Jersey, 2011.

Kostrhun, Petr. "90 let muzea Anthropos: Proměny muzea v období první Československé republiky." In *Transformations of the Anthropos Pavilion in Time (1928–2018)*, edited by Petr Kostrhun, Barbora Půtová and Zdeňka Nerudová, 11–104. Brno: Moravian Museum, 2018.

Kramerová, Daniela and Terezie Nekvindová. *Automat na výstavu: československý pavilon na Expo 67 v Montrealu = The automatic exhibition: the Czechoslovak pavilion at Expo 67 in Montreal.* Cheb: Galerie výtvarného umění, 2017.

Krch, Vojtěch. "Československý pavilon v Paříži." *Architekt SIA. Měsíčník pro architekturu, stavbu měst, bytovou péči a umění* 36, no. 9 (1937): 143.

Krejcar, Jaromír et al. "Československý pavilon na Mezinárodní výstavě umění a techniky v moderním životě." *Stavba* XIII (1936–37): 175.

Krejcar, Jaromír. "Poučení z pařížské výstavy." *Přítomnost* 14, no. 40, October 6, 1937, 633–635.

Krenn, Michael L., ed. *Race and U.S. Foreign Policy from 1900 Through World War II.* New York – London: Garland Pub., 1998.

Kříženecký, Jaroslav. "Zákonná úprava příbuzenských sňatků s hlediska biologicko-lékařského," *Právník* 57 (1918): 325–31.

Křížová, Markéta, "Aleš Hrdlička a transatlantický rozměr českého antropologického a eugenického myšlení první poloviny 20. století." In *Za rovnocennost evropských plemen. Československá antropologie tváří v tvář rasismu a nacismu*, 42–75. Prague: Ústav pro soudobé dějiny AV ČR, v. v. i.: Prague: NLN, 2023.

Křížová, Markéta. "Aleš Hrdlička a transatlantický rozměr českého antropologického a eugenického myšlení první poloviny 20. století." In *Za rovnocennost evropských plemen*

Československé antropologie tváří v tvář rasismu a nacismu, edited by Milan Ducháček et al., 42–75, Prague: Ústav pro soudobé dějiny AV ČR, v. v. i.: Prague: NLN, 2023.
Křížová, Markéta."Curators, Objects and the Indigenous Agency." *Reviews in Anthropology* 50, no. 1–2 (2021): 22–40, doi: 10.1080/00938157.2021.1962036.
Kroha, Jiří. "Několik poznámek k výstavní architektuře." *Horizont* (1927): 99–107.
Kubů, Eduard and Jaroslav Pátek. *Mýtus a realita hospodářské vyspělosti Československa mezi světovými válkami*. Prague: Karolinum, 2000.
Kybalová, Ludmila. *Emílie Paličková*. Prague: Nakladatelství československých výtvarných umělců, 1962.
Lagae, Johan. "Brussels 1935." In *Encyclopedia of World's Fairs and Expositions*, edited by John E. Findling and Kimberley D. Pelle, 277–281. Jefferson N.C.: McFarland, 2008.
Langhamer, Antonín. *Legenda o českém skle – The Legend of Bohemian Glass – Legende vom böhmischen Glas*. Zlín: Tigris, 1999.
Larson, Donald G. "San Francisco." In *Encyclopedia of World's Fairs and Expositions*, edited by John E. Findling and Kimberley D. Pelle, 305–309. Jefferson N.C.: McFarland, 2008.
Laughlin, Harry L. "The Eugenics exhibit at Chicago." *The Journal of Heredity* 26, no. 4 (1935): 155–162.
Leerssen, Joseph and Eric Storm, eds, *World Fairs and the Global Moulding of National Identities: International Exhibitions as Cultural Platforms, 1851–1958*. Leiden: Brill, 2022.
Lefebvre, Henri. *The Production of Space*, trans. Donald Nicholson-Smith. Oxford: Blackwell, 1991.
Leshne, Carla. "Progress is More than Just a Word: The Workers Film and Photo League at 1933 Chicago World's Fair." In *Meet Me at the Fair: A World's Fair Reader*, edited by Celia Pearce, et al., 371–378. Pittsburg: ETC Press, 2014.
Levey, George Collins. "Exhibition." *Encyclopedia Britannica*, vol. 9, 13th edition, 67–71. London: Encyclopedia Britannica, 1926.
Levy, Ruth. *A exposição do centenário e o meio arquitetônico carioca no início dos anos 1920*. Rio de Janeiro: EBA Publicações, 2010.
Lissitzky, El. "Exhibition Rooms (ca 1926)." In *El Lissitzky: Life, Letters, Texts*, edited by Sophie Lissitzky-Küppers, London: Thames & Hudson, 1968.
Lissitzky, El. *Russia: An Architecture for World Revolution*, translated by Eric Dluhosch. London: Lund Humphries Publishers, 1970.
Loos, Adolf. "Ornament and Crime (1908). In *Modernism: An Anthology of Sources and Documents*, edited by Vassiliki Kolocotroni, Jane Goldman and Olga Taxidou, 77–81. Chicago: University of Chicago Press, 1998.
Lugon, Olivier. "Dynamic Paths of Though: Exhibition Design, Photography and Circulation in the Work of Herbert Bayer." In *Cinema beyond Film: Media Epistemology in the Modern Era*, edited by François Albera and Maria Tortajada, 118–145. Amsterdam University Press, 2021.
Lukáč, Eduard Katarína Mayer, and Martina Lenhardtová. "Development of Adult Education in Interwar Slovakia (1918 – 1938)." *Ad Alta: Journal of Interdisciplinary Research* (2019): 173–180.
Macura, Vladimír. *Znamení zrodu: české obrození jako kulturní typ*. Prague: Československý spisovatel, 1983.
Makaryk, Irena R. *April in Paris. Theatricality, Modernism and Politics at the 1925 Art Deco Expo*. Toronto: Toronto University Press, 2018.
Marshall, Jennifer Jane. *Machine Art 1934*. Chicago, Ill. – London: The University of Chicago Press, 2012.

Masaryk, T. G. "Dodatek k Několika poznámkám o práci československých žen." In T. G. Masaryk, *Masaryk a ženy*, 22–24. Prague: Ženská národní rada, 1930,.

Masaryk, T. G. "Moderní názor na ženu." In T. G. Masaryk, *Masaryk a ženy*, 61–67. Prague: Ženská národni rada, 1930.

Matiegka, Jindřich. "Ethnogenie českého národa na základě archaeologickém a anthropologickém." *Památky archaeologické* XXIX, no. 2 (1917): 109–123.

Matiegka, Jindřich. "Tělesná povaha dnešního lidu československého." In *Československá vlastivěda II. Člověk*, 193–240. Prague: Sfinx, 1933.

Matiegka, Jindřich. "The origin and the beginnings of the Czechoslovak people." *Annual Report of the Board of Regents of the Smithsonian Institution 1919*, 471–486. Washington: Government Printing Office, 1921.

Matiegka, Jindřich. *Tělesné vlastnosti českého lidu*. Prague: Vilímek, 1917.

Matiegka, Jindřich. *Vznik a počátky národa československého*. Prague: Vilímek, 1917.

Maynez, Will. "Art and Danger at the Fair: Diego Rivera Escapes to the GGIE." A lecture presented at the Treasure Island Museum, November 16, 2019. Diego Rivera Mural Project at riveramural.org.

McLeary, Eric and Elizabeth Toon. "'Here Man learns about Himself': Visual Education and the Rise and Fall of the American Museum of Health," *American Journal of Public Health* (July 2012): 27-36, doi.org/10.2105/AJPH.2011.300560.

Meier, August and Elliott M. Rudwick. "Negro Protest at the Chicago World's Fair, 1933–1934." *Journal of the Illinois State Historical Society (1908–1984)* 59, no. 2 (Summer, 1966): 161–171.

Mergl, Jan and Lenka Merglová Pánková. *Moser Crystal Story: 1857–2017: 160 Years*. Karlovy Vary: Moser, 2017.

Merlínová, Lída. *Tatíček Masaryk*. Prague: Šolc and Šimáček, 1934.

Mertová, Martina and Rostislav Švácha. *Byli jsme světoví!: Expo Brusel, Montreal, Ósaka z archivu architekta Miroslava Řepy: pohledy do sbírek Muzea umění Olomouc = We were world-class! : Expo Brussels, Montreal and Ósaka from the archive of the architect Miroslav Řepa: collection insights*. Olomouc: Muzeum umění Olomouc, 2022.

Metelák, Alois. "Drobty o našem sklářství." *Národní listy* 73, no. 163, August 27, 1933, 5.

Michalová, Rea. *Karel Teige: Captain of the Avant-Garde*. Prague: KANT, 2018.

Mitchell, Timothy. "The World as Exhibition." *Comparative Studies in Society and History* 31, no. 2 (1989): 217–236.

Moholy-Nagy, László. "Constructivism and the Proletariat (1922)." In *Modernism: An Anthology of Sources and Documents*, edited by Vassiliki Kolocotroni, Jane Goldman and Olga Taxidou, 299–300. Chicago: University of Chicago Press, 1998.

Morelon, Claire. "State Legitimacy and Continuity." In *Embers of Empire: Continuity and Rupture in the Habsburg Successor States after 1918*, edited by Paul Miller and Claire Morelon, 43–63. New York–Oxford: Berghahn Books, 2017.

Moreno, María Pura. "Jaromir Krejcar, la apología de la técnica: el pabellón de Checoslovaquia en la Exposición Internacional de París, 1937." *Rita: Revista Indexada de Textos Académicos*, no.9 (2018): 134–143.

Morganová, Pavlína, Terezie Nekvindová, and Dagmar Svatošová. *Výstava jako médium: české umění 1957–1999*. Prague: AVU, 2020.

Morton, Patricia A. *Hybrid Modernities: Architecture and Representation at the 1931 Colonial Exposition, Paris*. Cambridge: MIT Press, 2000.

Mourey, Gabriel. "L'Exposition des arts décoratifs et industriel de 1925." *L'Amour de l'art* (August 15, 1925): 299–311.

Mueggenberg, Brent. *The Czecho-Slovak Struggle for Independence, 1914–1920.* Jefferson, North Carolina: McFarland, 2014.
Muñoz Torreblanca, Marina. "Barcelona's Universal Exhibition of 1888: An Atypical Case of a Great Exhibition." In *Cultures of International Exhibitions 1840–1940. Great Exhibitions in the Margins*, edited by Marta Filipová, 45–68. Farnham: Ashgate Publishing, 2015.
Nauman, Jaroslav. "Dojmy z Paříže poválečné." *Volná myšlenka*, May 30, 1920, 5
Nebeský, Václav. "Ženské umění." *Tribuna* 1, no. 23, February 26, 1919, 3.
Newhall, Marcus, *Sklo Union: Art before Industry: 20th Century Czech Pressed Glass.* Braintree: Hope Fountain, 2008.
Nochlin, Linda, "Malvina Hoffman: A Life in Sculpture." *Arts Magazine* (Nov. 1984): 106–110.
Nový, Petr, Milan Glaveš, and Patrik Illo. *Dva v jednom, 1918–2018. Design českého a slovenského skla.* Jablonec nad Nisou: Muzeum skla a bižuterie v Jablonci nad Nisou, 2018.
Nye, Joseph S. "Soft Power." *Foreign Policy* 80 (1990): 153–171.
Official Guide and Plan of the Brussels International Exhibition 1935. Brussels: Fack-Roussel, 1935.
Official Guide Book of the New York World's Fair 1939. New York: Exposition Publications, Inc. 1939.
Official Guide: Book of the Fair 1933. Chicago: Century of Progress Administration Building, 1933.
Official Guide: Book of the Fair, 1933 (Chicago: A Century of Progress Administration Building, 1933).
Official Guidebook. Golden Gate International Exposition. San Francisco: The Crocker Company, 1939.
Official Record of the Tasmanian International Exhibition, Held at Launceston, 1891–92 (Tasmania, 1893).
Official Souvenir Book, New York World's Fair Bulletin 1939. New York, Exposition Publications, 1939.
Okurka, Tomáš. "'Witness to the Momentous Significance of German Labour in Bohemia': Exhibitions in the German-Speaking Regions of Bohemia before the First World War." In *Cultures of International Exhibitions 1840–1940. Great Exhibitions in the Margins*, edited by Marta Filipová, 91–112. London: Ashgate, 2017.
Olejník, Milan. "The Development of the Slovak Industries during the First Ten Years of the Czechoslovak Republic." *Človek a spoločnosť: internetový časopis pre pôvodné teoretické a výskumné štúdie z oblasti spoločenských vied* 6, no. 1 (2003): 13–21.
Orzoff, Andrea. "Interwar Democracy and the League of Nations." In *The Oxford Handbook of European History, 1914–1945*, edited by Nicholas Doumanis, 261–281. Oxford: Oxford University Press 2014–2016.
Orzoff, Andrea. *Battle for the Castle: The Myth of Czechoslovakia in Europe, 1914–1948.* Oxford: Oxford University Press, 2009.
Oshikiri, Taka. "Selling Tea as Japanese History: Culture, Consumption and International Expositions, 1873–1910." In *World Fairs and The Global Moulding of National Identities: International Exhibitions as Cultural Platforms, 1851–1958*, edited by Joseph Leerssen and Eric Storm, 193–216. Leiden: Brill, 2022.
Pachmanová, Martina et al. "Výstavní praktiky státu: export československého umění." In *Budování státu: reprezentace Československa v umění, architektuře a designu = Building a State: The Representation of Czechoslovakia in Art, Architecture and Design*, edited by Milena Bartlová, 281–326. Prague: UMPRUM, 2015.
Pachmanová, Martina. *Civilizovaná žena: ideál i paradox prvorepublikové vizuální kultury.* Prague: VŠUP, 2021.

Pachmanová, Martina. *Z Prahy až do Buenos Aires: "ženské umění" a mezinárodní reprezentace meziválečného Československa*. Prague: VŠUP, 2014.

Pachmanová, Martina. *Zrození umělkyně z pěny limonády*. Prague: VŠUP, 2013.

Pacyga, Dominic A. *American Warsaw. The Rise, Fall and Rebirth of Polish Chicago*. Chicago: Chicago University Press, 2021.

Parezo, Nancy J. and Don D. Fowler. *Anthropology Goes to the Fair: The 1904 Louisiana Purchase Exposition*. Lincoln: University of Nebraska Press, 2007.

Pařízková, Jana and Martina Vlková. "Beer in the Czech Republic." In *Liquid Bread: Beer and Brewing in Cross-Cultural Perspective*, edited by Wulf Schiefenhovel and Helen Macbeth, 101–110. New York: Berghahn Books, 2011.

Pavitt, Jane. "The Baťa Project: A Social and Industrial Experiment." *Twentieth Century Architecture no. 1, Industrial Architecture* (1994): 31–44.

Pavlínek, Petr. "Regional Development and the Disintegration of Czechoslovakia." *Geoforum* 26, no. 4 (1995): 351–354.

Peabody, Rebecca. "Race and Literary Sculpture in Malvina Hoffman's 'Heads and Tales.'" *Getty Research Journal*, no. 5 (2013): 119–132.

Pečinková, Pavla. "Chapters from the History, 1895–1946." In *Vysoká škola uměleckoprůmyslová v Praze/Academy of Arts, Architecture and Design in Prague, 1885–2005*, edited by Martina Pachmanová and Markéta Pražáková. Prague: VŠUP, 2005.

Pekař, Josef. *Smysl českých dějin*. Prague: v.n., 1929.

Pěkný, Tomáš. *Historie Židů v Čechách a na Moravě*. Prague: Sefer, 2001.

Pemsel, Jutta. *Die Wiener Weltausstellung von 1873: das gründerzeitliche Wien am Wendepunkt*. Vienna and Cologne: Böhlau, 1989.

Pepchinski, Mary. "Woman's Buildings at European and American World's Fairs, 1893–1939." In *Gendering the Fair. Histories of Women and Gender at World's Fairs*, edited by Tracey Jean Boisseau and Abigail Markwyn, 187–207. Urbana: University of Illinois Press, 2010.

Perkins, Henry Farnham and Harry Hamilton Laughlin. *A Decade of Progress in Eugenics; Scientific Papers of The Third International Congress of Eugenics, Held at American Museum of Natural History, New York, August 21–23, 1932*. Baltimore: The Williams & Wilkins co, 1934.

Perman, David. *The Shaping of the Czechoslovak State: Diplomatic History of the Boundaries of Czechoslovakia 1914–1920*. Leiden: E. J. Brill, 1962.

Phillips, Christopher. "The Judgement Seat of Photography," *October* 22 (1982): 27–63.

Photo et Cinéma à l'Exposition de Paris 1937: Classe VI, Manifestations Cinématographiques. Classe XIV, Photographie et Cinématographie. Paris, s.n. 1937.

Pitlach, Milan and Ladislava Horňáková, *Le Corbusier, Chandigarh*. Ostrava: Kabinet architektury: Galerie výtvarného umění v Ostravě, 2014.

Pokluda, Zdeněk, *Baťovi muži*. Zlín: Kovárna VIVA, 2012.

Pokluda, Zdeněk. *Ze Zlína do světa - příběh Tomáše Bati = From Zlín into the World - The Story of Tomas Bata*. Zlín: Nadace Tomáše Bati, 2019.

Poskočil, Oldřich. "Mezinárodní knižní umění. Poznámky k lipské výstavě." *Typografia* 34, no. 10 (1927): 233–244.

Prasch, Thomas. "Eating the World: London in 1851." *Victorian Literature and Culture* 36, no. 2 (2008): 587–602.

Pravdová, Anna. "V centru pařížského dění: abstrakce 1924–1934." In František Foltýn et al., *František Foltýn 1891–1976, Košice–Paříž–Brno*, 61–96. Brno: Moravská galerie and Řevnice: Arbor vitae, 2007.

Pražák, Albert. *Československý národ*. Prague: Akademie, 1925.

President republiky, dr. T.G. Masaryk na návštěvě ve Zlíně v Baťových závodech v červnu 1928, film, *The National Film Archive*, https://www.youtube.com/watch?v=db1ajG9Uiew.

Przewodnik po Powszechnej Wystawie Krajowej. Poznań: Biuro Ogłoszeń "PAR," 1929, https://cyryl.poznan.pl/katalog.php?baza=obiekty&id_obiektu=83563.

Pynsent, Robert. *Questions of Identity: Czech and Slovak Ideas of Nationality and Personality.* Budapest: Central European University Press, 1994.

Qureshi, Sadiah. *Peoples on Parade: Exhibitions, Empire and Anthropology in Nineteenth-Century Britain.* Chicago: University of Chicago Press, 2011.

Rada Naczelna Ziemianek, eds. *Pawilon Ziemianek i Włościanek na Powszechnej Wystawie Krajowej w Poznaniu 1929 r.* Warszawa: Komitet Pawilonu Ziemianek i Włościanek, 1930.

Rádl, Emanuel. *O smysl našich dějin.* Předpoklady *k diskusi o této otázce.* Prague: Čin, 1925.

Rádl, Emanuel. *Rassové theorie a národ.* Prague: Kočí, 1918.

Raizman, David and Ethan Robey, eds. *Expanding Nationalisms at World's Fairs: Identity, Diversity and Exchange, 1851–1915.* London: Routledge, 2017.

Ramet, Sabrina P. and Carol Skalnik Leff. "Interwar Czechoslovakia – A National State for a Multiethnic Population." In *Interwar East-Central Europe, 1918–1941. The Failure of Democracy Building, the Fate of Minorities*, edited by Sabrina P. Ramet, 75–108. London – New York: Routledge, 2020.

Ramet, Sabrina P., ed. *Interwar East Central Europe, 1918–1941: The Failure of Democracy-Building, the Fate of Minorities.* London and New York, NY: Routledge, 2020.

Rampley, Matthew. "From Potemkin City to the Estrangement of Vision: Baroque Modernity in Austria, Before and After 1918." *Austrian History Yearbook* 47 (2016): 167–187.

Rampley, Matthew. "Peasants in Vienna. Ethnographic at the 1873 World's Fair," *Austrian History Yearbook* 42 (2011): 110–132.

Rampley, Matthew. *The Vienna School of Art History. Empire and the Politics of Scholarship, 1847–1918.* University Park: Pennsylvania University Press, 2013.

Rásky, Béla. "Artwork of the Month, May 2023: Photomontage at the Paris World's Fair by Robert Haas (1937)." *CRAACE*, https://craace.com/2023/05/30/artwork-of-the-month-may-2023-photomontage-at-the-paris-worlds-fair-by-robert-haas-1937.

Reichman, John J. *The Czechoslovaks of Chicago.* Chicago: Czechoslovak Historical Society of Illinois, 1937.

Rezende, Livia. "Nature and the Brazilian State at the Independence Centennial International Exhibition in Rio de Janeiro, 1922." In *Cultures of International Exhibitions 1840-1940. Great Exhibitions in the Margins*, edited by Marta Filipová, 163–182. Farnham: Ashgate Publishing, 2015.

Richards, Thomas. *The Commodity Culture of Victorian England: Advertising and Spectacle, 1851–1914.* Stanford University Press, 1990.

Ricke, Helmut, ed. *Czech Glass, 1945–1980: Design in an Age of Adversity.* Düsseldorf: Museum Kunst Palast and Stuttgart: Arnoldsche, 2005.

Robey, Ethan. "Kings, Peasants, Dragons, and Flowers. National Symbolism in Decorative Arts at the World's Fars." In *Inventing the Modern World: Decorative Arts at the World's Fairs, 1851–1939*, edited by Jason T. Busch and Catherine L. Futter, 162–189. New York: Skira Rizzoli, 2012.

Rocco, Vanessa. "Activist Photo Spaces: 'Situation Awareness' and the Exhibition of the Building Workers Unions." *The Journal of Curatorial Studies* 3, no. 1 (2014): 26–48.

Romlid, Christine. "Visualizing Sweden at the 1937 World Fair in Paris." In *Locating the Global Spaces, Networks and Interactions from the Seventeenth to the Twentieth Century*, edited by Holger Weiss. De Gruyter Oldenbourg, 2020.

Rosen, Jeff. "Of Monsters and Fossils: The Making of Racial Difference in Malvina Hoffman's Hall of the Races of Mankind." *History and Anthropology* 12, no. 2 (2001): 101–158.

Rothschild, Joseph. *East Central Europe Between the Two World Wars.* London: Routledge, 2015.

Rothschild, Joseph. *East Central Europe between the Two World Wars.* Seattle: University of Washington Press, 1974.

Rousová, Hana. "Art Deco." In *Design v Českých zemích 1900–2000: instituce moderního designu,* edited by Iva Knobloch and Radim Vondráček, 177–200. Prague: Academia, 2016.

Rusert, Brit and Whitney Battle-Batiste, *W. E. B. DuBois' Data Portraits.* Princeton Architectural Press, 2018.

Rydell, Robert W. *All the World's a Fair: Visions of Empire at American International Expositions, 1876–1916.* Chicago: University of Chicago Press, 1985.

Rydell, Robert W. *World of Fairs. The Century-of-Progress Expositions.* Chicago: University of Chicago Press, 1993.

Rydell, Robert W., and Laura Burd Schiavo, eds. *Designing Tomorrow: America's World's Fairs of the 1930s.* New Haven: Yale University Press, 2010.

Rydell, Robert W., John E. Finding, and Kimberly Pelle. *Fair America: World's Fairs in the United States.* Washington, DC: Smithsonian Institution Press, 2000.

S. Ú. "České sklo na Světové výstavě v Paříži." *Večerní Lidové noviny* 45, 31 July 1937, 3.

Samaan, A. E. *From a "Race of Masters" to a "Master Race': 1948 to 1848,* published at www.createspace.com, 2013.

Sandiford-Pellé, Isabelle. *Generál Pellé. obrázkový deník.* Prague: Ministerstvo obrany ČR - Prezentační a informační centrum MO, 2010.

Sawicki, Nicholas. "Czechoslovakia: Bauhaus Students and Associates." *Centropa* 3 no.1 (2003): 27–40.

Sayer, Derek. *Coasts of Bohemia: A Czech History.* Princeton, NJ: Princeton University Press, 1998.

Sayers, Paul T. "New York 1939–1940." In *Encyclopedia of World's Fairs and Expositions,* edited by John E. Findling and Kimberley D. Pelle, 299–305. Jefferson N.C.: McFarland, 2008.

Scarborough, Katherine. "Heart of America is Sound, Says Scientist." *The Sun,* November 21, 1926, 1.

Schlögel, Karl. *Moscow 1937.* Cambridge: Polity, 2013.

Schnapp, Jeffrey T. "Crystalline Bodies: Fragments of a Cultural History of Glass." *West 86th A Journal of Decorative Arts, Design, History and Material Culture* 20, no. 2 (2013): 173–194, https://doi.org/10.1086/674728.

Schuster, Sven. *Die Inszenierung der Nation. Das Kaiserreich Brasilienim Zeitalterder Weltausstellungen.* Frankfurt am Main: Peter Lang, 2015.

Schweiger, Werner J. *Wiener Werkstätte: Designs in Vienna 1903–1932.* London: Thames and Hudson 1984.

Scott. "Tales from the Swedish Café," *World's Fair Chicago, 1893,* 2019, https://worldsfairchicago1893.com/2019/07/20/swedish-cafe.

Secklehner, Julia. "Crossing Borders and Period Boundaries in Central European Art: The Work of Anna Lesznai (ca. 1910–1930)." In *Rethinking Period Boundaries: New Approaches to Continuity and Discontinuity in Modern European History and Culture,* edited by Lucian George and Jade McGlynn, 119–148. Oldenbourg: De Gruyter, 2022, doi:10.1515/9783110636000-005.

Secklehner, Julia. "'Feminine Horror' or 'Eminent Viennese Specialty'?: Vienna's Kunstgewerblerin in Paris, 1925." *Art East Central* 3, no. 1 (2023): 13–36, https://doi.org/10.5817/AEC2023-3-2.

Segawa, Hugo. *Architecture in Brazil. 1900–1990.* New York: Springer, 2013.

Ševeček, Ondřej and Martin Jemelka, eds. *Company Towns of the Baťa Concern. History – Cases – Architecture.* Stuttgart: Steiner, 2013.

Shanken, Andrew M. *Into the Void Pacific. Building the 1939 San Francisco World's Fair.* Berkley: University of California Press, 2015.

Shepherd, Deborah B. "The 1939–1940 New York World's Fair: Typical American Families Build Tomorrow." 2011, https://api.semanticscholar.org/CorpusID:111490811.

Sigel, Paul. *Exponiert. Deutsche Pavillons auf Weltausstellungen.* Berlin: Bauwesen Verlag,2000.

Šimůnek, Michal V. "Artur (Arthur) Brožek." In *The History of East-Central European Eugenics. Sources and Commentaries*, edited by Marius Turda, 147–148. London – New York: Bloomsbury Academic, 2015.

Šimůnek, Michal V. "Eugenics, Social Genetics and Racial Hygiene: Plans for the Scientific Regulation of Human Heredity in the Czech Lands, 1900–1925." In *Blood and Homeland: Eugenics and Racial Nationalism in Central and Southeastern Europe, 1900–1940*, edited by Marius Turda and Paul J. Weindling, 145–166. Budapest– New York: Central European University Press, 2007.

Šimůnek, Michal V. "Overview." In *The History of East-Central European Eugenics. Sources and Commentaries*, edited by Marius Turda, 131–132. London – New York: Bloomsbury Academic, 2015.

Šimůnek, Michal. "Eugenics, Social Genetics and Racial Hygiene: Plans for the Scientific Regulation of Human Heredity in the Czech Lands, 1900–1925." In *'Eugenics and Racial Nationalism in Central and Southeastern Europe, 1900–1940*, edited by Marius Turda and Paul J. Weindling, 145–166. Budapest– New York: Central European University Press, 2007.

Slačálek, Ondřej. "The Postcolonial Hypothesis. Notes on the Czech 'Central' European Identity." *Annual of Language & Politics & Politics of Identity* 10, no. 10 (2016): 27–44.

Šlapeta, Vladimír. *Benš: Adolf Benš (1894–1982): architektonické dílo Adolf Benš (1894–1982): architektonické dílo.* Prague: Nadace Charty 77, 2014.

Smith, Llewellyn. *Reports on the Present Position and Tendencies of the Industrial Arts as Indicated at the International Exhibition of Modern Decorative and Industrial Arts, Paris, 1925, with an Introductory Survey.* London: Harrow-HMSO Press, 1927.

Smith, Wilson. "Old London, Old Edinburg: Constructing Historic Cities." In *Cultures of International Exhibitions 1840–1940. Great Exhibitions in the Margins*, edited by Marta Filipová, 203–230. Farnham: Ashgate Publishing, 2015.

Sontag, Susan. *On Photography.* London: Penguin Books, 1977.

Soumar, Bohumil, ed. *Tchecoslovaquie à l'Exposition Internationale des Arts et Techniques dans la Vie Moderne Paris 1937.* Prague: Melantrich, 1937.

Stamenković, Aleksandra. "Yugoslav Pavilions at International Exhibitions in Artistic and Political Discourse 1918–1941." *Istorija 20. veka* 40, no. 2 (2022): 301–322.

Starita, Angela. "Brasilidade in Built Form: Tracing National Identity in Modernist Architecture in Brazil, 1922–1968." Dissertation submitted to the Faculty of New Jersey Institute of Technology at Rutgers, The State University of New Jersey – Newark, May 2020.

Stein, Claudia. "Organising the History of Hygiene at the *Internationale Hygiene-Ausstellung* in Dresden in 1911." *NTM Zeitschrift* für *Geschichte der Wissenschaften, Technik und Medizin* 21 (2013): 355–387, https://doi.org/10.1007/s00048-014-0109-5.

Sternstein, Malynne M. *Czechs of Chicagoland.* Arcadia Publishing, 2008.

Štícha, R. "Něco o Americe a výstavě v Chicagu." *Československá Paříž*, no. 28, March 4, 1934, 2.

Stirton, Paul. *Jan Tschichold and the New Typography: Graphic Design between the World Wars.* New York City: Bard Graduate Center – New Haven: Yale University Press, 2019.

Storm, Eric. "The Transnational Construction of National Identities: A Classification of National Pavilions at World Fairs." In *World Fairs and The Global Moulding of National Identities: International Exhibitions as Cultural Platforms, 1851–1958*, edited by Joseph Leerssen and Eric Storm, 53–83. Leiden: Brill, 2022.

Stratigakos, Despina. "Women and the Werkbund: Gender Politics and German Design Reform, 1907–14." *Journal of the Society of Architectural Historians* 62, no. 4 (2003): 490–511.

Šubert, F. A. "Národopisná výstava českoslovanská." *Osvěta, listy pro rozhled v umění, vědě a politice* 25 (1895): 30–45.

Sutnar, Ladislav, "Early modern design concepts." In *Ladislav Sutnar: Visual Design in Action*, edited by Reto Caduff and Steven Heller. Zurich: Lars Müller Publishers, 2015.

Sutnar, Ladislav. "Státní odborné školy na Mezinárodní výstavě v Barceloně 1929." *Horizont*, no. 25–26 (1930): 47–49.

Švéda, Josef. *Země zaslíbená. Antologie o obraze Ameriky v české literatuře a kultuře*. Příbram: Pistorius & Olšanská, 2016.

Svoboda, V. "Československý sklářský průmysl ve světě statistiky." *Sklářské rozhledy. Věstník sklářského ústavu v Hradci Králové* XII, no. 10 (1935): 153–154.

Svobodová, Markéta. *Bauhaus a Československo 1919–1938: studenti, koncepty, kontakty. The Bauhaus and Czechoslovakia 1919–1938: Students, Concepts, Contacts*. Prague: Kant, 2016.

Teige, Karel. "Konstruktivismus a likvidace umění." *Disk* II (1925): 4–8, reprinted in *Avantgarda známá a neznámá*, edited by Vladimír Vlašín. Prague: Svoboda, 1972.

Teige, Karel. "Výtvarné umění na brněnské Výstavě soudobé kultury." *Stavba* 7, no. 3 (1928–29): 44–45.

Tenorio-Trillo, Mauricio. *Mexico at the World's Fairs: Crafting a Modern Nation*. Berkeley: University of California Press, 1996.

Tenorio-Trillo, Mauricio. *Mexico at the World's Fairs. Crafting a Modern Nation*. Berkeley, Los Angeles – Oxford: University of California Press, 1996.

Teslow, Tracy. *Constructing Race. The Science of Bodies and Cultures in American Anthropology*. Cambridge: Cambridge University Press 2014.

The Basic Sciences and Medicine. A Century of Progress International Exposition Chicago 1933. Chicago, 1932.

Totzauer, David. "Pivovarnictví v Plzni a okolí v meziválečném období." MA Thesis at the University of West Bohemia Plzeň, 2020.

Turda, Marius. *Modernism and Eugenics*. Basingstoke: Palgrave Macmillan, 2010.

Turda, Marius. "Introduction." In *The History of East-Central European Eugenics, 1900–1945. Sources and Commentaries*, edited by Marius Turda, xi–xii. London – New York: Bloomsbury, 2015.

Turda, Marius. *The History of East-Central Eugenics, 1900–1945: Sources and Commentaries*. London and New York: Bloomsbury Academic, 2015.

Turnovský, Ladislav. "Československo na světové výstavě v Rio de Janeiro." *Národní listy*, February 10, 1923, 6.

Tyršová, Renata. *Svéráz v zemích československých*. Plzeň: Český deník, 1921.

Udovički-Selb, Danilo. "Facing Hitler's Pavilion: The Use of Modernity in the Soviet Pavilion at the 1937 Paris International Exhibition." *Journal of Contemporary History* 47, no. 1 (January 2012): 13–47.

Udovički-Selb, Danilo. "L'Exposition de 1937 n'aura pas lieu: The Invention of the Paris International Expo and the Soviet and German Pavilions." In *Architecture of Great Expositions 1937–1959: Messages of Peace, Images of War*, edited by Rika Devos, Alexander Ortenberg, and Vladimir Paperny, 23–50. Farnham: Ashgate Publishing, 2015.

Úlehla, Vladimír, Jaroslav B. Svrček, and František V. Vaníček. "Oddíl A. Člověk a příroda živá." In *Hlavní průvodce. Výstava soudobé kultury v Československu, Brno, 1928*, 91–98. Brno: VSK, 1928.

Ústavný zákon zo dňa 22.novembra 1938 o autonomii Slovenskej krajiny, no. 299, *Sbírka zákonů a nařízení státu československého* 1938, 1161–1164.

Varga, Bálint. *Monumental Nation: Magyar Nationalism and Symbolic Politics in Fin-de-siècle Hungary*. New York–Oxford: De Gruyter Berghahn 2016.

Vavrečka, Hugo. *Život je spíš roman*. Třebíč: Arca JiMfa; Ostrava: Jasmín, 1997.

Veszprémi, Nóra. "Universalizing The Local? Museums, Centres And Peripheries." *Revista da História da Arte. Instituto de História da Arte* 9, no. 2 (2021): 99–108.

Vlček, F. L. *Náš lid v Americe (Sebrané články a verše).* Týn nad Vltavou: O. Tunek, 1935.

Vlnas, Vít. "Mýty a kýče první republiky." *Nová Přítomnost* 8 (1991): 28–29.

Vogel, Klaus. "The Transparent Man—Some Comments on the History of a Symbol." In *Manifesting Medicine: Bodies and Machines*, eds. by Robert Bud, Bernard Finn, and Helmuth Trischler, 31–61. Amsterdam: Harwood Academic Publishers, 1999.

Vojan, Jaroslav E. S. "Budoucí forma češství v Americe." *Kalendář New Yorských listů* 18 (1939): 29.

Vojan, Jaroslav E. S. "How the Czechs Helped Build the American Civilisation." In *The Czechoslovaks of Chicago*, edited by John R. Reichman, 28–34. Chicago, 1937.

Vojan, Jaroslav E. S. "Naše účast na světové výstavě 1933." *Most* no. 5, October–November 1931, 4–5.

Vojan, Jaroslav E. S. "Století pokroku, světová výstava v Chicagu 1933." *Kalendář New Yorských listů na rok 1933*, vol. 12 (1933): 96.

Vs. "Vinníci ostudy s čs. Pavilonem na pařížské výstavě musí býti potrestáni." *Národní listy*, June 24, 1937, 3.

Výběr a výchova průmyslového člověka, instrukční příručka. Zlín: Baťa, osobní oddělení, 1938.

Vybíral, Jindřich. "What Is 'Czech' in Art in Bohemia? Alfred Woltmann and Defensive Mechanisms of Czech Artistic Historiography." *Kunstchronik* 59, no. 1 (January 2006): 1–7.

Webr, Otakar. "Sklo na mezinárodní výstavě moderních a dekorativních umění v Paříži." *Sklářské rozhledy* III, no. 1 (1926): 8–9.

Weigner, Leopold. *Lidové umění* československé. Prague: Josef Vilímek, 1917.

Wein, Martin. *History of the Jews in the Bohemian Lands.* Leiden – Boston: Brill, 2015.

Whalen, Philip. "Paris 1925." In *Encyclopedia of World's Fairs and Expositions*, edited by John E. Findling and Kimberley D. Pelle, 236–243. Jefferson N.C.: McFarland, 2008.

Wirth, Zdeněk, ed. *Československé umění.* Prague: Vesmír, 1926.

Wood, Andrew F. *New York's 1939–1940 World's Fair.* Charleston, SC: Arcadia, 2004.

World's Fair Memorial of the Czechoslovak Group (Czechs and Slovaks). Chicago: Czechoslovak Group, 1933.

Wörner, Martin. *Vergnügung und Belehrung: Volkskultur auf den Weltausstellungen, 1851–1900.* Münster–New York: Waxmann, 1999.

Wunderwald, Anke. "Das Poble Espanyol der Weltausstellung 1929. Barcelona als Imagination nationaler Einheit." In Cornelia Jöchner, Christin Nezik, Gáspár Salamon, Anke Wunderwald, *Museale Architekturdörfer 1880–1930 das Eigene in transnationalen Verflechtungen.* Dresden: Sondstein, 2023.

Wurts, Richard and Stanley Applebaum, et. al. *The New York World's Fair 1939/1940 in 155 photographs.* New York: Dover Publications, Inc., 1977.

Yugoslav Day, July 29, 1934, Lagoon Theatre. A Century of Progress International Exposition, Chicago (n.p., 1934).

Z. "Mezinárodní hygienická výstava." *Lidové noviny* 38, May 25, 1930, 7.

Zahra, Tara. *The Great Departure. Mass Migration from Eastern Europe and the Making of the Free World.* New York: W.W. Norton & Company, 2016.

Zemánek, Jiří. *Zdeněk Pešánek 1896–1965.* Prague: National Gallery – Gema Art, 1996.

Zintzen, Chris."Shifting Perspectives: Alpine Scenarios in the Work Complex Nach der Natur (Beyond Nature) by Austrian Architectural Photographer Margherita Spiluttini." In *The Draw of the Alps: Alpine Summits and Borderlands in Modern German-Speaking Culture*, edited by Richard McClelland, 103–122. Berlin and Boston: De Gruyter, 2023.

Index

Aalto, Alvar, 52, 88
Absolon, Karel, 167, 237
Africa, 48, 61, 110, 117, 172, 199, 200, 208, 221
African Americans, 195, 208, 225, 232
agency, 1, 3, 178–79, 256
Alsace, Alsatian village, 199
America. *See* United States of America
Americas, 25, 39, 57.
See also North America; South America
Angkor Wat, 57
Art Deco, 56, 105, 112, 133, 234
Asia, 10, 56, 221, 227
Association of the Czechoslovak Werkbund (Svaz československého díla), 26, 41, 79, 114, 122, 124, 136, 148, 150, 161, 253
Austria, 7, 12, 18, 38, 15, 125, 142, 153, 154, 159, 160, 166, 177, 243, 258
and exhibitions, 17, 20, 21, 28, 31, 35, 40, 48, 59, 78, 80n, 84, 85, 107, 125–26, 161–62, 163, 175, 236, 244.
See also Austria-Hungary
Austria-Hungary (Habsburg Monarchy), 2, 18–19, 154, 155, 159, 160, 162, 181, 192
and exhibitions, 12, 18, 20–21, 28, 139, 153, 208
avant-garde, 13, 74, 76, 113, 119, 123–24, 127, 137, 152, 169, 252, 253

Baťa (company), 13, 50, 95, 139, 166–76, 186, 247–52, 257
representations by, 64, 84, 93, 95–96, 100, 119, 140, 167–68, 175–76, 241–42 247–52, 258
visions of representation and social order, 14, 167, 168–69, 170–71, 172–75, 238–41, 248, 251
Baťa, Jan Antonín, 168, 170, 171, 239, 240, 248–49
Baťa, Tomáš, 166, 172, 239–40, 248–51
Bauhaus, 37, 73, 112
Bayer, Herbert, 73, 82, 87–88, 96
beer, 13, 34, 53, 142, 153–59, 176, 185, 197, 199, 212, 257
See also Pilsner
Behrens, Peter, 203, 243n
Belgium, 24, 59, 157, 160
and exhibitions, 24, 31n, 44, 47, 48, 61, 80n, 85, 125, 161
Beneš, Edvard, 5, 8, 28, 63, 169, 228, 252
Beneš, Vincenc, 127
Benka, Martin, 134–35
Benš, Adolf, 32, 112–13, 122
Bohemia, 5–8, 12, 20, 21, 39, 41, 55, 72, 107, 130, 142–43, 144, 150, 153, 155, 157, 159–61, 162, 164, 166, 176, 185, 244
representations of, 7, 9, 21, 28–29, 33, 64, 93, 94, 108, 140, 144, 145, 163, 182, 206, 209, 255
Bouda, Cyril, 64, 174, 247, 252
Bratislava, 70n, 91, 127, 134, 146, 150
See also School of Arts and Crafts

Braunerová, Zdenka, 108
Brazil, 23, 24, 26, 57, 110, 172, 179
Britain. *See* United Kingdom
Brno, 34, 70n, 92, 108n, 127, 146, 227
exhibitions in, 76, 77, 149, 167, 172, 203, 204, 237–38, 239
See also under world's fairs and exhibitions
Brožek, Artur, 244
Brůha, Jaroslav, 62
Bureau International des Expositions (BIE), 38, 55

Calder, Alexander, 52, 125
Čapek, Josef, 50, 119, 121, 127, 131
Čapek, Karel, 7–9, 18, 62, 193
ceramics, 27, 33, 39, 49, 58, 84, 91–92, 93, 94, 98, 108, 129, 132, 146, 163, 167, 191
Čermák, Antonín (Tony Cermak), 186–87
Cheb (Eger), 144
Chicago, 180, 181, 184–87, 206
See also under world's fairs and exhibitions
Čipera, Dominik, 172, 175
Čulen, Konštantín, 190
Czechoslovak Eugenics Society, 228
Czecho-Slovak Group, Chicago, 188
Czechoslovak Institute of National Eugenics, 228
Czechoslovakism, 6, 16, 30, 145, 187–89, 212, 218, 255
Czechs, 5–7, 8, 20, 26, 30, 51, 62, 75, 77, 144, 153, 154, 155, 160, 219–21, 223, 227, 252
diaspora in the US, 14, 45, 47, 63, 64, 180–82, 184–94, 199, 205–11, 221, 222, 256
representations of, 36, 37, 91, 116, 145, 154–55, 159, 200, 215, 219, 229, 247, 254, 255

Dawes, Rufus, 196
De Stijl, 31, 80, 112
Denis, Ernest, 108
Detva, 13, 144, 145–52, 166, 176, 254
Deutsche Werkbund, 79, 82, 122, 203
diaspora, émigrés, 14, 36, 179, 180–94, 212–13, 232, 256
involvement in world's fairs, 14, 43, 45, 47, 63, 64, 154, 182–84, 186–87, 188–89, 191–93, 198–99, 200, 206–12, 222, 256, 257
See also under Czechs; Slovaks; Ruthenians
Drahoňovský, Josef, 33, 92n, 114
Dreyfuss, Henry, 62
Drtikol, František, 113
Družstevní práce (The Cooperative Work), 79, 122
DuBois, W.E.B., 232
Duras, Mary, 51, 129, 201

Edgar, Emil, 72
Eitelberger, Rudolf von, 203
eugenics, 10–11, 15, 216–17, 226–32, 234–39, 244–45, 246, 249, 255
See also Czechoslovak Eugenics Society; Czechoslovak Institute of National Eugenics
exhibition elites, 3–4, 10, 16, 18, 27, 141, 149, 177–79, 253, 256, 257
exhibition villages, 11–14, 40, 44, 57, 154, 197, 199–200
exhibitions. *See* world's fairs and exhibitions
Field, Stanley, 232
Fierlinger, Zdeněk, 184
Filla, Emil, 50, 92n, 126

folk art, 13, 64, 80, 93, 107, 108–9, 114, 116, 127, 136, 145–46, 153, 190, 200, 223, 224, 254
as a commodity, 64, 93, 145–53, 205, 254
exoticization of, 11, 199–200, 206
gendered presentation of, 11, 153, 199–200, 205
juxtaposed to modernity, 65, 80, 94, 109, 132, 151–52, 191, 206, 254
representations of, 7, 13, 23, 26, 27, 33, 48, 53, 55, 58, 65, 93, 107–9, 110, 116, 125, 133, 135, 150, 159, 205, 208, 209–10, 211, 219
See also under race
food and drink at fairs, 11, 15, 34, 44, 153–56, 158, 159, 196, 197–98, 212, 255, 256
Ford (company), 44, 140, 245
Fordism, 166, 240
Foustka, Břetislav, 229

France, 24, 36, 48, 52, 59, 111, 133, 192
and exhibitions, 23, 24, 31, 47, 48, 52, 57, 69, 80n, 84, 88, 101, 105, 107, 108, 111–12, 113, 117, 139, 199, 236, 244
relations with Czechoslovakia, 2, 5, 24, 25, 107, 108, 144, 227
Franz Joseph I of Austria, 28–29
Fuchs, Bohuslav, 34, 50, 72, 77, 82–83, 94, 113

Galton, Francis, 216, 228
Geddes, Norman Bel, 62
Geddes, Patrick, 170
General Motors, 44, 61, 62, 140
Germans, 5–6, 7, 14, 51, 144, 145, 150, 153, 155, 156, 159–60, 161, 206, 219, 223, 255
Germany, 2, 6, 9, 18, 25, 31, 39, 55, 59, 60, 80, 111, 144, 150, 156, 158, 160, 163, 164–65, 166, 176, 177, 198, 228, 230, 258
and/at exhibitions, 18, 25, 31, 35, 40, 44, 51, 74, 82, 85, 87, 88, 99, 100, 110, 162, 165, 175, 236, 237, 243
glass, 50–55, 84–86, 195–65
as material used at exhibitions, 27, 28, 33, 36, 45, 49, 50, 51, 53, 54, 63, 70, 78, 84–87, 98, 110, 114, 125, 169, 174, 194, 242, 252, 253
exhibition of, 13, 27, 33, 46, 50, 53, 58, 64, 81, 91–92, 97, 98, 107, 130, 133, 142, 146, 159, 160–64, 167, 176, 191, 253, 248–52, 258
production of, 39, 47, 49, 92, 116, 140, 159–60, 164
See also Moser
Gočár, Josef, 32, 33, 50, 77, 100, 113, 115, 117, 126
Great Depression, 2, 13, 39, 42, 177, 194, 195, 230
Gutfreund, Otto, 33, 36, 50, 113

Haas, Robert, 125
Habsburg Monarchy. *See* Austria-Hungary
Hackenschmied, Alexandr, 169–70
Haerdtl, Oswald, 85, 125–26
Hamlin, Talbot Faulkner, 100
Haškovec, Ladislav, 228, 229
Heinz, 140–41, 154
Heythum, Antonín, 49, 50, 55, 57, 92, 97, 99, 133, 167, 199, 201, 253
Heythum, Charlotte, 201
Hlinka, Andrej, 125, 147
Hoffman, Josef, 85
Hoffman, Malvina, 232–34, 236
Hooton, Earnest A., 225–26, 246
Horejc, Jaroslav, 114
Hoyer, Egbert, 73
Hrdlička, Aleš, 192, 221–27, 228, 233, 246, 254
Hryniewicka-Piotrowska, Anatolia, 204
human zoo, 218–19
Hungary, 12, 18, 20, 38, 107, 125, 142, 151, 153, 160, 182, 187, 258
at exhibitions, 12, 18, 20, 38, 107, 125, 142, 151, 153, 160, 182, 187, 258
See also Austria-Hungary
Hurban, Vladimír, 60, 133, 176

Iofan, Boris, 51

Jablonec nad Nisou (Gablonz), 33, 46, 92, 92, 160, 165
Janák, Pavel, 25–27, 33, 51, 77, 79, 110, 114
Janeček, George, 60
Jirásková, Marta, 201
Johnová, Helena, 129, 201
Jurkovič, Dušan, 93, 147

Kafka, Bohumil, 107–8
Kafka, Břetislav, 91
Karfík, Vladimír, 173
Karlovy Vary (Karlsbad), 39, 46, 47, 92, 160, 162, 163, 165
Kars, Georges, 133
Kejř, Zdeněk, 54, 67, 70, 82, 89
Kelly, Edward J., 196
Kingdom of Serbs, Croats and Slovenes. *See* Yugoslavia.
Klenka, Richard, 41
Klos, Elmar, 169
Knüppelholz-Roeser, Margarethe, 203
Koch, Alois, 71
Komenský, Jan Amos, 62
Korbel, Mario, 64
Kościuszko, Tadeusz, 185
Kováčik, Andrej, 133
Krásná jizba (The Beautiful Room), 122
Krčmář, Jan, 53, 127
Krejcar, Jaromír, 50, 54, 67, 70, 71, 82, 83, 84–85, 87, 88, 89, 90, 94, 100, 113, 158, 168, 201

Kříženecký, Jaroslav, 229
Krofta, Kamil, 126
Kroha, Jiří, 71, 75–76, 97
Kupka, František, 107
Kysela, František, 27, 33, 91, 110, 114, 116–19, 121, 137, 252
labor, 3, 14, 43, 147, 148, 149, 160, 166, 172, 179, 193–95, 212–213, 238–40, 251, 256
  representations of, 95, 114, 119–21, 249, 251
  See also workers *under* women
Lalique, René, 31, 111, 161
Le Corbusier, 31, 78, 112, 113, 169
Léger, Fernand, 53, 112
Liberec (Reichenberg), 92, 144
Lissitzky, El, 73–75, 77, 81, 82, 96
London, 226
  See also under world's fairs and exhibitions
Loos, Adolf, 120
Ložek, Václav, 33

Machoň, Ladislav, 36, 37
Mánes (Association), 126, 127–28, 136
Mařatka, Josef, 58
Masaryk Academy of Labor, 228
Masaryk, Jan, 147
Masaryk, Tomáš Garrigue, 5, 8, 28, 29, 59, 107, 172, 181, 183, 185, 187–88, 202, 228, 239, 241
  statue/bust/painting of, 28, 36, 37, 41, 45, 49, 58, 65, 91, 182, 240–41, 244, 249, 252, 257
Masaryková, Frances Crane, 147
Matiegka, Jindřich, 219–21, 227, 237, 254
Meisner, Josef, 230
Melnikov, Konstantin, 78, 85, 112
Mendel, Gregor Johann, 216
Mendelianism, 216, 228, 229
Miró, Joan, 125
Mladá Boleslav, 72, 76
Moholy-Nagy, László, 73, 82, 887, 96, 120
Moravia, 5, 6, 90, 39, 55, 72, 107, 144, 159, 164, 166, 167, 176, 223–24
  representations of, 9, 108, 110, 145, 190–91, 224
Mościcka, Michalina, 204
Moser (Ludwig Moser & Sons, company), 13, 33, 141, 162–66, 255
Moser, Gustav, 165
Moser, Leo, 165
Moser, Ludwig, 162
Moser, Richard, 165
Mucha, Alfons, 107, 211
Munich Agreement, 6, 55, 60, 165
Museum of Decorative Arts, Prague, 130, 131
national days, 35n, 57, 182, 189, 200, 205–11, 255, 256
Nebeský, Václav, 127–28
New York, 38, 60, 133, 181, 191, 206, 227, 231
  See also under world's fairs and exhibitions
Niederle, Lubor, 108n, 219, 237
North America, 10, 17, 57n, 155, 221
Novotný, Otakar, 33, 91

Palacký, František, 91, 244
Paličková Milde, Emilie, 33, 34, 41n, 114, 201
Paris, 5, 38, 44, 107, 108, 127, 128, 172, 196, 197
  See also under world's fairs and exhibitions
Pellé, Maurice, 107–8
Pešánek, Zdeněk, 55, 94, 129–32, 137
Picasso, Pablo, 52, 112, 125
Pilsner, 13, 34, 141, 155–59, 166, 212, 257
Podhajská, Minka, 201
Poland, 12, 17, 31, 32, 35, 48, 49, 59, 61, 78, 88, 133, 177, 183, 185, 200, 204, 207, 258
Pospíšil, František, 227
Prague, 28, 79, 108, 112, 127, 146, 161, 188, 228, 244
  and/at exhibitions, 9, 28, 30, 33, 36–37, 41, 45, 47, 54, 64, 65, 72, 83, 92, 93, 94, 133, 150, 167, 171, 173, 174, 177, 200, 201, 256, 257
  See also Museum of Decorative Arts; School of Decorative Arts (UMPRUM); *and under* world's fairs and exhibitions
Procházková, Linka, 92n, 129, 201
Pulaski, Kazimierz, 185
Pytlík, Josef, 26

race, 216–17
  and folk culture, 109, 136, 223

as synonym for nation/ethnicity, 190, 208, 219
racial hierarchies, 15, 215, 217–18, 220, 222, 224–26, 227, 230, 233–34, 237–38, 246
racial ideas in Czechoslovakian thinking, 218–224, 230, 237, 244–45
representations of, 208, 217, 218, 222, 224–27, 230–34, 243, 246
Rádl, Emanuel, 8, 230
Rand, Sally, 43, 57, 197
Rašín, Alois, 23, 143
Reich, Lily, 40
Rieger, František Ladislav, 244
Roškot, Kamil, 44–45, 63–64
Ruszczyńska, Maria, 204
Ruthenia. *See* Sub-Carpathian Ruthenia
Ruthenians, 5, 14, 148
representations of, 145, 218, 255
diaspora in the US, 189, 193, 206, 211
Růžička, Vladislav, 227

School of Arts and Crafts, Bratislava, 93, 134
School of Decorative Arts (UMPRUM), Prague, 9n, 26, 27, 33, 91, 110, 114, 119–20, 122, 124, 136, 252, 253
Šetelík, Jaroslav, 37, 41, 45
shoes
exhibition of, 64, 95–96, 133, 139, 142, 167–68, 242
shoemaking, 93, 96, 119, 166–67, 168, 170–71, 242, 249–51
See also Baťa
Šíma, Josef, 112
Sirotek, Bedřich, 44, 49
Škoda, 50, 43, 89, 93, 95, 100
Slavíček, Antonín, 127
Slavíček, Jan, 54, 93
Slovak League of America, 133, 187, 188
Slovakia, 5–8, 19, 20, 21, 26, 39, 55, 107, 142, 144, 146–49, 153, 160, 175, 176, 187–88, 227n, 254
representations of, 7–9, 13, 33, 37, 41, 47, 93, 108–9, 116, 133–36, 145–46, 149–51, 176, 190–91, 199–200, 205–6, 215, 218, 254–55
Slovaks, 5–8, 51, 55, 109, 125, 145, 147–48, 190, 215, 220, 223–24
diaspora in the US, 14, 63, 64, 133, 180–82, 184–94, 198–99, 205–6, 209–11, 213, 222, 256
Smrčková, Ludvika, 163
Sokol, 9, 37, 48, 133, 181, 182, 185, 189, 191, 200, 206–7, 209, 211, 256
Soumar, Bohuslav, 45, 54, 67, 70, 82, 91, 92, 100n, 194, 253
South America, 10, 17, 24, 110, 221
Špaček, Stanislav, 182–83, 184
Spain, 31n, 35, 38, 40, 101, 125
Speer, Albert, 51
Štáfl, Otakar, 41, 45
Štech, Václav Vilém, 26, 33, 65, 114, 145–46, 150, 253
Štefánik, Milan Rastislav, 5, 91, 108
Strakonice, 92, 142
Štrimplová, Eva, 89, 90, 158, 201
Štursa, Jan Josef, 28, 33, 49, 50, 113, 127, 132, 244
Štýrský, Jindřich, 112, 126, 128
Sub-Carpathian Ruthenia, 5, 6, 7, 19, 20, 39, 142, 146, 193, 211, 254
representations of, 7, 9, 14, 145, 189, 206, 254–55
See also Ruthenians
Sucharda, Jan, 96, 97, 98, 100, 174
Sutnar, Ladislav, 41, 54, 55, 65, 67, 71, 78–84, 87, 88, 91, 94, 95, 97, 100, 122, 123, 124, 168, 244
Švabinský, Max, 107, 127
Švec, Otakar, 113, 115, 132, 184

tapestries, 114, 116–19, 121, 146, 252, 253
Tatra (company), 50
Tatra mountains, 37, 45, 47, 49, 65, 83, 92, 160, 244, 257
Teige, Karel, 76–78, 112, 119, 120, 121
Toyen, 112, 128
Turnovský, Ladislav, 24, 41, 45, 49, 53, 65, 70n, 91, 194, 253

Uherské Hradiště, 72, 167
Ukraine, 43, 183
United Kingdom, 31, 111, 163, 216, 228, 244
and/at exhibitions, 23, 24, 31, 44, 48, 61, 80n, 139, 244
relations with Czechoslovakia, 2, 25, 144–45, 160, 227

United States of America, 31, 36, 38, 39, 44, 111, 120, 121, 132, 150, 155, 165, 169, 193, 194, 207, 216, 221, 225–26, 246, 232, 245–46
and/at exhibitions, 10, 17, 23, 24, 35, 38, 40, 42–43, 55, 57–61, 85, 125, 132–33, 140, 141, 150, 175, 182, 184, 189, 196, 203, 210, 226, 232, 256
relations with Czechoslovakia, 2, 5, 25, 36, 55, 64, 107, 135–36, 145, 160, 175, 176, 181, 189, 190, 192, 205, 210, 211, 227, 250–51, 260–51
Czecho-Slovak diaspora in, 5, 14, 47, 64, 133, 180–82, 184–94, 198–99, 205–11, 212, 222, 256
Uprka, Joža, 108
Ústí nad Labem (Aussig), 144

van der Rohe, Mies, 40, 51, 73, 82
van der Velde, Henry, 203
Vaněk, Jan, 34
Vavrečka, Hugo, 172
Vienna, 7, 20, 71, 142–43, 172, 198
See also under world's fairs and exhibitions
Vítkovice Steels, 53, 63, 84, 89, 93, 95, 100, 162, 255, 257
Vydra, Josef, 93

Wachsman, Alois, 92
Whalen, Grover A., 60
Wiener Werkstätte, 85, 126, 163
Winkelmann, Emilie, 204
women, 201–2
at world's fairs, 3–4, 14, 128–29, 180, 194, 197, 200–205
absence in executive roles, 4, 128–29, 147, 257
as architects, 89, 201, 204
as artists and designers, 4, 51, 108, 128–29, 202–4, 256
as performers, 14, 128, 197, 201
as workers/service personnel, 11, 14, 147, 153, 179–80, 193–94, 197, 199, 200, 201, 202, 205, 212, 213, 256
associations of, 127, 180, 181, 186, 201–3, 256
representations of, 11, 14, 100, 113, 129, 130, 197, 199, 203–5, 233, 235, 236, 239, 249

Woodruff, Louise Lentz, 235
workers. *See* labor
workforce. *See* labor
World War I, 2, 10, 14, 36, 59, 107, 166, 175, 192, 218
World War II, 10, 55, 177, 221, 227n
world's fairs and exhibitions, 1–4, 10, 17, 21, 67, 103, 139, 180, 196, 216, 246
Barcelona
The International Exhibition of Barcelona (The Exposición International de Barcelona), 1929, 17, 18, 38–42, 51, 81, 82, 124, 149, 177, 257
Berlin
Great Berlin Art Exhibition, 1923, 74
Werkbund exhibition, 1930, 82
Brno
The Exhibition of Contemporary Culture, 1928, 77, 149, 167, 172, 204, 237–38, 239
The Exhibition of Modern Woman, 1929–30, 203
The Civilized Woman exhibition, 1929–30, 203
Brussels
The Universal and International Exposition (Exposition Universelle et Internationale), 1935, 17, 29, 47–51, 55, 57, 84, 149, 157, 161, 164, 167, 199, 241–42
Bucharest
The Exhibition of Czechoslovak Architecture, the Applied Arts and Building Industries Exhibition, 1930, 79n, 81, 122
Chicago
The World's Columbian Exposition, 1893, 42, 154, 181, 195, 203, 206, 208
Woman's World's Fairs, 1925–28, 204–5
Century of Progress, 1933, 28, 38, 42–47, 51, 57, 63, 123, 124, 132, 149, 156, 177–80, 182, 184, 186, 188, 189, 191, 194–99, 207, 208–11, 212, 222, 226, 230–31, 232, 233–36, 239, 243, 244, 249, 253, 256, 257
Cologne
The International Press Exhibition (Pressa), 1928, 75, 77

Werkbund Exhibition, 1914, 203
Dresden
International Hygiene Exhibition (Internationale Hygiene-Ausstellung), 1911, 242–43
Second International Hygiene Exhibition, 1930, 243–44
International Art Exhibition, 1926, 74
Düsseldorf
The Great Düsseldorf Exhibition of Healthcare, Social Welfare and Physical Exercise (Die Große Ausstellung Düsseldorf für Gesundheitspflege, soziale Fürsorge und Leibesübungen), 1926, 243
Johannesburg
Empire Exhibition, 1936, 149
Launceston, Tasmania
International Exhibition, 1891–92, 21, 163
Leipzig
Book Fair, 1927, 79n, 80, 122
International Book and Graphic Exhibition, 1914, 204
London
Great Exhibition, 1851, 1, 21, 51, 140, 153
Empire Exhibition, 1924–25, 11, 230
New York
The World of Tomorrow, 1939–40, 7, 9, 14, 17, 18, 38, 55, 57, 59, 60, 61–64, 68, 79, 90, 104, 119, 133–34, 140, 157, 158–59, 165, 170–73, 175, 177, 182, 212, 236–37, 245, 246, 247–48, 249, 251, 256, 258
Machine Art, 1934, 106
Paris
Exposition Universelle, 1855, 103
Exposition Universelle, 1867, 17, 154
Exposition Universelle, 1878, 153, 163n
Exposition Universelle, 1889, 51, 103, 163n
Exposition Universelle, 1900, 21, 107, 149, 153, 163n, 170, 232
Exhibition of Czechoslovak Folk Art, 1920, 23, 107–9, 110
The International Exhibition of Decorative Arts and Modern Industries (Exposition internationale des arts décoratifs et industriels modernes), 1925, 13, 17–18, 29–34, 77, 78, 85, 110, 111–120, 121–22, 124, 132, 150, 151, 161, 163, 164, 184, 201, 252, 257
Exhibition of Czech Female Painters, 1927, 202
The Colonial Exhibition, 1931, 110, 57, 230
International Exposition of Art and Technology in Modern Life (Exposition Internationale des Arts et Techniques dans la Vie Moderne), 1937, 7, 17–18, 51–55, 67–69, 77, 78, 79, 81, 82–83, 85–101, 110, 125, 132, 157–58, 161, 164, 167, 168, 170, 199, 201, 236, 253
Philadelphia
The Sesqui-Centennial International Exposition, 1926, 17, 34–38, 132, 149, 150, 156, 182–84, 191, 206, 256, 257
Poznań
International Hygiene Exhibition, 1933, 79n, 122
Polish General Exhibition (Powszechna wystawa krajowa), 1929, 204
Prague
Jubilee exhibition, 1891, 28–29, 72, 182
The Czechoslavic Ethnographic Exhibition, 1895, 126, 182, 219
National Exhibition of Glass, 1933, 50, 161
Rio de Janeiro
The Centennial International Exhibition (Exposição Internacional do Centenario), 1922, 23–29, 32, 33, 78, 80, 109, 110–11, 114, 143, 179, 245, 257
San Diego
The Panama California Exhibition, 1915, 221–22, 233
San Francisco
Golden Gate International Exposition, 1939–40, 17, 38, 55–59, 63, 133–35, 141, 177, 182, 212, 256
The Panama-Pacific International Exposition, 1915, 43, 175, 182
Seville
Ibero-American Exposition, 1929, 11, 39

St. Louis
Louisiana Purchase Exposition, 1904, 21, 43, 107, 149, 163n, 182, 221, 222, 243
Stockholm
The Stockholm International Exhibition, 1930, 122
Mother and Child exhibition, 1936, 236
Vienna
Weltausstellung, 1873, 20, 144, 155, 163, 203
Hygiene Exhibition, 1925, 243

Yugoslavia, Kingdom of (Kingdom of Serbs, Croats and Slovenes), 12, 38, 183
and exhibitions, 12, 17–18, 31, 43, 105, 207–8

Zlín, 93, 96, 166, 168–69, 171, 172, 173, 175, 176, 239, 240, 241, 242, 249, 250, 251
Zrzavý, Jan, 50, 92n, 127